HYPERSPECTRAL REMOTE SENSING OF TROPICAL AND SUBTROPICAL FORESTS

HYPERSPECTRAL REMOTE SENSING OF TROPICAL AND SUBTROPICAL FORESTS

Edited by

Margaret Kalacska
G. Arturo Sanchez-Azofeifa

CRC Press
Taylor & Francis Group
Boca Raton   London   New York

CRC Press is an imprint of the
Taylor & Francis Group, an **informa** business

CRC Press
Taylor & Francis Group
6000 Broken Sound Parkway NW, Suite 300
Boca Raton, FL 33487-2742

© 2008 by Taylor & Francis Group, LLC
CRC Press is an imprint of Taylor & Francis Group, an Informa business

First issued in paperback 2019

No claim to original U.S. Government works

ISBN-13: 978-0-367-45272-8 (pbk)
ISBN-13: 978-1-4200-5341-8 (hbk)

Library of Congress Cataloging-in-Publication Data

Hyperspectral remote sensing of tropical and sub-tropical forests / editor(s) Margaret Kalacska and G. Arturo Sanchez-Azofeifa.
    p. cm.
Includes bibliographical references and index.
ISBN 978-1-4200-5341-8 (alk. paper)
    1. Forests and forestry--Remote sensing. 2. Image processing--Digital techniques. 3. Multispectral photography. I. Kalacska, Margaret. II. Sánchez-Azofeifa, Gerardo-Arturo. III. Title.

SD387.R4H97 2008
577.34072--dc22
                                            2007045097

Visit the Taylor & Francis Web site at
http://www.taylorandfrancis.com

and the CRC Press Web site at
http://www.crcpress.com

Contents

Preface

Our main motivation for this book is to illustrate the potential for hyperspectral remote sensing to provide tools and information to infer and assess ecosystem characteristics at various spatial and temporal scales in the tropics and subtropics. The greater sensitivity and finer spectral resolution provided by these data offer unprecedented opportunities to study in detail even the most remote and inaccessible areas. These data also provide a means for retrospective or ongoing assessments. Examples from the chapters cover a range of ecosystem types, including mangroves, wooded savannas, rain forests, and dry forests, thus providing a nearly global perspective with study sites in Central and South America, Africa, Australia, and Hawaii.

In the broad field of remote sensing from the Earth and planetary sciences, hyperspectral sensors and data are a relatively new and untapped data source. Most frequently they have been applied in other environments for forestry, agricultural, mineral exploration, and geologic applications; in the tropics, however, they are still under-used. The diversity of tropical and subtropical ecosystems, while providing a complex and unique set of challenges to the use of hyperspectral data, also offers an ideal environment to develop, test, and apply new techniques. It is precisely the complex and dynamic nature of these ecosystems that makes them such interesting and exciting areas for research with hyperspectral data.

One of the most unique features of this book is the focus on the application of hyperspectral technology specifically addressing tropical and subtropical environments with real-world examples and actual data. The authors also integrate a range of analysis techniques, including hyperspectral reflectance indices, spectral mixture analysis, pattern classification, band selection, partial least squares, linear discriminant analysis, and radiative transfer models. The chapters present a comprehensive review of the current status and most innovative achievements of hyperspectral remote sensing in tropical and subtropical environments; over 500 separate studies are cited. And, as illustrated by the diverse backgrounds of the contributors, the most successful use of hyperspectral data in tropical or subtropical regions integrates a multidisciplinary approach spanning a wide range of expertise.

Several detailed volumes exist exploring the fundamentals of remote sensing, spectrometry, and image analysis in detail. Therefore, the purpose of this book is not to provide a tutorial in remote sensing; rather, its aim is to provide an illustration of the potential applications and analysis techniques that can be used, addressing the unique challenge of working in the tropics. Nevertheless, background information on hyperspectral remote sensing and the spectral characteristics of vegetation is provided in chapters 1, 2, 3, and 12.

This book is structured as a set of 13 contributed chapters addressing techniques and applications of hyperspectral remote sensing for tropical and subtropical forests with field spectrometry, airborne spectrometry and imagery, and satellite imagery. The following section provides an overview of each chapter.

OVERVIEW

CHAPTER 1: TROPICAL DRY FOREST PHENOLOGY AND DISCRIMINATION OF TROPICAL TREE SPECIES USING HYPERSPECTRAL DATA (K. L. CASTRO-ESAU AND M. KALACSKA)

Tropical tree species identification from remotely sensed imagery is a natural extension of previous research on their leaf optical properties and their discrimination using leaf spectra. Recently, efforts at mapping tree species in a species-rich tropical wet forest using airborne hyperspectral imagery have demonstrated promising results for distinguishing limited numbers of prominent tree species. These studies emphasize the importance of capturing the relevant information for species identification from within the large pool of data that comprises hyperspectral imagery.

In addition to reviewing the current status of remote sensing for species discrimination in tropical forests, this chapter highlights phenologic events in tropical dry forests that can provide unique opportunities for tree species identification using hyperspectral data. These opportunities will be most significant for species that exhibit a high degree of intraspecific synchronicity in leaf flush, leaf abscission, flowering, and/or fruiting events, which may be captured as predictable spectral responses at specific times during the year. Differences in the timing of these events among species will also facilitate discrimination.

CHAPTER 2: REMOTE SENSING AND PLANT FUNCTIONAL GROUPS: PHYSIOLOGY, ECOLOGY, AND SPECTROSCOPY IN TROPICAL SYSTEMS (M. ALVAREZ-AÑORVE, M. QUESADA, AND E. DE LA BARRERA)

Plant functional types may describe groups of plants with common responses to certain environmental influences and have been applied to several ecosystem functions such as biochemical cycles, fire resistance, invasion resistance, acquisition and use of resources, defense against herbivory, pollination, and seed dispersal, among others. In tropical systems, classifications have relied on ecophysiological characteristics and are based on species responses to different environmental factors. Most studies in tropical systems have analyzed single or a combination of a few ecophysiological characteristics for the discrimination of tropical functional groups. However, an analysis considering an assemblage of characteristics with ecological relevance that determines the establishment of species in a given habitat is more realistic.

This chapter considers the capability of hyperspectral sensors to detect characteristic chemical and anatomical properties of vegetative and reproductive tissues of plants. Exploration of the role of hyperspectral remote sensing in the assessment and determination of functional traits in the tropics is an important task in order to evaluate current and future applications of these technologies in ecological sciences. Four case studies in functional group detection are reviewed: (1) discrimination of successional stages, (2) discrimination of vegetation types, (3) discrimination of life forms, and (4) discrimination of biophysical properties. Broader applications of remote analysis of functional group studies are explored for species discrimination, carbon flux, the direct detection of plant functional types, and changes in functional properties of ecosystems.

CHAPTER 3: HYPERSPECTRAL DATA FOR ASSESSING CARBON DYNAMICS AND BIODIVERSITY OF FORESTS (R. LUCAS, A. MITCHELL, AND P. BUNTING)

The forests of the tropics and subtropics represent a diversity of habitats (e.g., rainforests, mangroves, and wooded savannas) that vary both spatially and temporally. A large proportion of forests is also secondary and exists at varying stages of degradation or regeneration. For remote sensing scientists, this spatial and temporal variation represents both an opportunity and a challenge for the use of hyperspectral data. In terms of opportunities, hyperspectral data have provided new options for assessing biological diversity and contributed to assessments of dead and live carbon, measures of forest health, and understanding of ecosystem processes (e.g., through retrieval of foliar biochemicals). The utility of these data in the tropics and subtropics has been limited by environmental conditions, sensor characteristics, and the complexity of the forested environment in terms of species diversity, multi-layering, and shadowing effects.

This chapter provides an overview of the use of hyperspectral data in tropical and subtropical forests and focuses primarily on forest types that are prevalent in these regions—namely, rainforests (evergreen and semi-evergreen), mangroves, and wooded savannas. The chapter conveys key features of hyperspectral data that allow different levels of information on forests to be extracted compared to multispectral counterparts. Using previously published research and case studies from Brazil and Australia, the use of hyperspectral data for assessing forest biodiversity, carbon dynamics, and health is demonstrated. Finally, the chapter provides some indication of the future directions for hyperspectral remote sensing of tropical and subtropical forests and outlines how existing and future sensors might be integrated to provide options for conserving, restoring, and sustainably utilizing forests.

FIELD SPECTROSCOPY

CHAPTER 4: EFFECT OF SOIL TYPE ON PLANT GROWTH, LEAF NUTRIENT/CHLOROPHYLL CONCENTRATION, AND LEAF REFLECTANCE OF TROPICAL TREE AND GRASS SPECIES (J. C. CALVO-ALVARADO, M. KALACSKA, G. A. SANCHEZ-AZOFEIFA, AND L. S. BELL)

An understanding of how leaf spectral reflectance changes as a function of soil type is key to interpreting intra- and interspecies spectral reflectance differences. Because leaf traits are influenced by leaf age, canopy position, and soil fertility, a comprehensive model of leaf spectral reflectance must include a clear description of soil–nutrient–plant interactions. However, because hyperspectral data present unique challenges in contrast to conventional data, this chapter begins with a review of their characteristics and introduces a form of classification based on the location of spectra in relation to their neighbors in multidimensional space. The overall objectives of the application section of this chapter are to (1) evaluate the effects of five soil types on plant growth and foliar nutrient concentration in seedlings of two tree species, (2) investigate the effect of plant genetic variation on plant growth as a response to soil type, (3) investigate the utility of the indirect leaf chlorophyll measurements from the Minolta SPAD–502 chlorophyll meter as a tool to evaluate plant vigor, and (4) assess the effect of soil type on the spectral response of tropical tree seedlings (from seeds and clones) and a tropical grass species. The grass is included as an exploration into natural modification of soil nutrients and toxicity through cadaveric decomposition and its effects on leaf spectral reflectance.

Cadaveric decomposition is an integral, though grossly overlooked, component of terrestrial ecology; very little is understood regarding its influence on biodiversity and ecosystem functioning (both above- and below-ground). It is shown that soil types with contrasting chemical and physical properties significantly affect plants' growth and in turn their spectral response. Specifically, in the cases where the photosynthetic/accessory leaf pigments are affected, the leaf spectra can be classified with high accuracy. Conversely, variations in soil nutrient concentration that either do not induce a change in the photosynthetic/accessory pigments or influence the vegetation in the same manner resulted in a classification problem that relied on the near-infrared wavelengths. Consequently, the spectra are not as readily separable. Importantly, the degree to which different soils affect leaf reflectance was found to be species specific.

Chapter 5: Spectral Expression of Gender: A Pilot Study with Two Dioecious Neotropical Tree Species (J. P. Arroyo-Mora, M. Kalacska, B. L. Caraballo, J. E. Trujillo, and O. Vargas)

Few studies have considered effects of dioecy on plant ecophysiological functioning, even though a large proportion of species in the tropics are dioecious. Maintaining a proper ratio of gender ensures species survival, an important consideration in natural forest management, reforestation, and conservation projects. Gender differences expressed in the spectral properties of leaves and leaf chemistry were examined for *Hyeronima alchorneoides* and *Virola koschnyi*, two neotropical dioecious tree species. Leaf-level spectra from an Analytical Spectral Devices, Inc. (ASD) FieldSpec handheld spectrometer were analyzed through spectral vegetation indices and pattern classification. Key spectral regions were identified as being the most promising regions warranting further exploration into differentiating individuals of different gender within species. Chlorophylls a and b and carotenoid concentrations were also examined, along with nitrogen concentration for both species, and were found to differ between males and females. The results also highlight the importance of considering the uniqueness of each species. While differences in the spectra were found for both species, *V. koschnyi* had a stronger expression of gender than *H. alchorneoides*. The most important spectral regions also differed somewhat between the two species. Results presented in this chapter show a definitive potential for the spectral differentiation of gender for dioecious species and possible biochemical explanations for these differences at the leaf level; in the future they may lead to a scaling up to the regional level and airborne imagery.

Chapter 6: Species Classification of Tropical Tree Leaf Reflectance and Dependence on Selection of Spectral Bands (B. Rivard, G. A. Sanchez-Azofeifa, S. Foley, and J. C. Calvo-Alvarado)

Few studies have focused on classification of tree species from spectral data despite the many potential applications in conservation biology and forest conservation/management. Fewer studies have examined classification of tropical tree species. Here the most important species for the survival of the great green macaw (*Ara ambigua*) are examined. Leaf reflectance of 20 species from Costa Rica can be classified accurately (79 to 97%) using linear discriminant analysis. In general, all of the species, except for *Dipteryx panamensis*, were identified correctly most of the time. *D. panamensis* was confused mostly with *Laetia procera* by all of the models. The species studied were classified best using bands or indices correlated with leaf water content, followed by pigmentation properties. The spectral range of input bands was manipulated in order to create classification models more appropriate for data affected by atmospheric absorption and sensor wavelength limitations. Models with only visible (VIS) and near-infrared (NIR) bands up to 1075 nm were insufficient to separate all 20 species studied, but despite the lack of strong water bands, some species were separated accurately. Models with only shortwave infrared (SWIR) and NIR bands, like models with only VIS and NIR bands, also had lower classification accuracies, but again not for all species. Overall accuracy was highest when the 350- to 2500-nm spectrum was exploited.

Chapter 7: Discriminating *Sirex noctilio* Attack in Pine Forest Plantations in South Africa Using High Spectral Resolution Data (R. Ismail, O. Mutanga, and F. Ahmed)

The wood-boring pest *Sirex noctilio* Fabricius (Hymenoptera: Siricidae) is causing mortality along the heavily afforested eastern regions of South Africa, with recent reports indicating that mortality might be as high as 30% in some forestry compartments. *S. noctilio* affects all commercial pine species in South Africa. Management strategies by South African forest companies now focus on the combined use of remote sensing, silvicultural treatments, and biological control to reduce *S. noctilio* population numbers and to minimize the potential economic threat to the industry. Remote sensing is a key component of the integrated management strategy and remains crucial for the detection and monitoring of the wasp and for the effective deployment of appropriate suppression activities. Limitations are primarily due to classification errors that arise because of the inability of broadband multispectral sensors to discriminate between the different damage classes associated with *S. noctilio* attack.

The question that then arises is whether, with the future availability and accessibility of high spectral (i.e., hyperspectral) resolution data in South Africa, there is potential to successfully discriminate between healthy and *S. noctilio*–attacked pine trees. The preliminary aim of this study was to use high spectral resolution data to identify diagnostic spectral features of *Pinus patula* needles showing varying degrees of *S. noctilio* attack. Specifically, the objectives were (1) to determine whether there is a significant difference between the mean reflectance at each measured band (from 400 to 1300 nm) for the three stages of *S. noctilio* attack (green, red, and gray), and (2) for the wavelengths that are spectrally different in this region, to test whether some bands have more discriminating power than others in the detection of *S. noctilio*–induced stress. Although no single band is capable of total separability, it is shown that spectral separability between all the classes (healthy, green, and red) is possible when using a four-band combination. Therefore, an important prerequisite (i.e., band selection) is established for the potential upscaling of results to either an airborne or space-borne platform.

AIRBORNE DATA

CHAPTER 8: HYPERSPECTRAL REMOTE SENSING OF EXPOSED WOOD AND DECIDUOUS TREES IN SEASONAL TROPICAL FORESTS (S. BOHLMAN)

Improvements in sensor technology and analytical methods in the past 10 years, in combination with an increased perception that tropical forests are not spectrally invariant, have provided important advances in quantifying biophysical information on mature tropical forest from remote sensing. Hyperspectral remote sensing, because of its greater spectral range and detail, provides an even better opportunity to detect subtle spectral differences relevant to species composition, biomass, or canopy functioning in mature tropical forests. Spectral mixture analysis (SMA) is one method used to gain ecological information from remotely sensed images that has been little used for mature tropical forests. SMA quantifies the proportion of different material present in a landscape (called "endmembers") rather than using an index or correlative approach. It can also be readily compared directly against field data and extrapolated to areas outside the study site.

Typical endmember spectra used in tropical land use studies are green vegetation, soil, shade/shadow, and, additionally in some cases, nonphotosynthetic vegetation (NPV), which refers to senescent vegetation and woody material. Seasonal, semideciduous tropical forests contain partially or fully deciduous trees in the dry season and thus have a large amount of wood exposed at the top of the canopy. This chapter addresses the question of whether NPV in the form of exposed canopy branches is an important component of seasonal tropical forests, and if it can be quantified with spectral mixture analysis. Hyperspectral data provide the best opportunity for detection because they provide narrow bands to possibly distinguish important absorption features associated with wood and include SWIR bands that have been shown to be important for quantifying NPV in other ecosystems. The linear mixture model worked well despite the high spectral heterogeneity within the green vegetation and NPV endmembers. SMA could provide important insights into temporal patterns of the landscape, showing the degree to which the species in the forest are synchronous in making the transition between deciduousness in the dry season and full leaf content in the wet season.

CHAPTER 9: ASSESSING RECOVERY FOLLOWING SELECTIVE LOGGING OF LOWLAND TROPICAL FORESTS BASED ON HYPERSPECTRAL IMAGERY (J. P. ARROYO-MORA, M. KALACSKA, R. L. CHAZDON, D. L. CIVCO, G. OBANDO-VARGAS, AND A. A. SANCHÚN HERNÁNDEZ)

Development of effective, long-term plans to conserve and manage biodiversity requires an understanding of the distribution, diversity, and abundance of species in both protected and unprotected areas. More than ever, due to extensive fragmentation and other anthropogenic effects, forest fragments in unprotected areas must serve a dual function: providing a sustainable source of timber and nontimber products and providing essential intact forest habitat for biodiversity conservation within the landscape matrix. Factors such as the initial structure of the forest and the scale and intensity of the logging operations, not to mention the short- and long-term effects of postharvesting activities, are important when forest management and biodiversity conservation goals converge. Therefore, monitoring of natural forest areas subject to forest management (selective logging) is necessary in order to assess forest recovery processes after disturbance. An ongoing major limitation for assessing forest recovery after selective logging is the lack of information representing a chronological sequence, thus precluding solid forecasting of regeneration trajectories. The use of hyperspectral data with multitemporal management information is a novel application of such data addressing the need for studies focusing on the assessment of forest recovery essential to monitoring efforts. The specific objectives for this chapter are to

- identify the most significant wavelengths for separating the spectra of natural forest area recovery following selective logging, thereby exploring the most likely forest characteristics (i.e., soil exposure, canopy structure, gap vegetation recovery) responsible for this separability
- determine the accuracy with which the spectral signature of different selectively logged natural forest areas can be classified
- determine how the spectral separability of the managed forest areas is affected by spatial resolution and in turn how this affects the accuracy of the different classifiers

The results indicate that it is possible to separate and accurately classify management units of various ages since selective logging with less than 5% error at both spatial scales (16 and 30 m). Nevertheless, at both scales examined, in order to discern the quantitative contribution of each canopy element to the separability of the four management units, an unmixing analysis of the spectra is required. Understanding of logging impacts at the landscape-scale level is necessary to enhance conservation practices outside protected areas.

CHAPTER 10: A TECHNIQUE FOR REFLECTANCE CALIBRATION OF AIRBORNE HYPERSPECTRAL SPECTROMETER DATA USING A BROAD, MULTIBAND RADIOMETER (T. MIURA, A. R. HUETE, L. G. FERREIRA, E. E. SANO, AND H. YOSHIOKA)

In this study, a new technique for calibrating airborne hyperspectral spectrometer data to reflectance factors is introduced. The technique acquires reference data over a white reference panel with a spectrometer before and after a flight and uses multi-band radiometer data obtained continuously over another panel during the flight to adjust the magnitude of the reference data for converting airborne data to reflectance spectra. This "continuous panel" method was evaluated and validated using an experimental data set obtained in a semi-arid grassland and used to acquire reflectance data in a tropical forest–savanna transitional region. Airborne spectral reflectance derived with this method was unbiased and showed high-quality spectral signatures that were not distorted due to atmospheric absorption features. Application results in the tropical region demonstrated the capability of the derived reflectance to depict well spatial variability of nonphotosynthetically active vegetation associated with land cover types, including tropical forests, savannas, pastures, and burned areas, with ligno-cellulose absorptions captured in their SWIR signatures. Absolute accuracy of the derived reflectance with the continuous panel method was estimated at 0.005 reflectance units with precision at the same level of 0.005 reflectance units (measurement uncertainty of 0.005 ± 0.005). This new technique provides valuable hyperspectral reflectance data sets that can be used for various applications in tropical and subtropical regions, including land cover assessments, scaling up and validation of satellite products, continuity studies, and creation of a reference hyperspectral reflectance database.

SATELLITE IMAGERY

CHAPTER 11: ASSESSMENT OF PHENOLOGIC VARIABILITY IN AMAZON TROPICAL RAINFORESTS USING HYPERSPECTRAL HYPERION AND MODIS SATELLITE DATA (A. R. HUETE, Y. KIM, P. RATANA, K. DIDAN, Y. E. SHIMABUKURO, AND T. MIURA)

Phenology represents the seasonal timing and annual repetition of biological life cycle events and is a characteristic property of ecosystem functioning and predictor of ecosystem processes. Despite its importance, the phenology of tropical rainforest ecosystems and the environmental conditions controlling its variability are not well understood. This is in part due to the complexity of tropical forest canopies, where a highly diverse tree species population can result in a wide variety of phenologic responses to the same or common environmental factors, such as rainfall, temperature, and photoperiod. This chapter examines patterns of optical-phenologic variability in an evergreen broadleaf tropical forest in the Amazon, using space-borne hyperspectral and moderate resolution satellite measurements. Moderate resolution satellite data provide high frequency but poor spatial resolution data of limited spectral content, while hyperspectral data offer finer resolution and spectral detail but at infrequent time intervals. Much of what is known about seasonal vegetation dynamics in the tropics comes from moderate and coarse resolution satellite measurements that are commonly used for large-scale vegetation monitoring and vegetation–climate studies; however, this often results in low spectral sensitivity for tracking temporal and spatial variability in tropical forest characteristics, including phenology. Hyperspectral remote sensing measurements add spectral fidelity to the extraction of optical-phenologic information from tropical forests.

The focus here is on the potential contributions of hyperspectral information in phenology studies and its scaling with high temporal frequency, moderate resolution satellite data for improved characterization of tropical rainforest phenology. The spectral and temporal variability in landscape phenology patterns of different physiognomic vegetation types in an evergreen broadleaf primary tropical forest is evaluated, including regenerating successional forests and pasture and agricultural areas. The goal is to investigate landscape phenology patterns in complex tropical rainforests and assess the extent, magnitude, and synchrony of phenology patterns. Wide variability and large seasonality were found in spectral signatures in the tropical forests analyzed here, with much of the variation occurring in forest conversion and regenerating areas of varying age classes and types of secondary forest regrowth. Unique phenology responses to seasonal drought periods were also observed across all vegetation types, both spectrally and temporally.

FUTURE DIRECTIONS

CHAPTER 12: HYPERSPECTRAL REMOTE SENSING OF CANOPY CHEMISTRY, PHYSIOLOGY, AND BIODIVERSITY IN TROPICAL RAINFORESTS (G. P. ASNER)

Hyperspectral remote sensing offers to revolutionize studies of tropical forest biochemistry, physiology, and biodiversity. However, the efficacy of imaging spectroscopy for forest research and monitoring rests in understanding the sources of variance in spectral signatures at different biophysical, taxonomic, community, and ecosystem levels of organization. Here, Asner discusses the issues affecting our ability to remotely explore humid tropical forests using airborne and space-based hyperspectral imaging. In doing so, the spectral properties of forests in Hawaii and the Brazilian Amazon are analyzed, building from the leaf to regional scales. Throughout the scaling process, three remote sensing approaches for exploring the functional properties of forests are highlighted: partial least squares (PLS) regression analysis, radiative transfer modeling, and hyperspectral reflectance indices. Asner's synthesis emphasizes that leaf spectral variability is driven by a constellation of biochemical and structural properties of foliage. No single wavelength band or region is uniquely sensitive to a particular biochemical. At the crown level, hyperspectral reflectance is driven by leaf optical properties, but is mediated by canopy structural and architectural characteristics.

These Hawaiian forest studies suggest that within-crown variability of hyperspectral signatures is low in comparison to microsite-, substrate-, or climate-mediated variability between crowns. Moreover, interspecific and regional hyperspectral studies using EO-1 Hyperion satellite data in Hawaii and Brazil show that imaging spectroscopy can be used to monitor forest–climate interactions in remote tropical regions. A conceptual framework for linking biochemical and spectral variability to canopy diversity of tropical forests is also presented. This framework is used to demonstrate biochemical diversity among Amazon forest species and to show how biochemical–spectral diversity translates to canopy species richness of lowland Hawaiian forests using airborne imaging spectroscopy. Finally, the interplay of sensor fidelity, measurement resolution, and analytical techniques for determining tropical forest biochemistry, physiology, and diversity using hyperspectral imagery is discussed.

CHAPTER 13: TROPICAL REMOTE SENSING—OPPORTUNITIES AND CHALLENGES (J. A. GAMON)

The fine spectral detail of hyperspectral remote sensing, or imaging spectrometry, presents a wealth of possibilities for expanding our understanding of tropical ecosystems. However, the potential of imaging spectrometry is limited by the shortage of hyperspectral data and the sheer volume and complexity of these data when they are available. This chapter focuses on two applications: (1) assessment of biodiversity, and (2) evaluation of biosphere–atmosphere interactions, which are both areas where imaging spectrometry can provide new opportunities. This chapter also describes the need for an informatics framework that encompasses the full complexity of hyperspectral data and outlines some basic considerations of the cyberinfrastructure needed to accommodate this task. Reflectance spectra provide a powerful, readily searchable proxy data set that can be related to a particular goal or ecological variable through proper calibration or validated models. Informatics for ecology and environmental science—or "eco-informatics," as it is sometimes called—is among the most complex informatics challenges that we face. Its complexity derives both from the nature and scale of the scientific questions addressed and from a variety of technical issues. From an informatics standpoint, this strength of hyperspectral remote sensing as a structured proxy data set has barely been tapped.

ACKNOWLEDGMENTS

The editors thank Mr. J. P. Arroyo-Mora, Mr. Ken Dutchak, and Ms. Eva Snirer for their help in editing and book cover design. This work was carried out with the aid of a grant from the Inter-American Institute for Global Change Research (IAI) CRN II #021, which is supported by the US National Science Foundation (Grant GEO-0452325) and the National Sciences and Engineering Research Council of Canada.

Contributors

Fethi Ahmed, Ph.D., is a senior lecturer at the School of Environmental Sciences, University of KwaZulu-Natal, Durban, South Africa, where he is the deputy head of the school. He both teaches and conducts research in environmental applications of geographical information systems (GIS), remote sensing, spatial analysis, and modeling. His main interests are in spatial databases, vegetation characteristics, hyperspectral remote sensing, and spatial statistics. His recent projects include the identification of invasive alien species and the modeling of their rate of spread; forest structural characterization using optical, radar, and LiDAR sensors and data fusion techniques; modeling forest productivity and water use; and use of hyperspectral sensors and foliar chemistry in characterizing vegetation health and vigor.

Mariana Alvarez-Añorve, B.S., obtained her B.S. in biology from the Faculty of Sciences at the Universidad Nacional Autonoma de Mexico (UNAM). She has collaborated on several research projects on the ecology of tropical tree species—specifically, on functional and morphological attributes of tropical trees, as well as on their response to nutrient stress. She has also studied the molecular and physiological responses of crop plants to chemical stress. Currently, she is a Ph.D. candidate in biological sciences at the Centro de Investigaciones en Ecosistemas, UNAM, under the guidance of Dr. Mauricio Quesada. Her research focuses on physiological, anatomical, and morphological traits of tropical tree species in different successional stages of tropical dry forests and on the identification and spectral characterization of plant functional groups with hyperspectral remote sensing techniques.

J. Pablo Arroyo-Mora, M.S., graduated from the School of Forestry Engineering at the Technological Institute of Costa Rica and completed his M.S. at the University of Alberta (Earth and Atmospheric Sciences Department) with a focus on GIS and remote sensing in tropical regions. Currently, he is a Ph.D. candidate in the Ecology and Evolutionary Biology Department at the University of Connecticut with a focus on forest logging practices related to the assessment of tree diversity and biodiversity database management in the San Juan Biological Corridor in Costa Rica. He is also interested in hyperspectral data analysis and machine learning techniques for ecological applications and has experience in secondary forest succession dynamics and forest restoration with native species plantations. His M.S. work focused on forest/nonforest mapping of tropical dry forest regions and the discrimination of tropical dry forest successional stages through a combination of ecology and remote sensing.

Gregory P. Asner, Ph.D., is a faculty member of the Carnegie Institution and Stanford University. His scientific research centers on how human activities alter the composition and functioning of terrestrial ecosystems at regional and global scales. Dr. Asner combines field work, airborne and satellite mapping, and computer simulation modeling to understand the response of ecosystems to land use and climate change. His recent work includes satellite monitoring of forest disturbance and selective logging throughout the Brazilian and Peruvian Amazon, invasive species and biodiversity in Hawaiian and Australian rainforests, El Niño effects on tropical forests, and impacts of livestock production on global biogeochemical cycles. Dr. Asner's remote sensing efforts focus on the use of new technologies for studies of ecosystem structure, chemistry, and biodiversity in the context of conservation and management.

Lynne S. Bell, Ph.D., is an associate professor in the School of Criminology at Simon Fraser University. Her scientific research interests include isotopic mass spectrometry to track human beings temporally and geographically. This type of work is important to forensic identification, monitoring human trafficking, and human security. She also has interests in the recovery and identification of clandestine graves using remote sensing technology. Other long-standing work includes the identification and recovery of molecular information from human skeletal material and understanding associated diagenetic processes that might affect preservation.

Stephanie Bohlman, Ph.D., is a tropical biologist whose main areas of research are forest canopy biology and physiology. One of the main tools she uses is remote sensing, striving to link plant physiology, canopy structure, and remote sensing. In particular, she is interested in landscape patterns of species and functional diversity in forests and their underlying physical and historical causes. She has a Ph.D. from the University of Washington, Seattle, and was a research associate at the Smithsonian Tropical Research Institute in Panama. She is currently a researcher at Princeton University, working with Steve Pacala to develop models of tropical forests that include remotely sensed data. Other current projects include mapping species distributions and tree mortality using high resolution images in Panama.

Peter Bunting, Ph.D., is a lecturer within the Institute of Geography and Earth Sciences at the University of Wales, Aberystwyth. He has a B.S. in computer science (2004) and a Ph.D. (2007) in the remote sensing of wooded savannas in Queensland using airborne and space-borne remote sensing data.

Julio C. Calvo-Alvarado, Ph.D., is a professor at the School of Forestry at the Costa Rican Technology Institute. His research and teaching activities are multidisciplinary in conservation and management of tropical natural resources such as soils, water, forests, and biodiversity. Dr. Calvo-Alvarado has worked with scientists from Europe, Canada, the United States, Germany, and Latin America in research topics such as climate change, forest hydrology, wetland restoration, watershed management, forest plantations, conservation biology, and the application of remote sensing for the monitoring of land use change and for the characterization of tropical forests. Many of his projects also include the human dimension as a strategy to understand interactions and to develop comprehensive conservation and management approaches that influence decision makers on the proper design and evaluation of environmental policies.

Benjamin L. Caraballo, B.A., graduated in 2006 from the Gallatin School of Individualized Study at New York University with a B.A. in ecology and photography. During those 4 years, Benjamin was the recipient of the Bill and Melinda Gates Millennium Scholarship award. At NYU, Benjamin worked closely with the Organization of Tropical Studies in Costa Rica and Duke University in an effort to develop spectral methods of identification of the neotropical species *Hyeronima alchorneoides*, an integral first step to future sustainable remote sensing techniques. He also worked closely with the School for International Training and spent 4 months in Tanzania studying wildlife ecology and the effects of Western ideals on the lives of the Maasai. He has worked with the Department of Vertebrate Paleontology at the American Museum of Natural History, creating Web modules that will grant public access to its collection of over 700,000 fossil specimens.

Karen L. Castro-Esau, Ph.D., graduated from the University of Alberta in 2006 with a Ph.D. in Earth and atmospheric sciences. Her thesis research focused on leaf spectra and leaf traits of tropical lianas and trees. The main objective of this research, which was carried out in Panama and Costa Rica, was to investigate the potential for species recognition using data gathered from remote sensors. Prior to her doctoral studies, Dr. Castro completed an M.S. degree in conservation biology and a B.S. degree in agriculture, also from the University of Alberta in Edmonton, Alberta, Canada. She currently is employed as a botanist with the Canadian Food Inspection Agency in Ottawa, Ontario, Canada, where she assesses the biosecurity risk of potentially invasive plants.

Robin L. Chazdon, Ph.D., is a professor in the Department of Ecology and Evolutionary Biology at the University of Connecticut. Her research focuses on tropical forest dynamics, succession, and regeneration. She is also actively studying aspects of biodiversity conservation in human-dominated tropical landscapes. Much of her work has been based in northeastern Costa Rica, at La Selva Biological Station and the La Selva–San Juan Biological Corridor. She has served as editor of the journal *Biotropica* (2003–2006) and president of the Association for Tropical Biology and Conservation (1998), and on the governing board of the Ecological Society of America (1998–2000) and the National Center for Ecological Analysis and Synthesis. She is currently a member of the founding board of the Costa Rica–USA Foundation and the La Selva Advisory Committee. In 2003 she was awarded the President's Medal from the British Ecological Society and in 2004 she received the Provost's Award for Excellence in Research. She is coeditor of two books and author of more than 70 journal articles and book chapters.

Daniel L. Civco, Ph.D. (1987, B.S. 1974, M.S. 1976) is an Earth resources scientist with more than 30 years' experience in remote sensing and GIS applications. He has been involved extensively in research addressing both inland and coastal wetland resources, land use mapping and change analysis, impervious surface detection, and natural resources inventory and analysis. Further, he has been involved in algorithm development and refinement for processing remote sensing and other geospatial data. Dr. Civco is director of the Center for Land Use Education and Research at the

University of Connecticut, and is the founder of its Laboratory for Earth Resources Information Systems (LERIS). He is a University of Connecticut teaching fellow and a fellow of the American Society for Photogrammetry and Remote Sensing, of which he is a former National Education Committee chairman and director of the Remote Sensing Applications Division.

Erick de la Barrera, Ph.D., is a researcher at Centro de Investigaciones en Eco-sistemas, Universidad Nacional Autónoma de México. He obtained his B.S. at the University of Guadalajara (1997) and his Ph.D. at the University of California, Los Angeles (2003). His research on cultivated plants has touched various aspects of plant physiological ecology, including gas exchange responses to various environmental factors and reproductive ecophysiology. Recent research interests in his lab, which has a biophysical focus, include studies on the evolutionary origin of nectar and plant responses to climate change scenarios. He is coauthor of over 15 peer-reviewed scientific articles and coeditor, with Dr. William K. Smith (Wake Forest University), of *Perspectives in Biophysical Plant Ecophysiology*.

Kamel Didan, Ph.D., received the Ph.D. degree in agricultural and biosystems engineering with a minor in industrial engineering from the University of Arizona in 1999. He is currently an associate research scientist and adjunct faculty in the Department of Soil Water and Environmental Science at the University of Arizona. He is an associate science team member of the EOS-MODIS instrument, where he leads the research and development of the MODIS vegetation index and surface reflectance spatial aggregation algorithms. His research interests include the development of algorithms, models, and knowledge extraction techniques for calibrated remotely sensed time series of land surface biophysical measurements. His research aims at assessing the short- and long-term climate-related and land use change influences on vegetation health and dynamics.

Laerte Guimaães Ferreira, Ph.D., received an M.S. degree in economic geology from the University of Brasília, Brazil, in 1993 and his Ph.D. degree in soil, water, and environmental science from the University of Arizona, Tucson, in 2001. He has been involved in the LBA Ecology project since 1999. Currently, besides being a co-investigator in one of the large-scale biosphere atmosphere (LBA) ecosynthesis and integration projects (TG 30), he is also a co-investigator in the project Impacts of Land Use Change on Water Resources in the Brazilian Cerrado (NASA LULCC). His research interests include applications of vegetation indices for biophysical monitoring and change detection, as well as the use of remote sensing and GIS approaches for environmental management and territorial planning.

Sheri Foley, M.S., received her master of science degree in Earth and atmospheric sciences from the University of Alberta, Edmonton, Alberta, Canada, in 2005. Her research was focused on remote detection of tropical tree species utilizing hyper-spectral imagery. Currently, she is employed by an environment consulting company, focusing her attention on geographic information systems and land-use management for forestry companies.

John A. Gamon, Ph.D., is primarily interested in the application of optical tools for studying plant ecophysiology, ecosystem gas exchange (photosynthesis, respiration, and evapotranspiration), and plant biodiversity. He also studies the impacts of disturbance (fire, drought, climate change) and species composition on ecosystem processes. Recent work has included the development of SpecNet (Spectral Network), an international network of collaborating sites and investigators combining optical and flux methods to understand ecosystem–atmosphere gas fluxes.

Alfredo R. Huete, Ph.D., is a professor in the Department of Soil, Water and Environmental Science at the University of Arizona. He received his M.S. in soil and plant biology from the University of California at Berkeley and Ph.D. in soil and water science from the University of Arizona in 1984. He is a MODIS science team member, responsible for the development of the vegetation index (VI) products and is involved in the large-scale biosphere atmosphere in the Amazon (LBA) experiment. His research includes satellite-based tropical phenology analysis, soil–vegetation–climate interactions, and the use of vegetation indices for land surface process studies.

Riyad Ismail, M.S., received a master's degree in geography from the University of Durban Westville, South Africa. He is currently a Ph.D. student at the University of KwaZulu-Natal, Pietermaritzburg. His research focuses on the use of high spatial and spectral resolution data for the accurate detection and monitoring of *Sirex noctilio* in commercial pine forests. Ultimately, the aim of his research is to develop an improved detection and monitoring framework that augments current pest management initiatives in South Africa. His Ph.D. promoters are Dr. Onisimo Mutanga and Professor Urmilla Bob. He is employed as a remote sensing analyst by Sappi Forest.

Margaret Kalacska, Ph.D., obtained both her M.S. and Ph.D. from the Earth and Atmospheric Sciences Department at the University of Alberta, Canada. Her doctoral research focused on the application of remote sensing to tropical dry forest ecology. She is a former TROPI-DRY postdoctoral fellow (University of Alberta) and National Sciences and Engineering Research Council of Canada postdoctoral fellow (Simon Fraser University). She is currently a research fellow in the Centre for Forensic Sciences and the School of Criminology at Simon Fraser University, Canada. Her research interests are the application of hyperspectral data/imagery, machine learning, and Bayesian networks for predictive models and the development of remote sensing techniques in the forensic and environmental sciences. She has experience with ecological, remote sensing, and spectrometry research in Costa Rica, Panama, Mexico, and Japan and has been involved in two airborne multi-/hyperspectral missions over Costa Rica (CARTA I and II).

Youngwook Kim, M.S., received his M.S. degree in urban planning from Seoul National University, Seoul, Korea. He is currently a Ph.D. candidate in the Department of Soil, Water, and Environmental Science at the University of Arizona. His research interests include hyperspectral remote sensing, its use in cross-sensor studies of vegetation time series, and the assessment of ecological parameters to environmental change through long-term remote sensing monitoring.

Richard Lucas, Ph.D., received his B.S. degree in biology and geography and a Ph.D. degree in the remote sensing of snow and vegetation using advanced, very high resolution radiometer (AVHRR) and Landsat sensor data from the University of Bristol, United Kingdom, in 1986 and 1989, respectively. He is currently a reader at the University of Wales, Aberystwyth (UWA), where he is involved primarily in the integration of single-date and time-series radar, optical (hyperspectral) and LiDAR data for retrieving the biomass, and structure and species/community composition of ecosystems ranging from subtropical woodlands to tropical forests and mangroves. His research focuses on the carbon dynamics of these ecosystems and their response to natural and human-induced change.

Anthea Mitchell, Ph.D., is a postdoctoral research fellow within the School of Biological, Earth and Environmental Sciences (BEES), University of New South Wales (UNSW). She obtained a B.S. in geography (Honors) in 1999 and a Ph.D. in the remote sensing of mangroves in 2004. Her current interests include hyperspectral remote sensing for crop and pasture characterization, habitat mapping and assessment, and the detection of vegetation stress and disturbance.

Tomoaki Miura, Ph.D., received his M.S. degree in resource management from the University of Nevada, Reno, in 1996 and Ph.D. degree in soil, water, and environmental science from the University of Arizona, Tucson, in 2000. He was involved in the LBA-Air-ECO project and is currently a science team member of the EOS-MODIS and EOS-ASTER instruments. His research interests include satellite data continuity, applications of vegetation indices for long-term trend analyses, and applications of remote sensing to natural resource and environmental management.

Onisimo Mutanga, Ph.D., is interested in the research areas of applied environmental GIS and remote sensing. He specializes in vegetation mapping and monitoring as well as wildlife habitat evaluation. His focus in recent years has been on the development of techniques for mapping tropical vegetation quality and quantity using hyperspectral remote sensing.

German Obando-Vargas, M.S., graduated from the School of Forest Engineering at the Technological Institute of Costa Rica and received his M.S. from the Tropical Agricultural Research and Higher Education Center (CATIE), where he specialized in natural resources management. He is currently director of the Department of Planning, Research and Development for FUNDECOR. His research interests are in the application of GIS to forestry and natural forest management. His main publications are "Trees of the Humid Tropics: Socioeconomic Importance," with Dr. Eugenia Flores, and "Electronic Use of Computers, Programs and Instruments in the Planning and Development of Management Plans of the Tropical Humid Forest," a case study for FAO. Presently, he is working with the government of

Costa Rica in the development of GIS tools and the use of GPS to control illegal logging in the country.

Mauricio Quesada, Ph.D., obtained his B.S. from the Department of Biology, University of Costa Rica, and M.S. and Ph.D. in ecology from Pennsylvania State University. Dr. Quesada has conducted research on pollination, plant reproductive systems, plant genetics, tree ecophysiology, and conservation biology. He has served as codirector of the Palo Verde Biological Station, Organization for Tropical Studies (OTS), Costa Rica, as an invited professor in the Department of Biology, University of Costa Rica, and as a research professor at the Chamela Biological Station, Institute of Biology and Ecology, Universidad Nacional Autónoma de México (UNAM). He is currently a research professor at the Centro de Investigaciones en Ecosistemas, UNAM. His current research interests include the effects of forest fragmentation on pollinators, plant reproductive systems, and the genetic structure of tropical trees of the dry forest. He is also investigating the effects of transgenic plants on their wild relatives. Finally, he is studying the relationship between plant functional groups in various stages of succession and remotely sensed data. He has over 50 publications in international peer-reviewed journals, book chapters, and an edited book.

Piyachat Ratana, Ph.D., received her M.S. degree in information technology for natural resources management, Bogor Agricultural University, Indonesia, and Ph.D. in soil, water, and environmental science at the University of Arizona in 2006. She is currently on the faculty in the Department of Geotechnology at Khon Kaen University, Thailand. Her research interests are in tropical phenology, climate–vegetation interactions, and drought detection in tropical environments.

Benoit Rivard, Ph.D., received his Ph.D. degree in Earth and planetary sciences from Washington University, St. Louis, Missouri, in 1990. He is currently a professor in the Department of Earth and Atmospheric Sciences of the University of Alberta. Dr. Rivard is a geologist with particular interest in the development of applied geological remote sensing. His key research preoccupation is to develop the analysis of hyperspectral sensing (field, airborne, and space borne) to improve the effectiveness of the oil/mining industry, and mapping agencies to delineate and manage their targeted resources. In this respect he has been working to (1) automate the hyperspectral analysis of rock cores and wall rock toward mineral mapping and rock type classification, (2) improve the analysis of hyperspectral imagery for northern regions that are remote and difficult to access, and (3) improve the analysis of hyperspectral data for boreal and tropical forests. Past research interests have included the use of radar remote sensing for lithologic and structural mapping and the development of methodologies for precise measurement of emissivity.

G. Arturo Sanchez-Azofeifa, Ph.D., conducts research related to the study of impacts of land use/cover change (LUCC) on biodiversity loss and habitat fragmentation in tropical dry forest environments. His research involves the study of theoretical linkages between remote sensing (multispectral and hyperspectral) and the spatial/temporal dynamics of leaf area index (LAI), primary productivity (PP), and photosynthetically active radiation (PAR). In addition, his research interests involve the development of techniques for the analysis and interpretation of the presence of non-self-supporting tropical systems (lianas) and tropical hardwood species (e.g., mahogany) at the leaf and canopy level. Dr. Sanchez-Azofeifa is one of the 2006 recipients of an Aldo Leopold Leadership Program fellowship.

Andrés A. Sanchún Hernández, B.S. Eng., graduated as a forest engineer from the Universidad Nacional, Costa Rica, and is currently pursuing a master's degree in natural resource management at UNED. Since 1998 he has worked in the planning, investigation, and development division of FUNDECOR (Foundation for the Development of the Central Volcanic Cordillera) as a researcher and parataxonomist and has participated in several research projects regarding the development of sustainable forest management. He is a coauthor of a digital dendrological key for tree species and a list of tree species and ecological guilds for the Central Volcanic Cordillera in Costa Rica.

Edson E. Sano, Ph.D., received his M.S. degree in remote sensing from the National Space Research Institute in Brazil in 1987 and his Ph.D. degree in soil, water, and environmental science from the University of Arizona, Tucson, in 1997. He was a science team member in the U.S.–Brazil joint projects in the LBA and SMEX03 programs. His research interests include land cover mapping over the Brazilian tropical savanna using optical and radar remotely sensed data and soil moisture estimation using both active and passive microwave sensors.

Yosio E. Shimabukuro is a researcher in vegetation and remote sensing at the Brazilian Institute for Space Research (INPE). His fields of interest include the use of satellite and ground-based remote sensing data for studying vegetation cover, and remote sensing and GIS techniques and models for environmental change detection over different biomes in Brazil.

Jolene E. Trujillo, B.S., completed a B.S. in biology at the University of New Mexico and is a former graduate of the Organization for Tropical Studies REU program in Costa Rica, where she investigated the impact of gender on the spectral properties of neotropical tree species. She is currently pursuing an M.S. in biology at Arizona State University, where she is investigating the impacts of ancient agriculture on modern landscapes. Her interests also include ecology with an emphasis on desert ecosystems and human impacts.

Orlando Vargas began working at the La Selva Biological Station in Costa Rica in 1982 and is currently the station naturalist and head of scientific operations. He is also part of the research team of the Digital Flora Project developing an Internet-accessible plant guide of the 1900 species found at La Selva. His former duties included working with the environmental education program and providing technical and botanical support at La Selva. In 1990 Orlando assisted the Biodiversity Institute (INBIO) on the *Flora of Costa Rica* manual, and worked as the naturalist at the Rara Avis Field Station. In 2002 he was a cofounder of the Reserva

Ecological Bijagual, a new conservation area in the Sarapiqui region of Costa Rica. Throughout his career he has collaborated on over 50 scientific papers, has participated directly in conservation projects in the Sarapiqui region of Costa Rica, and has served as a teaching assistant for various "Organization for Tropical Studies" courses and other workshops.

Hiroki Yoshioka, Ph.D., received his M.S. and Ph.D. degrees in nuclear engineering from the University of Arizona, Tucson, in 1993 and 1999, respectively. His research interests include cross-calibrations of satellite data products for continuity and compatibility of similar data sets and development of canopy radiative transfer models and their inversion techniques.

1 Tropical Dry Forest Phenology and Discrimination of Tropical Tree Species Using Hyperspectral Data

Karen L. Castro-Esau and Margaret Kalacska

CONTENTS

1.1 INTRODUCTION

Tropical tree species discrimination using hyperspectral data is an emerging field in which progress is closely linked to the availability and advancement of remote sensing technology. Initial interest in the optical properties of tropical plants was expressed in the latter half of the 1980s [1], and the potential for using these properties for tree classification was recognized at least a decade later [2,3]. While studies to that point focused on leaf-to-branch scale reflectance, the first attempt to classify tree crowns using hyperspectral imagery from airborne sensors was made only recently [4]. This small body of research to date has approached tree identification over multiple scales and has been largely exploratory in nature.

Understanding the facets of tree crown reflectance and the challenges involved in tree classification based on spectral reflectance begins at the leaf level. Incident light on a leaf is destined for one of three main pathways: reflectance, transmittance, or absorptance. A portion of the incident beam will undergo specular reflectance at the leaf's surface, or cuticle. Another portion will enter the leaf and be influenced by Beer's law of radiative transfer—its progress slowed according to the absorption coefficients of the cellular matter it encounters. At cellular interfaces, such as the boundary between cell walls and air spaces, rays may be redirected numerous times, governed by Snell's law of refraction. Red light intercepted by chlorophyll along the way is used to initiate the electron excitation process that drives photosynthesis. At slightly longer wavelengths (700–1300 nm), light is of little use to a leaf and is either reflected back out or transmitted on through the leaf, where it may engage with leaves lower in the canopy and contribute to leaf additive reflectance [5]. Still longer wavelengths (1300–2500 nm) are largely absorbed by water, leading to dominant absorption bands at 1400 and 1900 nm. Overall, spectral reflectance of leaves has been well studied in reviews [6–9].

Light interaction with a leaf is a complex phenomenon, and scaling out to branches and eventually tree crowns multiplies this complexity. The simpler issue of light travel through a single leaf becomes compounded by potential encounters with a multitude of leaves and additional vegetative (flowers, fruits, branches, stems, trunk, epiphytes, lianas, understory plants, and litter) and nonvegetative (soil) forest components. The expression of foliar chemistry in canopies is largely influenced by the amount of leafy material, measured as leaf area index (LAI), and the orientation of the leafy material, measured as leaf angle distribution (LAD) [10], which have bearing on the shape and amplitude, respectively, of the cumulative return. Whereas LAI and LAD are the dominant controls of canopy reflectance in full canopies, soil and understory vegetation have a major impact on reflectance of sparse canopies [10].

In tropical dry forests, foliar chemistry, LAI, and LAD can vary significantly— the former two seasonally and the latter diurnally. These variations are more distinct in partially to fully deciduous tropical dry forests than in evergreen wet forests that are by their very nature evergreen. Seasonal cycles in tropical dry forests that lead to phenomena such as leaf drop, leaf flush, and flowering events should provide unique opportunities for remote sensing of tree species or functional types. The objectives of this chapter, therefore, are to (1) review the current state of remote sensing for species identification in tropical forests in general, (2) discuss potential unique opportunities for tree species discrimination in tropical dry forests, and (3) consider the challenges and future directions of species identification using remotely sensed data in tropical forests.

1.2   LITERATURE REVIEW

1.2.1   LEAF OPTICAL PROPERTIES OF TROPICAL FOREST PLANTS: LOOKING FOR DIFFERENCES

Although leaf optical properties have traditionally been measured as diffuse reflectance and diffuse transmittance using a spectrometer with an attached integrating sphere, it is becoming increasingly commonplace to solely measure leaf bidirectional reflectance using a spectrometer. While the former permits the computation of absorptance (1-reflectance-transmittance), a measure of the radiation used by a leaf in photosynthesis, the latter is similar to the bidirectional reflectance recorded by an electro-optical airborne or satellite-borne sensor. Leaf spectral reflectance measurements are easily obtained in the field using a portable spectrometer, or in the lab using detached leaves, provided that the leaves are kept fresh [11].

By the very nature of the scientific method, hypothesis testing focuses on finding differences within or between groups. Research on the leaf optical properties of tropical plants is no different in this respect and has focused on finding differences within species, between species, or between groups of species, such as between life forms or between sun- and shade-adapted plants. Intraspecific differences in leaf optical properties are recognized to complicate the matter of finding interspecific or intergroup differences [12]. An example of variability in leaf optical properties within species occurs as leaves age. Roberts et al. observed typical changes in spectral reflectance of new leaves of an Amazon caatinga species, including decreased visible reflectance and increased NIR over the first 6 months, which relate to increasing chlorophyll content and the development of the mesophyll, respectively [13]. As leaves aged further, colonization by epiphylls (fungi, leafy liverworts, lichens, algae, bacteria) caused further spectral changes in the leaves. Older epiphyll-laden leaves reflected and transmitted less light and absorbed more, especially in the near-infrared region. Intraspecific differences in leaf spectra of tropical forest species—in this case, across multiple sites and multiple seasons—were also found by Castro-Esau et al. [12]. Classification across either boundary was considered futile due to the high degree of variability within species. Within these boundaries (i.e., spectra gathered from a single site within a single season), however, classification of tropical plant spectra was much more promising. Indeed, efforts at classifying tropical trees based on their leaf spectra have been largely successful, with overall accuracies consistently exceeding 85% for up to 27 species [3,4,12,14].

While interspecific differences in leaf optics can be used to discriminate between species, differences in leaf optics between groups of species can provide insight into the ecology and physiology of those groups. At Parque Natural Metropolitano, Panama, differences in leaf spectral reflectance between two life forms, lianas and trees, were found to be related to differences in their chlorophyll contents and potentially linked to different strategies of phenology and coping with drought stress [15,16]. These two groups could be readily classified using hyperspectral leaf reflectance data during the dry season. The same could not be said for lianas and trees at a tropical wet forest site, where the two life forms could not be reliably distinguished and did not manifest differences in chlorophyll content. In an earlier study [17], another difference between the leaf optical properties of lianas and trees at Parque

Natural Metropolitano was observed—that is, that lianas consistently showed lower leaf transmittance than trees. Higher transmittance in trees may be an adaptive strategy to allow a greater proportion of light to reach lower layers of leaves, whereas lianas tend to be monolayered.

Also comparing leaf optical properties of two species groups, Lee and Graham [1] and Lee et al. [18] compared tropical forest sun and shade plants. They found their leaf optical properties to be highly similar, despite distinct anatomical differences between the two groups, including, for shade-adapted taxa, lower specific weights and equidiameter palisade cells (vs. columnar for sun-adapted taxa). Similarly, within species, Poorter, Oberbauer, and Clark [19] did not observe significant differences among reflectance values (400–700 nm) of understory, mid-canopy, and canopy leaves for four tropical rain forest trees in Costa Rica, although for two species small but significant differences in absorptance (400–700 nm) were observed. Leaf optical properties were also found to be similar among sun leaves of pioneer and climax tree species in the same study.

The few studies mentioned here represent the bulk of research done to date on leaf optical properties of tropical plants. They also lead to additional questions. For instance, why do leaf optical properties of sun and shade leaves not differ, considering the magnitude of differences in their structural attributes? The results of studies by Lee and Graham [1], Lee et al. [18], and Poorter et al. [19] are surprising. Although, in each case, the entire photosynthetic range (400–700 nm) was considered as a whole, a reexamination of this issue may be warranted to determine if there are significant differences between light acclimation groups at individual narrow-width wavebands, both in the photosynthetic range as well as the near-infrared range. Differences in near-infrared reflectance could be expected due to differences in the arrangement, shape, and number of layers of mesophyll cells in shade-adapted plants as compared to sun-adapted plants.

Questions also remain surrounding the physiological differences between lianas and trees in tropical dry forests that lead to differences in their leaf spectral reflectance, as well as the degree of intragroup spectral variation that could potentially prevent the discrimination of the two life forms at the canopy level. Furthermore, epiphytes such as bromeliads are another tropical life form whose optical properties remain completely unstudied, although limited research has focused on epiphytic lichens in tropical forests. Additional studies that explore the variability in leaf spectral reflectance of tropical plant species due to factors such as leaf maturation and senescence, leaf angle, herbivory, and mineral nutrition, as well as differences among species, functional groups, or life forms, will be valuable not only for better understanding their leaf optical properties, but also for the goal of interpreting canopy-level spectra.

1.2.2 TROPICAL TREE SPECIES DISCRIMINATION AT THE CANOPY LEVEL

1.2.2.1 Aerial Photography

Identification of tropical tree species using data gathered with an airborne hyperspectral sensor is a relatively new endeavor. Earlier attempts were made to visually recognize tropical trees from aerial photographs. Some of the most primitive involved medium- to small-scale black and white prints, from which only a very few species could be identified (Nyyssonen [20], cited in Myers and Benson [21]). Later attempts have met with greater, albeit still limited, success using large-scale color aerial photographs [21–23]. These studies rely on the development of crown typology schemes to standardize crown descriptions and to serve as an aid for classifying the tree crowns. Features included in crown typologies include crown size and dominance, crown shape and structure, crown contour, crown texture, branching habit, branch visibility, foliage density, presence or absence of fruits or flowers, and color [22–24].

By carefully examining crown typological features in aerial photographs, interpreters can classify a portion of the total number of tree crowns. This process, done manually, is not entirely unlike the automated process of feature selection and supervised classification using modern image-processing software. In a tropical rain forest in Queensland, Australia, 24 of 111 tree species were identified with greater than 75% accuracy using large-scale aerial photographs [21]. In French Guiana, 12 tree categories, mainly of commercial interest, were identified using stereoscopic analysis of photographic prints over an area of 5 ha, although the success rate was not quantified [23].

Lack of a greater body of research on tropical tree identification from aerial photographs could be due to a number of factors, including the high level of expertise required by interpreters, in terms of the knowledge of features and habits of a large number of tree species; the laborious nature of visual crown identification; and limited success rates. Photo interpretation is described as an art as well as a science [25]. The human eye is effective at identifying distinct patterns and shapes (e.g., palm trees, conifers) and at demarcating individual trees, especially if they are well spaced [25,26]. The eye is less effective at distinguishing between color tones or infrared wavelengths [26].

1.2.2.2 Mapping of Tropical Mangroves

Mapping of tropical mangrove species has been attempted with a wide range of sensor capabilities, including, in recent years, hyperspectral remote sensors. This range includes both moderate spatial resolution multispectral sensors (e.g., Landsat Thematic Mapper (TM), 30 m, in Rasolofoharinoro et al. [27]) and high spatial resolution multispectral sensors (e.g., Quickbird, 2.8 m, in Wang et al. [28]. In terms of spectral resolution, it also includes broadband sensors (e.g., Landsat TM, in Rasolofoharinoro et al. [27]; IKONOS; and Quickbird, in Wang et al. [28]—all less than or equal to seven bands); and narrowband sensors (e.g., Hyperion, in Demuro and Chisholm [29], cited in Vaiphasa et al. [30], with 220 bands; and airborne visible/infrared imaging spectrometer (AVIRIS), in Hirano, Madden, and Welch [31], with 224 bands). In addition to studies involving airborne and satellite-borne electro-optical sensors, synthetic aperture radar has also been employed in mapping of tropical mangroves (e.g., Proisy et al. [32] and Held et al. [33]). Green et al. [34] summarized mangrove mapping studies prior to 1998. These studies were largely based on aerial photos, Landsat and SPOT (Satellite Pour l'Observation de la Terre) technology, or radar (ERS-1 [European Remote Sensing Satellite], JERS-1 [Japanese Earth Resources Satellite]), and involved the labeling of few mangrove classes (mostly 2–4, none > 10).

Compared to other tropical forests, mangroves are recognized as "desirable test beds" for remote sensing research, due to variable composition in terms of species, structure, and canopy closure over relatively short distances, while exhibiting relatively low species diversity [33]. In an exploratory study to determine the feasibility of discriminating mangroves to the species level using hyperspectral data of leaves under laboratory conditions, Vaiphasa et al. [30] found that 16 species of Thai mangroves were statistically different at most spectral locations. The result held promise for further investigations using onboard hyperspectral sensors. However, spectral similarities among members of the Rhizophoraceae family (six species) were noted as a limitation. In addition to mangrove mapping studies using moderate spatial resolution hyperspectral data (e.g., AVIRIS, 20 m, 224 bands in Hirano et al. [31]; and Hyperion, 30-m, 220 bands in Demuro and Chisholm [29], cited in Vaiphasa et al. [30]). The studies that most closely approach the use of high spatial resolution hyperspectral data (known to the authors) are those by Green et al. [34] and Held et al. [33], both using compact airborne spectrographic imager (CASI) imagery 1-m and 8 bands; 2.5 m and 14 bands, respectively, for the two studies.

However, even the CASI sensor, which offers programmable bands, is restrictive in the amount of spectral data it provides relative to a hyperspectral sensor that covers the full visible and near-infrared ranges. Green et al. [34] demonstrated the superiority of CASI over Landsat TM for discriminating between mangrove and nonmangrove vegetation as well as for discriminating more mangrove classes (nine) from the Turks and Caicos Islands. Five of these classes were species- or genus-level classes (*Conocarpus erectus*, *Avicennia germinans*, short and tall *Rhizophora*, *Salicornia* spp.); others included short and tall mixed mangrove habitats, sand, and mud crust. Held et al. [33] explored the complexities involved in integrating CASI and airborne synthetic aperture radar (AIRSAR) (C, L, and P bands) data for classifying 10 mangrove classes in Queensland, Australia. Classification based on combined bands from the two sensors proved more accurate than with either alone; the CASI data provided information on spectral quality of mangrove classes and the AIRSAR data offered information on structure and texture. Both studies resulted in classification accuracies ranging between 80 and 85%.

1.2.2.3 Mapping Species-Rich Tropical Forests

Unlike the growing body of research for mapping tropical mangrove species, research on tree identification in more species-rich tropical forests is extremely limited. The complexity of this task is obvious considering that tree species numbers in tropical forests can exceed 300 per hectare [35]. Coupled with high species diversity is an overall lack of airborne hyperspectral imagery over tropical forests to date, which is evident in the fact that the two most recent papers published on this topic use the same data set. Although not yet achieved, automated tropical tree recognition, unlike visual interpretation of large-scale aerial photographs, is a highly desirable goal that would facilitate ecological studies of tropical forests, including demographic studies, ecosystem characterization, mapping of endangered or endemic species, quantifying carbon pools and sequestration rates, and identifying important food sources for wildlife [4,12].

Research by Clark, Roberts, and Clark [4] and Zhang et al. [36] represents the pioneering work on tropical tree species recognition using hyperspectral data thus far.

Both studies focus on airborne hyperspectral digital imagery collection experiment (HYDICE) hyperspectral images acquired on March 30, 1998, over tropical wet forest old growth and surrounding anthropogenic land cover classes at La Selva, Costa Rica. The tree species of interest, as well as the data analysis approaches, differ between the two studies. Clark et al. focused on seven species of canopy emergents, which were well represented in the imagery (*Balizia elegans, Ceiba pentandra, Dipteryx panamensis, Hyeronima alchorneoides, Hymenolobium mesoamericanum, Lecythis ampla,* and *Terminalia oblonga*). These species are a combination of deciduous and evergreen trees. Crowns were considered to have either low or high leaf cover at the time of image acquisition. Only one species, *T. oblonga,* overlaps with the study by Zhang et al. Four other species were considered by this second research group, chosen based on large size, apparent lack of lianas and epiphytes, and their unambiguous location in the images. The species were *Castilla elastica, Ficus insipida, Luehea seemannii,* and *Zygia longifolia* (fig. 1.1).

In the case of Clark et al. [4], "crown-scale" and "object-based" classification approaches were explored. In the first, a class was assigned based on the average of all crown or sunlit-only crown pixel spectra from each individual tree crown. The object-based classification involved assigning a class based on the majority value of the classified pixels. Using linear discriminant analysis and 30 selected bands, 92% accuracy was achieved using the crown-scale classification, in comparison to 86% for the object-based classification.

The study by Zhang et al. [36] differed from that of Clark et al. [4] in that its primary focus was to compare derivative and wavelet analysis for tree species discrimination. A secondary aim was to quantify intra- and interspecies spectral variability in the tree crown spectra. The capacity for wavelet-transformed spectra for reducing within-crown and within-species variation and capturing between-species variations was demonstrated over that of the derivative-transformed spectra. Key to the studies by Zhang et al. and Clark et al. was the ability to capture the relevant information for species discrimination from the larger pool of data that comprises the hyperspectral image—in the first study, through feature selection, and in the second, in recognizing the value of specific wavelet coefficients and the wavelet energy feature.

Other initiatives that involve mapping tropical forest species with airborne hyperspectral (or quasi-hyperspectral) imagery include the classification of *Eucalyptus* forests in Queensland, Australia, and the discrimination of lianas and trees in Panama. In the first case, Goodwin et al. [8a] (2005), using CASI-2 data (80 cm, 10 spectral bands), were unable to discriminate between *Eucalyptus* species, although they were successful at classifying an amalgamated *Eucalyptus* class, *Syncarpia glomulifera,* mesic vegetation, soil, and shadow, with 95% overall accuracy. From this study, it is clear that while species-level classifications may not always be possible, there may still be opportunities to map species to genus or family level. While 10 bands is much fewer than hyperspectral sensors provide, this study is included here because the sensor collects a larger number of narrow-width wavebands than traditional multispectral sensors (≤7 bands) do.

In the second case, Sanchez-Azofeifa and Castro-Esau [37] and Kalacska et al. [16] are exploring the potential for mapping the presence or absence of liana communities in a tropical dry forest of Panama. Compared to tree leaves, higher reflectance

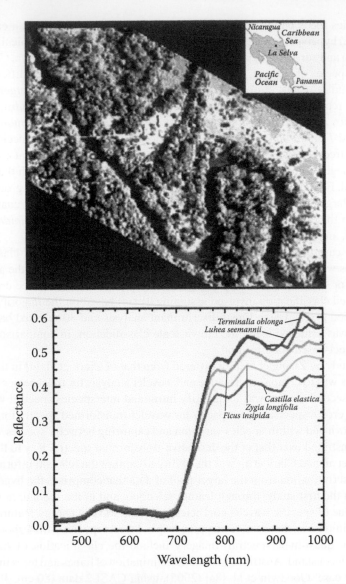

FIG. 1.1 Color composite of HYDICE bands (red = 641 nm, green = 552 nm, blue = 460 nm) showing the location of the 17 tree crowns sampled. Crown colors are for the following species: green = *Ficus insipida*, yellow = *Zygia longifolia*, red = *Castilla elastica*, blue = *Luehea seemannii*, magenta = *Terminalia oblonga*. (Zhang, J. et al., *Remote Sensing of Environment*, 105, 129, 2006. Reproduced with permission.) See CD for color image.

of liana leaves at 550 nm is an effect that appears to be transferred to the crown level. This effect has been observed from in situ crown spectra, gathered from a canopy crane, as well as from HYDICE images of the area [15,16,37]. Additional research over a greater variety of host and nonhost trees as well as over different life zones will further elucidate the potential for mapping the extent and spread of lianas in tropical forests.

1.2.2.4   Research on Tree Crown Delineation Using Multispectral Imagery and LiDAR

As part of this discussion on tropical tree species identification from hyperspectral imagery, the contribution of high spatial resolution (<5 m) multispectral imagery and, more recently, LiDAR (light detection and ranging) for tree crown delineation should not be ignored, although neither has been used extensively in the tropics. Satellite-borne multispectral sensors such as IKONOS (4-m multispectral, 1-m panchromatic), Quickbird (2.8-m multispectral, 0.7-m panchromatic), and Orbview-3 (4-m multispectral, 1-m panchromatic) have yet to be equaled by any satellite-borne hyperspectral sensor in terms of spatial resolution. To date, similar spatial resolutions can only be matched by airborne hyperspectral sensors flown at low altitudes, such as CASI (variable number of programmable bands; see, for example, Bunting and Lucas [38]), HYDICE (210 bands; see, for example, Clark et al. [4] and Zhang et al. [36]), AVIRIS (224 bands; see, for example, Xiao, Ustin, and McPherson [39]), Probe (128 bands), and HyMap (100–200 bands). LiDAR, in turn, is not a traditional passive electro-optical sensor at all, but rather an active remote sensor that sends out laser pulses and measures return time. In addition to determining tree locations and height, structural features that can also be resolved using LiDAR include canopy height, canopy topography, forest volume, aboveground biomass, basal area, and number of forest strata, as well as forest successional status [26,40].

Although multispectral sensors with high spatial resolution may have limited capacity to identify tropical tree species, due to a lack of fine spectral detail, it is evident that some proportion of canopy and emergent trees can be located from these images. A significant amount of research has been devoted to tree crown delineation in temperate forests. Numerous approaches have been developed, which are categorized by Bunting and Lucas [38] as "those that either detect crown centroids and boundaries, follow valleys (also referred to as contouring), or match templates" [38,41–44]. These methods are most successful in simple forests, such as orchards and conifer forests, where crowns are symmetrical, and less so in hardwood forests, which have a more variable geometry [25,38]. Circumscribing crowns is often viewed as a first step in mapping floristic composition, after which spectral information can be used to identify species [25,26].

With the exception of Bunting and Lucas [38], who developed a tree crown delineation algorithm for Australian mixed-species forests in Queensland, the potential for applying crown delineation algorithms to images of more species-rich tropical forests has not been tested to date. In a tropical moist forest in Amazonas, Brazil, however, Read et al. [45] recognized many individual trees from merged 1-m panchromatic and 4-m multispectral IKONOS imagery of logged and old-growth forest in the Brazilian Amazon. They demonstrated the potential of this type of imagery for studies of tree demography, forest gap formation and recovery, and logging assessment and monitoring. They also indicated that species-level research may be possible for crowns with distinctive spectral reflectance characteristics and/or architectural attributes [45]. However, locating the corresponding trees at ground level was a significant challenge without a unique spatial context and/or highly accurate global positioning satellite (GPS) positions.

Similarly, crown detection and delineation algorithms for LiDAR data are in development [38,46] but have not yet been applied to tree species identification in tropical forests. LiDAR location and structural data offer new opportunities to enhance species classification and the overall study of tropical floristic composition. Gillespie, Brock, and Wright [26] suggest a three-step sequence to use LiDAR and spectral sensors to their greatest advantage in mapping floristic composition in tropical forests. First, LiDAR data are used to locate tree crown centers; second, spectral signatures from these centers are compared with those in a spectral library; and, third, additional LiDAR (e.g., crown height, crown diameter) and spectral (e.g., biochemistry indices, phenology) information is used to improve the species classifications. Clark, Clark, and Roberts [47] explored the use of small-footprint LiDAR for estimating tree height and subcanopy elevation in a tropical wet forest and surrounding secondary forests, plantations, pastures, and swamps at La Selva, Costa Rica. In a later study, Clark et al. [4] used information from a LiDAR digital canopy model to co-locate crown centers in HYDICE images. The crowns were then classified to species using the hyperspectral data, demonstrating the complementary nature of the two data types. Further exploration of the potential of LiDAR data usage in conjunction with hyperspectral data for tree identification is warranted.

1.3 TROPICAL DRY FOREST PHENOLOGICAL EVENTS AS REMOTE SENSING OPPORTUNITIES FOR SPECIES DISCRIMINATION

1.3.1 TROPICAL DRY FOREST CHARACTERISTICS

Within the range of climatic descriptors that delimit the tropical dry forest life zone (mean annual biotemperature higher than 17°C, mean annual precipitation of 250–2000 mm) [35,48], in which rainfall alone can vary eightfold, considerable latitude can be observed in terms of species richness, composition, structure, and productivity. From one dry forest to the next, variation is also evident in the expression of phenological events. Despite the large gradient in tropical dry forest characteristics, this life zone is noted for its pronounced rainfall seasonality, a factor that distinguishes it from either tropical moist or wet forest. It is this rainfall seasonality that is associated with marked phenological events that should favor species discrimination in these forests. The focus of this section, therefore, is to describe the tropical dry forest ecosystem, with particular attention to these phenological events, and to consider how these events may be detected by electro-optical sensors.

Tropical dry forest is described as seasonally semideciduous forest [35]. Compared with tropical moist or wet forests, tropical dry forests not only have a higher proportion of dry-season deciduous trees, but they are shorter and have smaller, more open crowns, which rarely make lateral contact [35]. It follows that leaf area index and biomass are lower in tropical dry forests. Tree density and species density are also lower [49]. On the other hand, tropical dry forests tend to have high structural and physiological diversity, and high floristic endemism [50]. The distribution of rare species is more often clumped than that of common species [51]. With respect to its physical structure, canopy trees generally reach 20–25 m; smaller trees, 10–20 m; and the shrubby layer beneath, 2–3 m. Species with compound leaves are commonly found in the canopy. The shrub layer beneath is typically dense and thorny, and often characterized by many-stemmed plants [35].

1.3.1.1   Seasonality of Precipitation

In a tropical dry forest, both the annual quantity of precipitation, mentioned before, and the seasonal pattern of precipitation affect ecosystem functioning. Holdridge et al. [35] observed that life zone vegetation boundaries corresponded with total amounts of precipitation—more so than its seasonal distribution—and gave the former priority as an ecological factor in the life zone system. In the interest of observing phenological events, however, it is the periodicity of annual rainfall that is paramount. Its importance with respect to the ecology of tropical dry forests is stressed by Murphy and Lugo [49]: "Rainfall seasonality becomes a dominant ecological force when temporal patterns of biological activity such as growth or reproduction become synchronized with the availability of water or when the geographic distributions of plant or animal taxa are constrained by moisture limitations during certain times of the year." A trend of increasing dry season length and severity with increasing latitudinal distance from the equator has been noted. Two dry periods (one short, one long) or one extended dry period also tended to occur with increasing distance from the equator (Richards [52], cited in [49]).

1.3.1.2   Functional Groups of Trees

Within the phenological complexity of a tropical dry forest is a near continuum in the expression of deciduousness among tree types (fig. 1.2). These functional groups of trees vary in their adaptive strategies to drought, such as in rooting depth and the maintenance of stem water potential, which in turn influence their response to internal environmental cues that stimulate phenological processes such as leaf shedding and flushing [53]. Trees described as deciduous or stem succulent trees tend to be leafless for several months of the year, while brevideciduous trees may be leafless for only a few days or weeks. Leaf drop is rapid and leaf flush is highly synchronous in conspecific trees of stem succulent trees, more so than in other functional types [53]. Leaf phenological events are not as conspicuous in leaf exchanging or evergreen species within tropical dry forests. In leaf-exchanging species, abscission of old leaves and emergence of new leaves are highly synchronized, such that the latter occurs immediately after the former [53]. The two processes tend to overlap for evergreen species.

A gradient of deciduousness can be observed not only among functional tree types (fig. 1.2), but also along the precipitation gradient that describes increasingly dry forests. In Panama, for instance, the percentages of species that exhibited deciduousness in wet (2830 mm rainfall/year), medium (2570 mm rainfall/year), and dry (2060 mm rainfall/year) forests were 14, 28, and 41%, respectively [54]. In another study, 75% of species were partially or completely deciduous in a tropical dry forest in Guanacaste, Costa Rica (1533 mm rainfall/year) [55].

The stimuli for phenological events in tropical dry forest trees, and for the specific timing of those events, include tree water status [56], day length [53], internal growth factors, or a combination of both biotic and abiotic factors [49]. Furthermore, the tree functional types may respond differently to those stimuli. Leaf fall is often

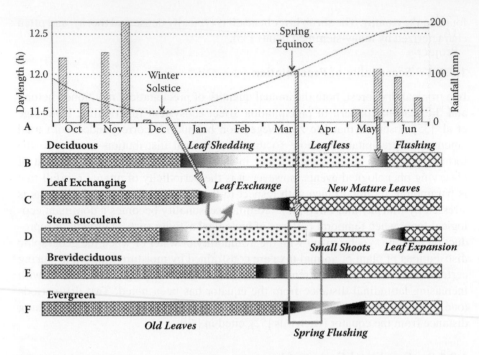

FIG. 1.2 Seasonal variation in environmental factors (A: rainfall, day length) for Guanacaste, Costa Rica (10°30′ N) and vegetative phenology in tropical dry forest trees of various functional types (B–F) in tropical semideciduous forests. Arrows indicate causes of bud break during the dry season. B: the first heavy rains of the rainy season; C: drought-induced leaf shedding; D–F: increasing day length around the spring equinox. (Rivera, G. et al., *Trees— Structure and Function*, 16, 445, 2002. Reproduced with permission.)

hastened by dry weather and delayed by wet weather. In a study by Borchert [57], it was found that seasonal variability in leaf fall was less for deep-rooted species that were able to tap lower soil water resources. Leaf emergence, oddly enough, often tends to precede the rainy season [49,56]. Rivera et al. [53] observed that, for tropical forests with a 4- to 6-month dry season, spring flushing coincided with an increased photoperiod of 30 minutes or less, weeks prior to the onset of the rainy season. Borchert [56] and Reich and Borchert [58] also found a strong correlation between phenological events and tree water status, the latter of which was determined by rainfall and soil moisture availability.

1.3.2 REMOTE SENSING OF TROPICAL DRY FORESTS

From a remote sensing perspective, observable, predictable phenological events are the most desirable for tree identification studies. Ideally, the events should be temporally staggered among species, yet exhibit a high degree of synchronicity within species. This is not always the case. In fact, Condit et al. [54] noted that for nonevergreen species in wet, medium, and dry forests in Panama, individuals of a species were often deciduous at different times and to different degrees. Recognizing the potential for detecting predictable phenological events at a site will be

TABLE 1.1

Projected Seasonal Variation in Crown Spectral Reflectance of Tropical Dry Forest Trees of Different Functional Groups

Relative crown reflectance	Tree functional group[a]				
	Deciduous	Leaf exchanging	Stem succulent	Brevideciduous	Evergreen
Red[b]					
Rainy season	Low	Low	Low	Low	Low
Winter solstice	Med[c]	Low–med	Med–high	Low	Low
Dry season	High	Low–med	High	Low–med	Low
Spring equinox	High	Low	Med	High	Med
Near infrared[d]					
Rainy season	High	High	High	High	High
Winter solstice	Med–high	Med–high	Low	High	High
Dry season	Low	Med	Low	Low–med	High
Spring equinox	Low	High	Med	Low	Med
Shortwave infrared[e]					
Rainy season	Low	Low	Low	Low	Low
Winter solstice	Low–med	Low	Low–med	Low	Low
Dry season	High	Med	High	Med	Low
Spring equinox	High	Low	Med	High	Low–med

[a] See functional groups given in figure 1.2 (after Rivera et al. [53]); rainy season ca. June–October; winter solstice ca. December; dry season ca. February; spring equinox ca. March.
[b] ca. 630–690 nm.
[c] Med = medium.
[d] ca. 760–900 nm.
[e] ca. 1550–1750 nm; 2080–2350 nm.

facilitated by knowledge of the ecology of the site and, in the larger picture, the ecology of tropical dry forests over the broad range of climatic conditions that define this life zone. The literature does provide numerous examples of synchronous phenological events that can offer windows of opportunity for tree species or functional group discrimination. Frankie, Baker, and Opler [55], for example, provide a list of phenological records for 113 tree species at a tropical dry forest at Comelco, Guanacaste, Costa Rica. As another example, Borchert [56,59] accumulated similar data for 37 species at a second site: Hacienda La Pacifica, also in Guanacaste, Costa Rica. The same author also demonstrated the usefulness of herbarium specimens for studying the phenology and flowering periodicity of tropical dry forest species [60]. Rivera et al. [53] charted out the vegetative phenology for approximately 50 tropical dry forest tree species spanning five countries and both hemispheres. These records, which provide the months of leaf drop/leaflessness, leaf flushing and/or flowering, and mature fruit periods for these species, can provide an important resource for remote sensing studies at these sites (table 1.1).

In this section, while it is recognized that operational, satellite-borne high spatial resolution hyperspectral imagery is not yet available, opportunities for detecting phenological events as an aid to species discrimination are described with the assumption that, with continued advances in technology, such technology will be available in the future.

1.3.2.1   Leaf Flush

Many tropical dry forest species flush their leaves in the late dry season to early wet season. Frankie et al. [55], for example, noted a large flushing period between late April and June, with the peak in May, at a site in Guanacaste, Costa Rica (fig. 1.3). Different functional groups vary widely in the timing and degree of synchronicity of this event [53] (fig. 1.2). If the chronological sequence of flushers were consistent from year to year, comparisons of images taken during the dry season and at early and mid-flush would be useful in delimiting the evergreen species as well as the early and late flushers. Indeed, many species are consistent in the timing of their flushing events. For over 50 tropical dry forest tree species from Argentina, Costa Rica, Java, Thailand, and tropical savannas of Brazil, intraspecific bud break was remarkably synchronous and interannual variation in bud break was remarkably low, suggesting photoperiod control of these events [53]. There may also be opportunities to detect species with unusual flushing behavior. For example, Frankie et al. [55] noted that a number of the riparian species at the tropical dry forest site flushed much earlier (January–March) than the majority of the tree species. Another species, *Lysiloma seemannii*, was the only one of 113 species to flush 1 month into the rainy season. In a study by Devall, Parresol, and Wright [61] in central Panama, *Cordia alliodora* was distinct from other species in remaining leafless for the first 2 months of the rainy season.

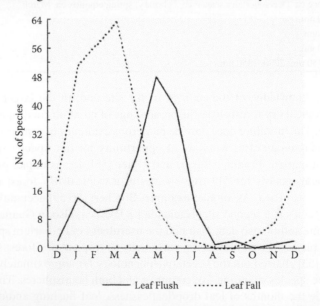

FIG. 1.3   Leaf fall and leaf flushing periodicities of dry forest tree species from Guanacaste, Costa Rica. (Frankie, G.W. et al., *Journal of Ecology*, 62, 881, 1974. Reproduced with permission.)

1.3.2.2   Leaf Abscission

The observation of periodic and synchronous leaf drop events follows the same logic as that for leaf flushing events. It may be possible to distinguish the early dry season deciduous species from the late leaf dry season deciduous, after which only evergreen species remain in leaf. Differences between species in time of leaf drop may afford additional opportunities to distinguish between species that flushed their leaves at the same time. For some species, however, immediate or nearly immediate flushing of leaves after leaf drop by evergreen, leaf-exchanging, and brevideciduous trees may reduce or eliminate the opportunities to observe such phenological events (e.g., *Cassia grandis* in Rivera et al. [53]; fig. 1.2). As well, synchronicity within species and between years is notably less for leaf abscission than for bud break, with some exceptions [53].

1.3.2.3   Flowering

Synchronous intraspecific flowering events may also provide important windows for discrimination of canopy species when individuals have large numbers of non-green flowers that render a unique spectral signature (fig. 1.4). In tropical dry forests,

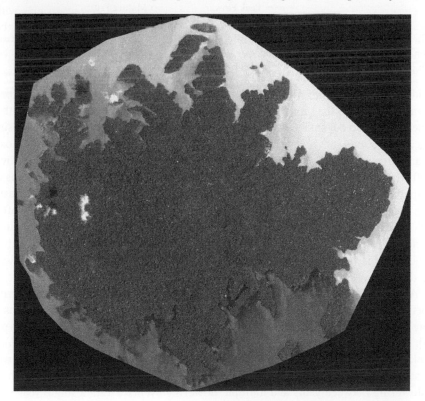

FIG. 1.4   Quickbird image (2.8 m) of Barro Colorado Island, Panama, showing flowering *Tabebuia* spp. (mostly *Tabebuia ochracea*) trees. True color composite with bands 3:2:1 displayed in red–green–blue. See CD for color image.

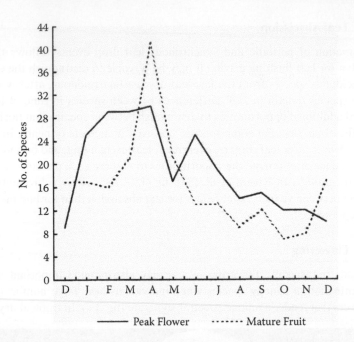

FIG. 1.5 Flowering periodicity and fruiting periodicities of dry-forest tree species from Guanacaste, Costa Rica. (Frankie, G.W. et al., *Journal of Ecology*, 62, 881, 1974. Reproduced with permission.)

deciduous trees often flower when the tree is leafless or as leaves begin to emerge, preceding the onset of the rainy season [49,55] (fig. 1.5). Synchronous flowering within a species can be advantageous in terms of increasing the potential for outcrossing [62]. *Pachira quinata* (Bombacaceae) trees flower in synchrony in Guanacaste, Costa Rica, with a peak flowering time in early March [62]. These trees produce well over 1500 white flowers at this time, with larger numbers of flowers on isolated trees as compared to those in continuous stands [62]. Tree species at Comelco, Guanacaste, were considered to be either "dry-season-flowering" species, which were often leafless and produced "conspicuous masses of flowers," or "wet-season-flowering" species, whose flowering events were less conspicuous due to the more closed nature of the fully leaved canopy, as well as factors such as smaller flower size, paler colors, and the relative rarity of the species [55].

Among both the dry- and wet-season flowering types, species ranged from having short flowering periods (less than a week) to long flowering periods (e.g., 2 months or more). Mean flowering time was approximately 5½ weeks for both flowering types [55]. In the same study, only two species (*Sterculia apetala* and *Crescentia alata*) displayed unsynchronized flowering among individuals of the same species during the study period [55]. Other minor flowering patterns included differences in timing of flowering between years and multiple flowering events within a single year [55]. It should be noted here that, although likely less so, conspicuous fruiting events could also be of interest from a remote sensing perspective, offering similar opportunities to the flowering events (fig. 1.5).

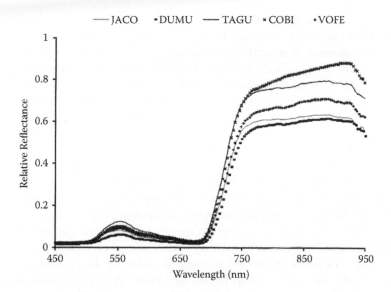

——JACO   •DUMU   —— TAGU   × COBI   •VOFE

FIG. 1.6   Average crown reflectance spectra (field of view ca. 3.6 m) for five species at Fort Sherman, Panama. Species codes: JACO = *Jacaranda copaia*, DUMU = *Dussia mundu*, TAGU = *Tapirira guianensis*, COBI = *Cordia bicolor*, and VOFE = *Vochysia ferruginea*. (Castro-Esau, K. et al., *American Journal of Botany*, 93, 517, 2006. Reproduced with permission.)

1.3.2.4   Single-Date Imagery

There may be little advantage from single-date imagery for tropical tree identification in dry forests as compared to wet forests, with the exceptions of generally lower species diversity (i.e., fewer species to consider) and less continuous canopy, which would facilitate the delimitation of single crowns. Rather, the opportunities may be similar to those observed by Clark et al. [4] and Zhang et al. [36] in a tropical wet forest in Costa Rica, which favor some of the more common emergent species. Timely single-date images may also be important for capturing important phenological events in tropical dry forests, such as the massive flowering events of *Tabebuia* spp. in Central America (fig. 1.4). Species with unique leaf spectral signatures should be detectable as well, such as those with relatively high leaf anthocyanin content or those with relatively high or low leaf area index, exhibited as correspondingly high or low near-infrared reflectance. Single-date crown spectra of five species from Fort Sherman, Panama (a tropical wet forest), for example, show differences in spectral reflectance in both the visible and near-infrared regions (fig. 1.6).

1.3.2.5   Multitemporal Imagery

In terms of tree species discrimination, the advantages afforded by tropical dry forest phenology over the more constant, leaf-on condition of evergreen forests will be more evident in remote sensing studies using multitemporal imagery. It is not surprising that for many remote sensing studies, classification results are higher when using multitemporal imagery than when using single-date imagery [63]. For instance, differences in timing of emergence, anthesis, and seed maturation among crops may be captured at key times during the growing season and are essential to mapping crop species (e.g., Collins [64]). Although hyperspectral imagery provides a powerful tool for tree identification in general, the daunting complexity of many tropical forests may necessitate the combined advantage of multitemporal hyperspectral imagery.

While satellite-borne high spatial resolution hyperspectral data are not currently available for tropical forests, they are anticipated in the future as remote sensing technology continues to develop. Spatial resolution of the hyperspectral imager currently in space (EO-1 Hyperion), at 30 m, is too coarse for tree crown delineation. The failed OrbView-4, had it achieved orbit, would have offered hyperspectral imagery (250 bands, 450–2500 nm) at an 8-m spatial resolution and provides testimony to satellite hyperspectral capabilities of the future. Anticipated sensors with high spectral and spatial resolution capabilities will invariably provide new opportunities for mapping tropical tree species. A potential approach to start with is the delineation of deciduous vs. evergreen species using dry and wet season images. With early-dry, mid-dry, late-dry, and rainy season images, even more tree functional types and species could potentially be mapped, based on the phenological events observed. (Note that both Borchert [59,60] and Rivera et al. [53] provide means of describing dry forest tree functional types.) This approach will inherently provide valuable information on the proportion of tree functional types in an area as well as on forest photosynthetic capacity over the two seasons.

1.4   CHALLENGES FOR THE FUTURE

The "art and science" of tropical tree identification from hyperspectral imagery faces numerous challenges, although initial studies using airborne hyperspectral imagery by Clark et al. [4] and Zhang et al. [36] indicate that mapping at least a portion of tree species in tropical forests is feasible. Tree crown delineation in tropical forests is an area of research that needs to be explored because it enables the extraction of appropriate spectral data from crowns and can be considered a first step in classifying tree species [43]. Spectral mixing is another major challenge. The presence of lianas in the canopy is one of many examples of spectral mixing that can occur. Lianas tend to form monolayers of leaves over their host tree, thereby masking the tree spectral signature [15,36]. Leaf angles are another issue, as these can change significantly throughout the day. Some tropical dry forest trees experience mid-day wilt [65], resulting in a large decrease in the near-infrared contribution of those trees. On windy days, furthermore, the impact of leaf angle and leaf additive reflectance could be impossible to predict (Sanchez-Azofeifa, 2006, pers. comm.).

In addition to the complexities surrounding hyperspectral data analysis, there is a continuing need for developing a sophisticated understanding of tropical tree ecology and phenology and their implications for tree discrimination. For example, a recent study showed earlier flowering (15–20 days) of a tropical dry forest tree in disturbed habitat as compared to undisturbed habitat, as well as a lower proportion of trees that flowered in the disturbed habitat [66]. Successional stage, furthermore, has been shown to be an important factor in determining the number of species and individuals exhibiting drought deciduousness in tropical dry forests [50]. Understanding unique patterns of deciduousness should provide a significant advantage for unraveling seasonal changes in spectral signatures when multitemporal images allow the privilege of such comparisons.

An overall lack of hyperspectral imagery of sufficient spatial resolution for tree discrimination studies remains a significant limitation to advancement in this area. We await the time when operational, satellite-borne high spatial resolution hyperspectral imagery is available and, in the meantime, anticipate additional opportunities through the use of airborne hyperspectral imagery. On that note, new opportunities for tropical tree discrimination studies are available from a recent mission in Costa Rica. Through a cooperative effort between El Centro Nacional de Alta Tecnologia (CENAT—Costa Rican Centre for High Technology) and El Programa Nacional de Investigaciones Aerotransportadas y Sensores Remotos (National Airborne and Remote Sensing Program), hyperspectral imagery with the HyMap II sensor (HyVista Corporation, Castle Hill, New South Wales, Australia) in conjunction with the MASTER stimulator (moderate resolution imaging spectroradiometer (MODIS)/ advanced space-borne thermal emission and reflection radiometer (ASTER)) [67] was collected for nearly 90% of the country at approximately 16-m spatial resolution in March through April 2005 (fig. 1.7). Higher spatial resolution data (~5 m) were also collected for the Santa Rosa National Park, the Maritza Biological Station, and the Los Inocentes Lodge and Center for Conservation, all in northwestern Costa Rica. In addition, the aircraft, a WB-57 (fig. 1.8) from the NASA Johnson Space Center, also carried an infrared camera.

During an earlier Costa Rican airborne research and technology (CARTA) mission in 2003 only MASTER imagery and high resolution infrared photographs were taken. The Hymap II instrument records data in 128 bands (table 1.2), of which 125 are delivered to the users (3 bands in the visible range are deleted during preprocessing) in the 450- to 2500-nm range. Data products from this mission were radiance ($\mu W/cm^2$ *nm*sr) and surface reflectance and a geocorrection file that may be applied by the user to either the radiance or reflectance images. Atmospheric correction was done with the HyCor package and included the application of the *at*mosphere *rem*oval (ATREM) algorithm followed by an EFFORT (Empirical Flat Field Optimal Reflectance Transformation) polishing to remove any systematic ATREM errors. In the aircraft, the HyMap sensor is mounted on a gyro-stabilized platform and is connected to a differential GPS (DGPS) integrated with a Boeing CMIGITS II GPS/INS inertial monitoring unit to geocode the raw data. Chapter 9 utilizes hyperspectral data from the CARTA II mission.

Additional advances in the area of leaf optical properties of tropical trees and tropical tree identification are already evident from this anthology:

Alvarez-Añorve et al. (chapter 2) examine the potential for applying hyperspectral data to identifying plant functional groups as a means to characterize successional stages.

Lucas et al. (chapter 3) explore the use of hyperspectral data for assessing forest biodiversity, carbon dynamics, and health, and also present research on tree species classification in wooded savannas and mangroves. Both chapter 3 and chapter 12 also present the utility of combining information from other sources such as LiDAR with hyperspectral imagery.

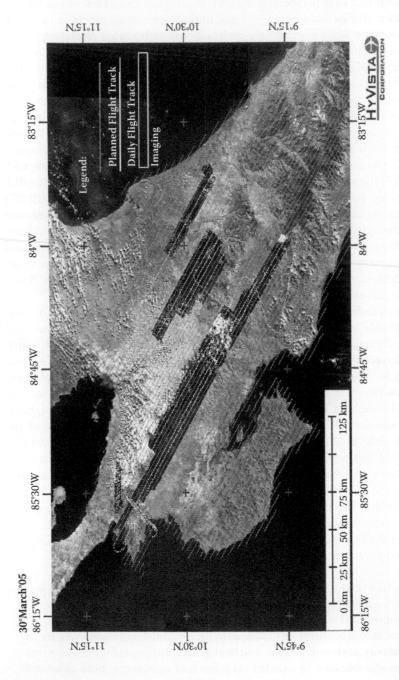

FIG. 1.7 Flight plan of CARTA II mission flown March–April 2005 over Costa Rica. Hyperspectral imagery was gathered with the HyMap II sensor in conjunction with the MASTER sensor (MODIS/ASTER airborne simulator). Lines indicate planned flight lines. Black boxes indicate data acquired March 30, 2005. See CD for color image.

FIG. 1.8   NASA WB-57 aircraft used to carry the HyMap II, MASTER sensor, and infrared camera for the CARTA II mission, March–April 2005.

TABLE 1.2

Typical Spectral Configuration for the HyMap II Airborne Hyperspectral Sensor

Module	Spectral range	Average spectral sampling interval
Visible	450–890 nm	15 nm
NIR	890–1350 nm	15 nm
SWIR 1	1400–1800 nm	13 nm
SWIR 2	1950–2480 nm	17 nm

An exploratory analysis into the effect of soil type on the spectral properties of tree seedling and grass species is carried out by Calvo-Alvarado et al. (chapter 4).

The previously unconsidered possibility of gender differentiation in tropical dioecious trees using spectral reflectance is investigated by Arroyo-Mora et al. (chapter 5).

Rivard et al. (chapter 6) take a new look at band selection for discrimination of trees important to an endangered bird species.

Ismail et al. (chapter 7) use a new approach to discriminate pine forest that has undergone pest infestation in South Africa.

Bohlman (chapter 8) examines the application of spectral unmixing to explore the contribution of various canopy elements to the signature of the canopy.

Arroyo-Mora et al. (chapter 9) assess recovery following selective logging from airborne (HyMap) data and pattern classification techniques.

Miura et al. (chapter 10) present a new technique for calibrating airborne hyperspectral spectrometer data to reflectance factors.

Huete et al. (chapter 11) compare patterns of optical-phenologic variability in an evergreen broadleaf tropical forest in the Amazon from space-borne hyperspectral imagery (Hyperion) and MODIS.

Asner (chapter 12) delves further into the understanding of intercrown vs. intracrown variability in reflectance and the spectrobiochemical properties of the canopy.

Gamon (chapter 13) reexamines the potential of hyperspectral remote sensing for assessing biodiversity and the evaluation of biosphere–atmosphere interactions. He also explores the challenges of establishing an informatics framework in which to link remotely sensed and ecological data.

It is likely that these studies will be some of the first among many in a burgeoning field of remote sensing research in tropical forests, spurred on by the desire to understand, characterize, and monitor tropical forests in a more adequate manner.

Despite the challenges, the prospects for tropical tree identification are no doubt intriguing. Until such time that satellite-borne high spatial resolution hyperspectral imagery is available, studies using airborne hyperspectral imagery should continue to shed light on the potential for species discrimination in tropical forests as well as the feasibility of different approaches to data analysis. A furtherance of leaf-level studies will also be highly valuable for interpreting images as well as understanding seasonal changes in spectral reflectance that tie in with events such as leaf flushing, leaf maturation, and leaf senescence, as well as flowering and fruiting.

REFERENCES

1. Lee, D.W. and Graham, R., Leaf optical-properties of rainforest sun and extreme shade plants, *American Journal of Botany*, 73, 1100, 1986.
2. Fung, T. and Siu, W.L., Hyperspectral data analysis for subtropical tree species recognition, in *IGARSS '98 Sensing and Managing the Environment, IEEE International Geoscience and Remote Sensing*, Seattle, WA, 1998.
3. Cochrane, M.A., Using vegetation reflectance variability for species-level classification of hyperspectral data, *International Journal of Remote Sensing*, 21, 2075, 2000.
4. Clark, D.A., Roberts, D.A., and Clark, D.A., Hyperspectral discrimination of tropical rain forest tree species at leaf to crown scales, *Remote Sensing of Environment*, 96, 375, 2005.
5. Jensen, J.R., *Remote sensing of the environment: An earth resource perspective*. Prentice Hall, Upper Saddle River, NJ, 2000.
6. Knipling, E.B., Physical and physiological basis for the reflectance of visible and near-infrared radiation from vegetation, *Remote Sensing of Environment*, 1, 155, 1970.
7. Gates, D.M., Keegan, H.J., Schleter, J.C., et al., Spectral properties of plants, *Applied Optics*, 4, 11, 1965.
8. Gausman, H.W., *Plant leaf optical properties in visible and near-infrared light*. Graduate Studies No. 29, Texas Tech Press, Lubbock, TX, 1985.
8a. Goodwin, N., Turner, R., and Merton, R., Classifying eucalyptus forests with high spatial and spectral resolution imagery: An investigation of individual species and vegetation communities, *Australian Journal of Botany*, 53, 337, 2005.
9. Grant, L., Diffuse and specular characteristics of leaf reflectance, *Remote Sensing of Environment*, 22, 309, 1987.
10. Asner, G.P., Biophysical and biochemical sources of variability in canopy reflectance, *Remote Sensing of Environment*, 64, 234, 1998.
11. Foley, S., Rivard, B., Sanchez-Azofeifa, G.A., et al., Foliar spectral properties following leaf clipping and implications for handling techniques, *Remote Sensing of Environment*, 103, 265, 2006.

12. Castro-Esau, K., Sanchez-Azofeifa, G.A., Rivard, B., et al., Variability in leaf optical properties of mesoamerican trees and the potential for species classification, *American Journal of Botany*, 93, 517, 2006.

13. Roberts, D., Nelson, B.W., Adams, J.B., et al., Spectral changes with leaf aging in Amazon caatinga, *Trees*, 12, 315, 1998.

14. Fung, T., Ma, F.Y., and Siu, W.L., *Band selection using hyperspectral data of subtropical tree species*, http://www.gisdevelopment.net/aars/acrs/1999/ps3/ps3055pf.htm, 1999.

15. Castro-Esau, K.L., Sanchez-Azofeifa, G.A., and Caelli, T., Discrimination of lianas and trees with leaf-level hyperspectral data, *Remote Sensing of Environment*, 90, 353, 2004.

16. Kalacska, M., Bohlman, S.A., Sanchez-Azofeifa, G.A., et al., Hyperspectral discrimination of tropical dry forest lianas and trees: Comparative data reduction approaches at the leaf and canopy levels, *Remote Sensing of Environment*, 109, 406, 2007.

17. Avalos, G., Mulkey, S.S., and Kitajima, K., Leaf optical properties of trees and lianas in the outer canopy of a tropical dry forest, *Biotropica*, 31, 517, 1999.

18. Lee, D.W., Bone, R.A., Tarsis, S.L., et al., Correlates of leaf optical-properties in tropical forest sun and extreme shade plants, *American Journal of Botany*, 77, 370, 1990.

19. Poorter, L., Oberbauer, S.F., and Clark, D.B., Leaf optical properties along a vertical gradient in a tropical rain-forest canopy in Costa Rica, *American Journal of Botany*, 82, 1257, 1995.

20. Nyyssonen, A., Aerial photographs in tropical forests, *Unasylva*, 16, 3, 1962.

21. Myers, B.J. and Benson, M.L., Rain forest species on large-scale color photos, *Photogrammetric Engineering and Remote Sensing*, 41, 505, 1981.

22. Myers, B.J., Guide to the identification of some tropical rain forest species from large-scale color aerial photographs, *Australian Forestry*, 45, 28, 1982.

23. Trichon, V., Crown typology and the identification of rain forest trees on large-scale aerial photographs, *Plant Ecology*, 153, 301, 2001.

24. Sayn-Wittgenstein, L., Milde, R., and Inglis, C.J., Identification of tropical trees on aerial photographs, in Information Report FMR-X-113, 1978, Forest Management Institution, Canada.

25. McGraw, J.B., Warner, T.A., Key, T.L., et al., High spatial resolution remote sensing of forest trees, *Trends in Ecology and Evolution*, 13, 300, 1998.

26. Gillespie, T.W., Brock, J., and Wright, C.W., Prospects for quantifying structure, floristic composition and species richness of tropical forests, *International Journal of Remote Sensing*, 25, 707, 2004.

27. Rasolofoharinoro, M., Blasco, F., Bellan, M.F., et al., A remote sensing based methodology for mangrove studies in Madagascar, *International Journal of Remote Sensing*, 19, 1873, 1998.

28. Wang, L., Sousa, W.P., Gong, P., et al., Comparison of IKONOS and Quickbird images for mapping mangrove species on the Caribbean coast of Panama, *Remote Sensing of Environment*, 91, 432, 2004.

29. Demuro, M. and Chisholm, L., Assessment of Hyperion for characterizing mangrove communities, in *Proceedings of the International Conference AVIIRIS 2003 Workshop*, 2003.

30. Vaiphasa, C., Ongsomwang, S., Waiphasa, T., et al., Tropical mangrove species discrimination using hyperspectral data: A laboratory study, *Estuarine, Coastal, and Shelf Science*, 65, 371, 2005.

31. Hirano, A., Madden, M., and Welch, R., Hyperspectral image data for mapping wetland vegetation, *Wetlands*, 23, 436, 2003.

32. Proisy, C., Mougin, E., Fromard, F., et al., Interpretation of polarimetric radar signatures of mangrove forests, *Remote Sensing of Environment*, 71, 56, 2000.

33. Held, A., Ticehurst, C., Lymburner, L., et al., High resolution mapping of tropical mangrove ecosystems using hyperspectral and radar remote sensing, *International Journal of Remote Sensing*, 24, 2739, 2003.

34. Green, E.P., Clark, C.D., Mumby, P.J., et al., Remote sensing techniques for mangrove mapping, *International Journal of Remote Sensing*, 19, 935, 1998.

35. Holdridge, L.R., Grenke, W.C., Hatheway, W.H., et al., *Forest environments in tropical life zones: A pilot study*. Pergamon Press, New York, 1971.

36. Zhang, J., Rivard, B., Sanchez-Azofeifa, G.A., et al., Intra- and interclass spectral variability of tropical tree species at La Selva, Costa Rica: Implications for species identification using HYDICE imagery, *Remote Sensing of Environment*, 105, 129, 2006.

37. Sanchez-Azofeifa, G.A. and Castro, K.L., Canopy observations on the hyperspectral properties of a community of tropical dry forest lianas and their host trees, *International Journal of Remote Sensing*, 27, 2101, 2006.

38. Bunting, P. and Lucas, R.M., The delineation of tree crowns in Australian mixed species forests using hyperspectral compact airborne spectrographic imager (CASI) data, *Remote Sensing of Environment*, 101, 230, 2006.

39. Xiao, Q., Ustin, S.L., and McPherson, E.G., Using AVIRIS data and multiple-masking techniques to map urban forest tree species, *International Journal of Remote Sensing*, 25, 5637, 2004.

40. Lefsky, M.A., Cohen, W.B., Parker, G.G., et al., Lidar remote sensing for ecosystem studies, *Bioscience*, 52, 19, 2002.

41. Leckie, D.G., Geougeon, F.A., Walsworth, N., et al., Stand delineation and composition estimation using semiautomated individual tree crow analysis, *Remote Sensing of Environment*, 85, 355, 2003.

42. Bai, Y., Walsworth, N., Roddan, B., et al., Quantifying tree cover in the forest-grassland ecotone of British Columbia using crown delineation and pattern detection, *Forest Ecology and Management*, 212, 92, 2005.

43. Leckie, D.G., Cloney, E., and Joyce, S., Automated detection and mapping of crown discoloration caused by jack pine budworm with 2.5-m resolution multispectral imagery, *International Journal Earth Observation and Geoinformation*, 7, 61, 2005.

44. Gougeon, F.A. and Leckie, D.G., The individual tree crown approach applied to IKONOS images of a coniferous plantation area, *Photogrammetric Engineering and Remote Sensing*, 72, 1287, 2006.

45. Read, J.M., Clark, D.B., Venticinque, E.M., et al., Application of merged 1-m and 4-m resolution satellite data to research and management in tropical forests, *Journal of Applied Ecology*, 40, 592, 2003.

46. Leckie, D., Gougeon, F., Hill, D., et al., Combined high density LiDar and multispectral imagery for individual tree crown analysis, *Canadian Journal of Remote Sensing*, 29, 633, 2003.

47. Clark, M.L., Clark, D.B., and Roberts, D.A., Small-footprint LiDar estimation of sub-canopy elevation and tree height in tropical rain forest landscape, *Remote Sensing of Environment*, 91, 68, 2004.

48. Holdridge, L.R., *Life zone ecology*. Tropical Science Center, San Jose, Costa Rica, 1967.

49. Murphy, P.G. and Lugo, A.E., Ecology of tropical dry forest, *Annual Review of Ecology and Systematics*, 17, 67, 1986.

50. Kalacska, M., Sanchez-Azofeifa, G.A., Calvo-Alvarado, J., et al., Species composition, similarity and diversity in three successional stages of a seasonally dry tropical forest, *Forest Ecology and Management*, 200, 227, 2004.

51. Hubbell, S.P., Tree dispersion, abundance, and diversity in a tropical dry forest, *Science*, 203, 1299, 1979.

52. Richards, P.W., *The tropical rain forest, an ecological study*. Cambridge University Press, Cambridge, 1952.

53. Rivera, G., Elliot, S., Caldas, L.S., et al., Increasing day-length induces spring flushing of tropical dry forest trees in the absence of rain, *Trees—Structure and Function*, 16, 445, 2002.

54. Condit, R.C., Watts, K., Bohlman, S.A., et al., Quantifying the deciduousness of tropical forest canopies under varying climates, *Journal of Vegetation Science*, 11, 649, 2000.

55. Frankie, G.W., Baker, H.G., and Opler, P.A., Comparative phenological studies of trees in tropical wet and dry forests in the lowlands of Costa Rica, *Journal of Ecology*, 62, 881, 1974.

56. Borchert, R., Soil and stem water storage determine phenology and distribution of tropical dry forest trees, *Ecology*, 75, 1437, 1994.

57. Meinzer, F.C., Andrade, J.L., Goldstein, G., et al., Partitioning of soil water among canopy trees in a seasonally dry tropical forest, *Oecologia*, 121, 293, 1999.

58. Reich, P.B. and Borchert, R., Water stress and tree phenology in a tropical dry forest in the lowlands of Costa Rica, *Journal of Ecology*, 72, 61, 1984.

59. Borchert, R., Water status and development of tropical trees during seasonal drought, *Trees—Structure and Function*, 8, 115, 1994.

60. Borchert, R., Phenology and flowering periodicity of neotropical dry forest species: Evidence from herbarium collections, *Journal of Tropical Ecology*, 12, 65, 1996.

61. Devall, M.S., Parresol, B.R., and Wright, S.J., Dendroecological analysis of *Cordia alliodora, Pseudobombax septenatum* and *Annona spraguei* in Central Panama, *IAWA Journal*, 16, 411, 1995.

62. Fuchs, E.J., Lobo, J.A., and Quesada, M., Effects of forest fragmentation and flowering phenology on the reproductive success and mating patterns of the tropical dry forest tree *Pachira quinata, Conservtion Biology*, 17, 149, 2003.

63. Castro-Esau, K., Sanchez-Azofeifa, G.A., and Rivard, B., Monitoring secondary tropical forests using space-borne data: Implications for Central America, *International Journal of Remote Sensing*, 24, 1853, 2003.

64. Collins, W., Remote-sensing of crop type and maturity, *Photogrammetric Engineering and Remote Sensing*, 44, 43, 1978.

65. Gamon, J.A., Kitajima, K., Mulkey, S.S., et al., Diverse optical and photosynthetic properties in a neotropical dry forest during the dry season: Implications for remote estimation of photosynthesis, *Biotropica*, 37, 547, 2005.

66. Herrerias-Diego, Y., Quesada, M., and Stoner, K.E., Fragmentation on phenological patterns and reproductive success of the tropical dry forest tree *Ceiba aesculifolia, Conservation Biology*, 20, 1111, 2006.

67. Hooke, S.J., Myers, J.J., Thome, K.J., et al., MODIS/ASTER airborne simulator (MASTER)—A new instrument for Earth science studies, *Remote Sensing of Environment*, 76, 93, 2001.

2 Remote Sensing and Plant Functional Groups
Physiology, Ecology, and Spectroscopy in Tropical Systems

Mariana Alvarez-Añorve, Mauricio Quesada, and Erick de la Barrera

CONTENTS

2.1 PLANT FUNCTIONAL GROUPS IN TROPICAL SYSTEMS

The term "functional groups" was proposed by Cummins [1] to classify species playing similar roles or performing analogous processes in the ecosystem. Plant functional types may describe groups of plants with common responses to certain environmental influences [2,3] and have been applied to several ecosystem functions such as biochemical cycles, fire resistance, invasion resistance, acquisition and use of resources, defense against herbivory, pollination, and seed dispersal, among others [4].

In general, functional grouping of species allows us to simplify biodiversity into components capable of explaining patterns or processes in a certain system [5]. This concept has been useful to predict the types of responses of vegetation to environmental changes even without detailed information about each species [6]. Thus, functional groups are often used in global models of vegetation [7,8] and climatic change [9].

The most important approaches to plant functional groupings are based on the use of functional characteristics and have been used by ecologists for decades [10–14]. According to Reich et al. [6], there are four main kinds of functional groupings. The first one is based on categorical qualitative approaches that classify groups of plants based on certain characteristics such as life form and type of photosynthesis, among others. The second one groups species along a continuum of quantitative characteristics such as growth rate, specific leaf area, maximum photosynthetic capacity, etc. [15,16]. The third model is based on a combination of quantitative characteristics that may influence each other such as leaves, seeds, and tree height [17]. Finally, a fourth model groups plant species based on their responses to specific environmental factors; a good example of this is the classification of tropical plant species proposed by Mulkey, Wright, and Smith [18] on the basis of their shade tolerance or the C-S-R scheme of plant strategies proposed by Grime [10].

2.1.1 Studies on Functional Groups

In general, the number of studies regarding functional groups has increased over the last two decades. An important number of these studies have focused on plant groups associated to successional stages after anthropogenic disturbance; another set of studies evaluated the response of predefined functional groups to global increments of CO_2 and temperature due to climate change; and a third group of studies evaluated the importance of functional diversity in synthetic or natural communities. Most of these studies, however, have been conducted in temperate regions rather than in tropical forests where biodiversity is higher and ecosystems are fragile [19]. The few studies in the tropics suggest that the quantity of plant functional groups is potentially very high and more complex [20]. Most of the studies in the tropics have been conducted in tropical rain forests with few in tropical dry and cloud forests [21–24]. Tropical systems other than tropical rain forests have been largely overlooked in the scientific literature [25].

In an extensive review, we surveyed the literature from different databases using a combination of the following keywords: "plant functional group," "plant functional type," and "tropical." Searches were conducted in the Science Citation Index and Biological Abstracts databases as well as in the main editorials (Blackwell Science, Springer–Verlag, Elsevier) and scientific societies of the most important indexed journals of ecology, physiology, and conservation biology. In this revision we found a total of 50 studies that analyze plant functional groups in tropical systems. Fifty-four percent of the studies use data of tropical plants in combination with data from other ecosystems to analyze three main aspects of functional groups related to (1) theoretical generalizations of ecological processes [2,6,26–30]; (2) performance of morphological and physiological characters and of their interrelationships [31–36], among others; and (3) the role of functional groups on the dynamics of communities and ecosystems [2,37–40]. The remaining 46% of these studies exclusively analyze tropical species to identify functional groups de novo or to evaluate the performance and consistency of functional groups previously defined for the tropics.

Ecophysiological characteristics are among the most used plant traits to classify functional groups. In tropical systems these traits have been mainly used for grouping plant species on the basis of (1) shade tolerance [41–43]; (2) maximum potential height [44-46]; (3) increment in diameter used as an indicator of growing rate [20,44,47–50]; (4) elements related to photosynthetic capacity [15,51]; and (5) water status and/or water use efficiency (i.e., water storage stems [21]), stomatal conductance [51], C isotopic composition ($\delta^{13}C$) [52,53], or the difference between the air and leaf temperatures (ΔT) (i.e., indicator of heat dispersion capacity [23]).

Some other classifications including ecophysiological characteristics are based on species responses to different environmental factors such as plant response to abnormal patterns of drought and rain [22] or the response to changes in the landscape (i.e., structure of forest fragments [54]). Other studies have identified a combination of various other characteristics to define functional groups. These include reproductive traits, life forms and patterns of distribution in forests in different successional stages [55], demographic characteristics such as mortality rate [44], and, in the case of invasive species, the impact on native species [56]. In summary, we can classify the studies of functional groups of tropical plants in the following groups: (1) studies that characterized plant species associated to different successional stages, (2) studies that classify plants based on growth patterns, and (3) studies that classify plants based on water stress, water storage capacity, and heat dissipation.

The analysis of functional plants under succession has been emphasized due to the increasing importance of tropical secondary forests. There is increasing knowledge that recognizes different strategies of groups of plants specialized to regenerate in different successional stages; these studies have become crucial for the conservation of tropical systems [57,58]. Another set of studies emphasized the use of growth patterns to determine functional groups because growth rate is a trait that is correlated with ecophysiological characteristics of great importance for species performance [59]. Finally, some studies classify plants based on water use mechanisms because this resource plays a decisive role in species distribution and diversity gradients in tropical forests [60]. Water availability is an important determinant of species distribution in tropical dry forests [61].

As shown thus far in this review, most studies in tropical systems have analyzed single or a combination of a few ecophysiological characters for the discrimination of tropical functional groups. However, an analysis considering an assemblage of characteristics with ecological relevance that determine the establishment of species in a given habitat would be more realistic. Thus, we consider that a possible combination of plant traits should include plant growth rate, leaf longevity, specific leaf area, photosynthetic capacity, leaf water content, water use efficiency, and certain reproductive parameters. The use of this kind of assemblage would generate more consistent information about a great number of species in order to detect tropical functional groups. The information generated in most studies until now did not consider a multivariate approach of both vegetative and reproductive parameters. Therefore, there is a lack of comprehensive analyses of functional groups in tropical systems—communities that are highly diverse and complex.

2.2 SOLAR IRRADIATION AND LEAF OPTICAL PROPERTIES

Solar electromagnetic radiation reaches the Earth's outer atmosphere with an energy of 1366 W m^{-2}. During its trajectory to the planet's surface, approximately 40% of the energy is attenuated, with an essentially absolute filtering of wavelengths below 200 nm and above 10,000 nm [62,63]. Considering that such radiation can be absorbed, transmitted, or reflected at different wavelengths, remote sensing takes advantage of the net reflected radiation by the various objects on the surface of the Earth, allowing for applications, for instance, in tropical ecology, as discussed later in this chapter. In the present section we will consider some of the factors that result in the range of reflected wavelengths registered by remote sensing satellites, with a special focus on leaf-level properties.

2.2.1 SOLAR IRRADIANCE AND RADIATION SOURCES

The wavelengths of the solar radiation reaching the outer layers of the atmosphere approximately range from 200 to 1600 nm, with most of the incident energy within the range of visible light [62,63]. Various components of the atmosphere attenuate such solar irradiance. For instance, the stratospheric ozone layer filters out wavelengths below 350 nm [64]. Water vapor, in turn, absorbs infrared radiation with major bands at 900, 1100, and above 1200 nm [63]. CO_2 has narrow absorption bands at 2700, 4300, and 15,000 nm [65]. In fact, the infrared absorption properties of water vapor and CO_2 are widely utilized for the measurement of real-time gas exchange (i.e., CO_2 uptake and transpiration by plants).

In addition to the aforementioned solar irradiance attenuation by absorption, air and suspended particles further attenuate it by scattering, a phenomenon of special relevance for visible light. Indeed, when a light beam's trajectory is intercepted by an object, some of its energy is lost on impact and the rest is re-irradiated concentrically [66]. In fact, the sky's brightness and blue color during clear days are due to the scattering of light by air molecules—a phenomenon known as Rayleigh scattering—while larger particles, such as dust, further reduce the light's energy, as can be observed during red-sky sunsets due to Mie scattering [63,66].

In any case, six sources of radiation can reach an object (e.g., a leaf) on Earth's surface. First is direct solar irradiation or *sunlight*, for which most of its energy comes from visible light and whose range largely coincides with photosynthetically active radiation (wavelengths of 400–700 nm), with a special enrichment in yellow-orange (approximately 560–640 nm) wavelengths [62,63]. Due to scattering and absorption by the atmosphere, *skylight* also reaches the surface of the Earth during clear days, with energy of only 10% of that of direct solar irradiation with a peak wavelength near 400 nm [63,67]. Finally, visible *cloudlight*, with a peak around 500 nm, results from the transmission through clouds of irradiance [67]. The optical properties of the objects encountered by direct sunlight, skylight, and cloudlight result in reflected long-wave radiation that also reaches objects such as leaves. Thus, the three remaining radiation sources can be identified as *reflected sunlight*, *reflected skylight*, and *reflected cloudlight*.

2.2.2 Leaf Optical Properties

In the previous section we considered the six sources of radiation that can reach a leaf. Now we will discuss some leaf optical properties that result in the wavelength ranges, or bands, that are actually registered by remote sensors and some of their ecophysiological implications.

Pigments are highly conjugated biological molecules that absorb light at certain wavelengths. The most obvious pigment for studying plants is chlorophyll, which has absorption peaks in the blue and, especially, red regions of the visible spectrum, while light absorption of intermediate wavelengths is substantially reduced, especially in the green region. As a result, most of the red and blue radiation is absorbed by leaves, while most of the green light is reflected, conferring the familiar color to plants and vegetation in general. In this respect, remote sensors are able to detect a depletion in the red region that can be correlated with chlorophyll content (e.g., Castro-Esau, Sanchez-Azofeifa, and Caelli [68]). Higher chlorophyll contents, as suggested by remotely measured red-light depletion, can be indicators of higher canopy density or a more complex community structure. Another possibility is that such depletion indicates higher nitrogen content in the plant tissue. This is due to the fact that the most abundant protein in plants and on Earth is responsible for CO_2 fixation. Thus, chlorophyll content can be used as a proxy for determining protein and nitrogen content for plant tissue, as well as soil nitrogen levels [69,70].

While chlorophyll's maximum absorption occurs in the red region, accessory pigments absorb light of shorter wavelengths and, in consequence, of higher energy [63,70]. Of particular importance for tropical and subtropical forests, where the solar angle leads to higher irradiances than at higher latitudes, pigments that absorb in the blue-green region can be mentioned. In addition to funneling energy toward photosynthesis they also double as photoprotective pigments. First, as a result of photochemistry, a very reactive form of oxygen can result from an interaction with chlorophyll molecules. Carotenoids can quickly absorb the energy from such *free radicals*, thus preventing cellular damage. The second photoprotective function of carotenoids, specifically linked with the xanthophyll cycle, is nonphotochemical quenching. In this case, excited chlorophyll molecules can return to a basal state either by fluorescence (i.e., emitting light) or by transferring the excitation energy directly to other molecules. When exposed to high-light environments, a finite number of chlorophyll molecules become saturated, as illustrated by the numerous light response curves of net CO_2 uptake available in the literature, and such excess energy can inhibit or damage the photosynthetic machinery. In this respect, under high light the xanthophyll violaxanthin (absorption occurs in the blue region) is converted to zeaxanthin (absorption occurs in the green region), which in turn is converted back to violaxanthin when the light decreases [70].

Leaves also absorb infrared radiation, mainly from reflected light, which can increase their temperature. As a response, transpiration rates may also increase, taking advantage of the cooling effect resulting from evaporation. Nevertheless, because all objects with a temperature higher than absolute zero emit radiation, in addition to reflected radiation, leaves irradiate in the infrared [63]. The wavelength in which they maximally irradiate can be predicted, as a function of their surface temperature, by the Wien displacement law, $\lambda_{max} T = 3.67 \times 10^6$ nm K, on a photon basis, where λ_{max} is the wavelength of maximum photon flux density, and T is the surface temperature of an object. For instance, the sun's surface temperature is 5800 K; according to Wien's displacement law, it maximally irradiates at 630 nm, while a leaf at 30°C would maximally irradiate at 35,631 nm. The relative importance of emitted vs. reflected infrared radiation in terms of the spectral signature of vegetation registered by remote sensors is not yet known. Yet, because infrared is absorbed by water, such bands can be utilized for assessing the water status of vegetation (e.g., Castro-Esau et al. [68] and Hunt, Rock, and Nobel [71].

Plant anatomy also influences the optical properties of leaves. For instance, leaves from xeric environments may be more reflective of shorter wavelengths due to the higher contents of silicates in their leaves. In addition, some species present calcium oxalacetate crystals, which are believed to dissipate excess energy [63,72]. Also, the thickness of a leaf's mesophyll influences the amount of absorbed light as a consequence of the multiple layers of cells per leaf unit area [63,72]. Studies about the particular influences of cuticle composition, trichomes, and mesophyll thickness on leaf reflectance are recent and scarce, so further studies characterizing various functional groups may improve our understanding of the biological implications of remotely sensed spectra.

2.3 HYPERSPECTRAL DATA APPLICATIONS ON FUNCTIONAL GROUP DETECTION IN TROPICAL FORESTS: CASE STUDIES

Hyperspectral data are narrowband information on the reflectance of an object on Earth gathered by remote sensing analysis from in situ, airborne, or satellite sensors. This state-of-the-art technology allows for a detailed analysis of objects in the landscape from air and space; an example is provided by two hyperspectral satellites, Hyperion and Proba, now in orbit [73].

In hyperspectral imagery it is possible to subdivide the spectral range into over 200 intervals, each approximately 10–20 nm in width. If a radiance value is obtained for each interval, then a spectral curve of the wavelength intensity can be generated from the reflectance of each object in the landscape. The area covered by each hyperspectral image (and spatial resolution) varies by sensor; for example, Hyperion's ground coverage is 7.5 by 100 km with a resolution of 30 m per pixel [73]. Hyperspectral imaging is, then, a powerful and versatile means for continuous sampling of broad intervals of the spectrum.

The capability of hyperspectral sensors to detect numerous narrow bands can be applied to detect from space characteristic chemical and anatomical properties of vegetative and reproductive tissues of plants. A number of recent studies have indicated the advantages of using discrete narrowband data (i.e., hyperspectral data) from specific portions of the spectrum, rather than broadband data (i.e., multispectral data), to obtain the most sensitive quantitative or qualitative information on vegetation or crop characteristics (i.e., references 74 through 77). For example, Thenkabail et al. [76] established the advantages of using narrowband Hyperion data over broadband IKONOS, ETM+, and ALI data in studying rainforest vegetation. When compared to broadband data from IKONOS, ETM+, and ALI sensors, Hyperion's narrow bands explained 36–83% more of the variability in biomass and increased by 45–52% land use/land cover (LULC) classification accuracies as verified by ground truthing. The overall accuracy in classifying nine rainforest LULC classes was 96% and was achieved by using 23 Hyperion wavebands. In comparison, the overall accuracies were only 48% for IKONOS (four bands), 42% for ETM+ (six nonthermal bands), and 51% for ALI (nine multispectral bands). Similarly, Lee et al. [78] and Kalacska et al. [79] indicated that the large number of narrow bands of hyperspectral data is an advantage for the estimation of structural and functional canopy characteristics.

Given the fact that 47% of the global forest cover is in the tropics [80], where the most biodiversity can be found, it is necessary to apply modern techniques to describe and study functional attributes of tropical plant communities using a large-scale landscape approach that can be corroborated with ground truth data at the species level. Thus, exploration of the role of hyperspectral remote sensing in the assessment and determination of functional traits in the tropics is an important task in order to evaluate current and future applications of these technologies in ecological sciences. At present, however, hyperspectral imaging techniques have been poorly applied in tropical zones, with only a few efforts to detect and study plant functional groups. In order to exemplify applications of hyperspectral remote sensing in this respect, in the following sections we will describe some case studies regarding identification of different kinds of vegetation based on functional attributes of species.

2.3.1 DISCRIMINATION OF SUCCESSIONAL STAGES

The area covered by secondary forests has increased over the last decades worldwide, encouraging the interest in successional studies and the development of new techniques oriented to detect these habitats using remote sensing images (e.g., Landsat, SAR, MODIS, AVIRIS). Thus, studies in this respect could involve the detection and characterization of plant successional groups, which would be of crucial importance for conservation purposes [25].

In general, recent forest clearings are spectrally distinct as they have higher reflectance than mature forest in visible, near, and middle infrared wavebands used by satellite sensors. In forest succession, red reflection exhibits a slight decrease as increasing leaf area absorbs rising amounts of radiation in this range; meanwhile, near-infrared reflectance (NIR) increases as additional leaf layers are added to a canopy as a result of the increasing reflectance from the spongy mesophyll. Nevertheless, subsequent canopy maturation, characterized by the acquisition of more layers and complexity, reduces reflectance, given that shadowing traps incoming energy. Shadowing also depresses shortwave infrared reflectance (SWIR). Provided that the SWIR is influenced by water absorption, increasing canopy moisture content also leads to a decrease in SWIR through secondary forest succession. In fact, several studies have indicated that SWIR bands contain most of the information relevant to plant regeneration [81].

At present, a few attempts have been made to discriminate tropical forest successional stages by using hyperspectral remote sensing. For example, Thenkabail et al. [76] used hyperspectral imaging in order to detect different types of LULC in several ecoregions of West Africa, including humid forests. Specifically, they classified primary forests without evidence of anthropogenic disturbance, degraded primary forest with some evidence of anthropogenic disturbance, young secondary forest between 9 and 15 years old, mature secondary forest between 15 and 40 years old, and mixed secondary forest with significant anthropogenic disturbance. They also attempted to identify LULC classes of agricultural lands recently abandoned with regrowth vegetation between 1 and 8 years old. Only seven to nine Hyperion bands were required to separate pristine vs. degraded primary forest, young vs. mature vs. mixed secondary forest, and fallows of 1–3, 3–5, or 5–8 years old. When all rainforest vegetation was pooled, approximately 23 Hyperion bands were required to achieve adequate separability.

Indeed, examination of average reflectance spectra for the different vegetation types indicated that the Hyperion data provided many possibilities for separating vegetation categories using specific narrow bands throughout the 600- to 2350-nm spectral range. The most important wavebands were early mid-infrared (EMIR; 1300–1900 nm) bands followed by far near-infrared (FNIR; 1100–1300 nm), far mid-infrared (FMIR; 1900–2350 nm), and red (600–700 nm) wavebands. The results of this study showed that the two most frequently occurring wavebands sensitive to predicting forest biomass were centered at 682 and 1710 nm. Consequently, this study reaffirms the importance of using bands near 680 nm, as previously established by Thenkabail, Smith, and De Pauw [82], that are within a maximal absorption region for crops and vegetation. Some of the most important bands useful for detecting different vegetation types in this study were those related to leaf biochemical and physical traits such as content of water, chlorophyll, starch, lignin, cellulose, and proteins. Biophysical characteristics of vegetation such as biomass, vegetation growth, and leaf types were also important to obtain this segregation. Provided that most of these traits do differ among species in tropical forests [72], the potential use of hyperspectral remote sensing in the tropics to identify groups of species displaying different functional attributes associated to specific successional stages is clear.

2.3.2 Discrimination of Vegetation Types

Hyperspectral remote sensing has already shown important capabilities to differentiate vegetation types, especially when combined with data from other sensors such as radar that provide concurrent information about forest structure. An example for high-diversity mangrove systems is provided by Held et al. [83]. The study emphasizes the potential of hyperspectral scanners for identifying groups of species occurring under different grades of environmental stress. Provided that every group of species shares an assembly of physiological, anatomical, and ecological characteristics that allows survival under particular conditions, the differences among such groups should be detected by hyperspectral sensors facilitating the discrimination of plant functional groups.

In mangrove ecosystems, gradients in salinity, tide action, and drainage often cause major differentiation in species composition and structure across a linear spatial arrangement from the water edge to inland. Thus, high-diversity mangrove systems can contain up to 30 different species, broadly segmented into "mangrove zones" [83]. In order to describe these ecosystems accurately and objectively in terms of their zonation, productivity, and diversity patterns, Held et al. [83] conducted an analysis combining high spatial (3-m pixels or less) and spectral resolutions by using SAR (synthetic aperture radar) and the airborne hyperspectral scanner CASI (compact airborne spectrographic imager).

Although SAR data separated the vegetation into its general structural groups quite well, it had difficulty discriminating any further detail; thus, the SAR-only analysis correctly classified 57.9% of the mangrove types. The CASI-only data, on the other hand, provided finer detail but exhibited considerable confusion between structurally different vegetation classes—specifically, between the sand-dune vegetation and mixed stands of the plant *Bruguiera* sp. The CASI-only analysis classified 71% of the mangrove types present in the subset. However, when the data from both sensors were considered together, the classification accuracy increased to 76%. There are only a few cases where mangrove types have been classified at this level of detail within an individual estuary. This study, therefore, showed that there is considerable scope in use of high-resolution hyperspectral data for detecting, mapping, and monitoring mangroves at the necessary level of detail for mangrove diversity, ecological, and even ecophysiological studies. This study also shows the potential of a combination of remote sensing techniques for increasing the level of accuracy in the detection of vegetation types.

2.3.3 Life Form Discrimination

A specific case attempting to discriminate between species with different biochemical properties and ecological traits with hyperspectral data is the study of Castro-Esau et al. [68]. The objective of this study was to determine if it is possible to distinguish between lianas and supporting trees, at the leaf level, using hyperspectral reflectance measurements taken for two communities of tropical liana/tree species from a tropical dry forest (Parque Natural Metropolitano, Panama) and from a tropical wet forest (Fort Sherman, Panama). The study showed that lianas and trees from the tropical dry forest are distinguishable based on their spectral reflectance at the leaf level with the use of pattern recognition techniques. It is suggested that the chlorophyll concentration of liana leaves is lower than for tree leaves and that this difference is highly significant, resulting in an increase in reflectance at 550 nm as liana quantity/coverage increases.

Differing levels of water or nutrient stress (i.e., nitrogen) between lianas and trees could also have induced the differences observed in leaf reflectance between lianas and trees, and/or possibly differences in photosynthetic capacity between the two structural groups. Indeed, lianas typically were forming monolayers above tree crowns, favoring high light interception and low light transmission. Trees, in contrast, favor greater light transmission [68]. According to the authors, further study is required to clarify the physiological mechanisms between the two groups and whether such differences are maintained throughout the year. It must be emphasized, however, that in this case hyperspectral data effectively reflected general physiological differences between species with different ecological traits. Information produced in this respect could be useful for mapping species or communities with applications in biodiversity assessment studies. Indeed, mapping of lianas would be helpful for carbon budgets' estimation, because carbon sequestration is impeded in areas where liana proliferation obstructs tree regeneration to the point, in some cases, that a net release of CO_2 can be measured for some tropical forests [84,85].

2.3.4 DISCRIMINATION OF BIOPHYSICAL PROPERTIES

Hyperspectral remote sensing has proven to be useful in distinguishing biophysical properties of tropical forests, which usually differ among vegetation types. Kalacska et al. [79], for instance, examined forest structure and biodiversity of tropical dry forest from satellite imagery. They addressed the inference of neotropical dry forest biophysical characteristics (i.e., structure), biomass, and species richness directly from hyperspectral remote sensing imagery acquired over three seasons: wet, transition, and dry. They also examined six narrowband spectral vegetation indices that were sensitive to canopy characteristics in other ecosystems: normalized difference vegetation index (ND705), canopy normalized difference vegetation index (NDcanopy), single ratio (SR705), canopy single ratio (SRcanopy), modified single ratio (MSR), and canopy structure index (CSI). The results of this study showed that all canopy characteristics share similarities in shortwave infrared, except for biomass, which had important spectral regions in the visible, near infrared, and shortwave infrared. In addition, for species richness, the shortwave infrared was also heavily favored with the exception of one wavelength from the visible.

In contrast, for canopy height, basal area, and the Holdridge complexity index (HCI), none of the wavelengths were from the visible range, but instead all wavelengths were from the near infrared and shortwave infrared (743–2257 nm). Differences in canopy openness and structure among seasons were considered the most important factor in predicting biomass. According to the authors, low-canopy leaf area index (LAI) observed in the dry season exposed woody material, leaf litter, and soil with minimal to no contribution from green leaves, accounting then for the importance of the shortwave- and near-infrared regions. The intermediate stage had an important contribution of woody material and dry leaves, of less importance of the soil and green leaves; while in the late stage the green leaf was an important contribution to the spectral response in the visible region of the spectrum.

These studies provide new techniques to identify groups of species with functional differences by their particular biophysical properties, especially when such differences are reflected at the phenological level. In this sense, hyperspectral technology would allow for more accurate quantification of forest biophysical and biochemical attributes, which is essential for biodiversity assessment, land cover characterization, biomass modeling, and carbon flux estimation [86].

2.4 HYPERSPECTRAL REMOTE SENSING AND FUNCTIONAL GROUP STUDIES: PRESENT AND FUTURE APPROACHES

The capability of hyperspectral remote sensing to discriminate among plant functional groups has already been explored directly or indirectly, as mentioned in the previous section. However, such a discriminatory capability has several potential applications that have not been properly explored at present in the tropics: for example, differentiation of land vegetation/forest types, carbon flux estimates, description of ecosystem status, and assessment of vegetation functional changes, among others. Now we will briefly explain some of these potential applications in order to present a general view of future approaches for studies on tropical vegetation.

2.4.1 Vegetation Type Discrimination on Tropical Land Vegetation

Vegetation surveys in tropical forests are difficult and time consuming because plant species diversity is extremely high, their taxonomy is known only by a few specialists, and remote areas are logistically difficult to work in. As a result, it is difficult to collect field data that cover the area of interest sufficiently [87]. Remote sensing data are then an alternative technique that could be applied for separating vegetation types [88].

Provided that tree species composition is related to soil differences [89], spectral characteristics may behave similarly because they are mostly determined by the forest canopy. For example, Salovaara et al. [87] were able to separate floristically defined terra firma forest classes in Amazonia from Landsat ETM+ images with a reasonable accuracy. On the other hand, inundated and noninundated forests have been mapped and discriminated with high accuracy by Hess et al. [90]. Hyperspectral data, however, could help discern unique spectral patterns of different vegetation types and could be useful in extracting biophysical information such as biomass. These data will then be extremely useful for vegetation studies since they can contribute to tropical functional group differentiation. Indeed, hyperspectral remote sensing would be especially helpful for this purpose in complex tropical rain forests, where the distribution patterns of individual species are poorly known and a high level of accuracy is required. This type of datum could additionally offer valuable information for sustainable resource use and biodiversity conservation, where vegetation types can be used as surrogates for modeling the distributions of species and communities [91].

2.4.2 Carbon Flux Estimations

Globally, terrestrial vegetation sequesters some 100 Pg of carbon from the atmosphere each year for the production of organic matter through photosynthesis, half of which occurs in the tropics. The role of the various major forest ecosystems in the carbon cycle must therefore be assessed, particularly as carbon sinks that may be managed to reduce the atmospheric carbon load. Currently, attention is focused on tropical forests, which cover 7% of the Earth's land surface, playing a major but poorly understood role in the cycling of carbon [92].

Many studies have used remote sensing analysis to discriminate between mature tropical forest and nonforest areas [93] because monitoring of secondary forest regrowth may be important in the carbon balance of the tropics. Tropical secondary forests that follow nonforested stages have the facility to decrease atmospheric carbon concentrations to some degree in relatively short periods of time [94]. Recent special attention to tropical secondary forests may then be attributed to their capacity to act as carbon sinks and their potential role to serve as regulators of climate change. A secondary forest may actually have higher net primary production than a mature forest and may rapidly sequester carbon from the atmosphere, converting it to biomass. The strength of this carbon sink and the size of the resulting carbon pool depend on a range of factors, such as species composition and, remarkably, on the age of the regenerating forest [95]. Thus, to understand the role of regenerating forests as carbon sinks, information about their age, species composition, location, and extent is required. The most feasible way to derive this information is through remote sensing [92].

In order to analyze the progress and future potential of research to monitor carbon sequestration, Castro-Esau, Sanchez-Azofeifa, and Rivard [96] reviewed the attempts for estimating secondary forest biomass from space-borne data in the neotropics. This review states that considerable progress has been made in classifying neotropical secondary forests according to age using Landsat TM data. Currently, most Landsat TM studies of secondary forests have separated few broad regrowth classes of fairly young age (usually up to 20 years old or less) and with varying degrees of accuracy. The majority of studies analyzed have involved multitemporal images of sites in the Brazilian Amazon. Foody et al. [92], for instance, investigated ability of Landsat TM data to identify different successional stages of tropical rain forest in Amazonia. A range of forest classes (11) varying in strength as carbon sinks was identified accurately from these data. Their results also indicated that the youngest age class may be more variable in composition and spectral response than the older forest. This could be a function of a range of successional pathways being followed.

These results demonstrated that it is possible to use image classifications to scale up point measurements of carbon flux between regenerating forest classes and atmosphere over large areas. Moreover, the dynamics of the forest succession was to some extent manifested in the remotely sensed data. Castro-Esau et al. [96] stated that, although much research is required, it is possible that accurate classification of secondary forests in tropical areas will necessitate the use of imagery with higher spectral resolution (i.e., hyperspectral data) from which unique spectral signatures (i.e., for certain common crops) might be determined for their distinction from secondary forests. Furthermore, the capability of hyperspectral remote sensing to discriminate the dominant species within a regenerating forest would refine carbon accounting models, increasing classification accuracy of regeneration stages. Hyperspectral remote sensing could then be used to classify regenerating tropical forest classes accurately and even to identify different successional pathways.

Hyperspectral data will also be especially useful in highly heterogeneous sites, where pixels with mixed classes would be abundant and an entire area of interest with secondary forests of varying ages might occupy only a few pixels [96]. However, we must consider that the use of only hyperspectral data could eventually fail when discriminating different secondary forests, especially when dealing with highly homogeneous areas. An example of such limitation is the study performed by

Lucas et al. [97], who made the first attempt at mapping tropical forest regeneration stages using only hyperspectral data in the Brazilian Amazon and failed to obtain an acceptable level of accuracy for further estimation of biomass and carbon accumulation rates. Thus, the combination of hyperspectral imagery with data from other sensors is recommended. The integration of Landsat TM data with selected Hyperion scenes, for instance, may be useful for the separation of secondary forests from land cover types that have appeared spectrally similar [96]. Nonetheless, the determination of secondary forest biomass content from remote sensing data with greater precision would provide a better understanding of the role of secondary forests in global biogeochemical cycles as well as of their potential for mitigating atmospheric carbon [96].

2.4.3 Direct Detection of Plant Functional Types

Hyperspectral data have directly been used to discriminate plant functional types in nontropical zones. Studies carried out by Schmidtlein and Sassin [98] used hyperspectral remote sensing to analyze gradual floristic differences difficult to assess by conventional field surveys. They ranked species according to their functional responses and successfully modeled gradients in the appearance of plant functional response groups. These modeled gradients served to map species distributions. The results from this study indicate the potential of hyperspectral remote sensing and gradient analysis for mapping of continuous gradients in species assemblages. Of course, longer-term work with multitemporal data is needed to determine whether the approach can become a useful supplement to ground surveys. Such investigation must also be performed in the tropics in order to analyze functional diversity.

A specific attempt to discriminate tropical plant functional groups is currently being performed in a Mexican tropical dry forest by the authors. In order to identify and characterize secondary forest we are determining functional groups of plant species from different successional stages in the tropical dry forest of Chamela, a highly diverse tropical system located on the Pacific coast of Mexico. To define functional groups we decided to evaluate different morphological and physiological attributes highly involved in resource acquisition mechanisms, such as maximum photosynthesis, leaf dynamics, relative growth rate, and various leaf traits (specific leaf area, blade shape, thickness, water content, and chlorophyll content).

Preliminary results suggest that attributes such as water content, photosynthesis, specific leaf area, and chlorophyll content account for a substantial part of the functional differences observed among species from different successional stages, as well as among individuals of the same species growing in different stages. The phylogenetic origin of species appears to be an important factor for the variability of functional traits among evaluated stages. However, species growing in the different successional stages showed important differences in such traits, providing evidence of the response of certain functional traits to specific environmental factors associated with successional change. Variability of evaluated functional traits must be reflected at the spectral level, as demonstrated by Castro-Esau et al. [72]. Thus, by relating these attributes to the spectral reflectance of each species, we will be able to recognize the different types of secondary tropical dry forests (successional stages) by hyperspectral remote sensing. This kind of information is crucial for the understanding of the natural regeneration process as well as to determine the area covered by secondary vegetation in the tropics.

2.4.4 CHANGES IN FUNCTIONAL PROPERTIES OF ECOSYSTEMS

As mentioned previously, remote sensing can provide accurate estimates of functional features of tropical forests. Studies assessing changes in functional properties of the tropics by using remote sensing have been performed with coarse-resolution imagery by Koltunov et al. [99]. Provided that changes in forest function can be expressed as changes in forest phenology, these changes can be detected via remote sensing. Koltunov et al. used MODIS imagery to show that removal of timber species during selective logging changes forest composition and structure. Imagery analysis showed changes sufficiently large in magnitude to alter biosphere–atmosphere exchange of CO_2, water vapor, and energy in the logged regions of the Amazon basin. These changes could, in turn, alter a range of biogeochemical processes in the region and may have cascading effects on the regional climate system.

In the same sense, in order to understand canopy gap dynamics following selective logging, Asner et al. [100] used a spectral mixture analysis of the Landsat ETM data to estimate damage as well as to monitor intensity and canopy closure following timber harvests in eastern Amazon forests. They evaluated the impacts of different kinds of logging in terms of the canopy gap formations and showed that approximately one-half of the canopy opening caused by logging is closed within 1 year of regrowth following timber harvests. As stated by Asner et al., forest canopy damage monitored by remote sensing has several applications at the regional level—for example, to predict the location of fire-prone sites and respiration hotspots likely to result from coarse woody debris and damaged roots. Spatial and temporal dynamics of faunal species can also be linked to forest disturbance. Thus, an understanding of the components of canopy recovery will be important both for carbon balance as well as for other ecological and biogeochemical functions of the system [100]. These findings highlight the need for using higher resolution data to carry out a detailed analysis of the consequences of selective logging or other anthropogenic disturbances that may cause changes in the functional properties of tropical systems.

2.5 CONCLUSIONS

The current conservation status of tropical forests demands systematic approaches to study functional diversity. Such approaches must focus on the discrimination of coherent (consistent) plant functional groups as representative as possible of the great diversity of these systems. Assemblages of ecophysiological characteristics are considered of great utility to characterize tropical functional groups. These characteristics are susceptible to being detected and analyzed with modern techniques such as hyperspectral remote sensing, which allows for studying vegetation at a landscape level. At present, the capability of hyperspectral remote sensing to detect anatomical, biochemical, and biophysical properties of vegetation has been applied to tropical forests to discriminate successional stages, structural characteristics, life forms, and vegetation types, among others. Thus, the high accuracy and discriminatory capability of this technique on the detection of information of ecological relevance could be used for the discrimination of plant functional groups in the tropics. Such an approach would generate valuable knowledge about tropical functional diversity, would constitute a powerful tool for the study of forest dynamics, and would contribute to our understanding of the responses of tropical vegetation to human disturbance and climatic change.

REFERENCES

1. Cummins, K.W., Structure and function of stream ecosystems, *Bioscience*, 24, 631, 1974.
2. Lavorel, S. and Garnier, E., Predicting changes in community composition and ecosystem functioning from plant traits: Revisiting the Holy Grail, *Functional Ecology*, 16, 545, 2002.
3. Lavorel, S., McIntyre, S., Landsberg, J., et al., Plant functional classifications: From general groups to specific groups based on response to disturbance, *Trends in Ecology and Evolution*, 12, 474, 1997.
4. Blondel, J., Guilds or functional groups: Does it matter? *Oikos*, 100, 223, 2003.
5. Naeem, S., Species redundancy and ecosystem reliability, *Conservation Biology*, 12, 39, 1998.
6. Reich, P.B., Wright, I.J., Cavender-Bares, J., et al., The evolution of plant functional variations: Traits, spectra and strategies, *International Journal of Plant Science*, 164, S143, 2003.
7. Cramer, W., Using plant functional types in a global vegetation model, in *Plant functional types: their relevance to ecosystem properties and global change*, T.M. Smith, H.H. Shugart, and F.I. Woodward, eds. Cambridge University Press, Cambridge, 1997, 271.
8. Leemans, R., The use of plant functional type classifications to model global land cover and simulate the interactions between the terrestrial biosphere and the atmosphere, in *Plant functional types*, T. Smith, H. Shugart, and F. Woodward, eds. Cambridge University Press, Cambridge, 1997, 289.
9. Bugmann, H., Functional types of trees in temperate and boreal forests: Classification ad testing, *Journal of Vegetation Science*, 7, 359, 1996.
10. Grime, J.P., Evidence for the existence of three primary strategies in plants and its relevance to ecological and evolutionary theory, *American Naturalist*, 111, 1169, 1977.
11. Pearcy, R.W. and Ehleringer, J., Comparative ecophysiology of C3 and C4 plants, *Plant Cell and Environment*, 7, 1, 1984.
12. Garnier, E., Growth analysis of congeneric annual and perennial grass species, *Journal of Ecology*, 80, 665, 1992.
13. Craine, J.M., Tilman, D., Wedin, D., et al., Functional traits, productivity and effects on nitrogen cycling of 33 grassland species, *Functional Ecology*, 16, 563, 2002.
14. Ackerly, D.D. and Monson, R.K., Waking the sleeping giant: The evolutionary foundations of plant function, *International Journal of Plant Sciences*, 164(suppl), S1, 2003.
15. Ellis, A.R., Hubbell, S.P., and Potvin, C., In situ field measurements of photosynthetic rates of tropical tree species: A test of the functional group hypothesis, *Canadian Journal of Botany*, 78, 1336, 2000.
16. Foster, T.E. and Brooks, J.R., Functional groups based on leaf physiology: Are they spatially and temporally robust? *Oecologia*, 144, 337, 2005.
17. Westoby, M., A leaf-height-seed (Lhs) plant ecology strategy scheme, *Plant Soil*, 199, 213, 1998.

18. Mulkey, S.S., Wright, S.J., and Smith, A.P., Comparative physiology and demography of three neotropical shrubs: Alternative shade-adaptive character syndromes, *Oecologia*, 96, 526, 1993.
19. Groombridge, B. et al., *Global biodiversity: Status of the Earth's living resources: a report*. Chapman & Hall, London, 1992.
20. Clark, D.B. and Clark, D.B., Assessing the growth of tropical rain forest trees: Issues for forest modeling and management, *Ecological Applications*, 9, 981, 1999.
21. Borchert, R., Soil and stem water storage determine phenology and distribution of tropical dry forest trees, *Ecology*, 75, 1437, 1994.
22. Borchert, R., Rivera, G., and Hagnauer, W., Modification of vegetative phenology in a tropical semideciduous forest by abnormal drought and rain, *Biotropica*, 34, 27, 2002.
23. Souza, G., Ribeiro, R., Santos, M., et al., Functional groups of forest succession as dissipative structures: An applied study, *Brazilian Journal of Biology*, 64, 3, 2004.
24. Velaquez-Rosas, N., Meave, J., and Vazquez-Santana, S., Elevational variation of leaf traits in montane rain forest tree species at La Chinantla, Southern Mexico, *Biotropica*, 34, 534, 2002.
25. Sanchez-Azofeifa, G.A., Quesada, M., Rodriguez, J., et al., Research priorities for neotropical dry forests, *Biotropica*, 37, 477, 2005.
26. Davis, A.J., Liu, W.C., Perner, J. et al., Reliability characteristics of natural functional group interaction webs, *Evolutionary Ecology Research*, 6, 1145, 2004.
27. Fonseca, C.R. and Ganade, G., Species functional redundancy, random extinctions and the stability of ecosystems, *The Journal of Ecology*, 89, 118, 2001.
28. Gillison, A.N. and Carpenter, G., A generic plant functional attribute set and grammar for dynamic vegetation description and analysis, *Functional Ecology*, 111, 1169, 1997.
29. Colassanti, R.L., Hunt, R., and Askew, A.P., A self-assembling model of resource dynamics and plant growth incorporating plant functional types, *Functional Ecology*, 15, 676, 2001.
30. Eviner, V.T. and Chapin, F.S., Functional matrix: A conceptual framework for predicting multiple plant effects on ecosystem processes, *Annual Review of Ecology and Systematics*, 34, 455, 2003.
31. Coomes, D.A. and Grubb, P.J., Colonization, tolerance, competition and seed-size variation within functional groups, *Trends in Ecology and Evolution*, 18, 283, 2003.
32. Reich, P.B., Ellsworth, D.S., and Walters, M.B., Leaf structure (specific leaf area) modulates photosynthesis-nitrogen relations: Evidence from within and across species and functional groups, *Functional Ecology*, 12, 948, 1998.
33. Reich, P.B., Walters, M.B., Ellsworth, D.S., et al., Relationships of leaf dark respiration to leaf nitrogen, specific leaf area and leaf life-span: A test across biomes and functional groups, *Oecologia*, 114, 471, 1998.
34. Wright, I.J., Clifford, H.T., Kidson, R., et al., A survey of seed and seedling characters in 1744 Australian dicotyledon species: Cross-species trait correlations and correlated trait-shifts within evolutionary lineages, *Biological Journal of the Linnean Society*, 69, 521, 2000.
35. Poorter, H. and Navas, M.L., Plant growth and competition at elevated CO_2: On winners, losers and functional groups, *New Phytologist*, 157, 175, 2003.
36. Diaz, S., McIntyre, S., Lavorel, S., et al., Does hairiness matter in Harare? Resolving controversy in global comparisons of plant trait responses to ecosystem disturbance, *New Phytologist*, 154, 7, 2002.
37. Wardle, D.A., Bonner, K.I., and Barker, G.M., Stability of ecosystem properties in response to aboveground functional group richness and composition, *Oikos*, 89, 11, 2000.
38. Walker, B.H., Biodiversity and ecological redundancy, *Conservation Biology*, 6, 18, 1992.
39. Petchey, O.L. and Gaston, K.J., Extinction and the loss of functional diversity, *Proceedings of the Royal Society of London*, 269, 1721, 2002.

40. Petchey, O.L. and Gaston, K.J., Functional diversity (FD), species richness and community composition, Proceedings of the Royal Society of London Series B-Biological Sciences, *Ecology Letters*, 5, 402, 2002.
41. Shugart, H.H., Plant and ecosystem functional types, in *Plant functional types: their relevance to ecosystem properties and global change*. T.M. Smith, H.H. Shugart, and F.I. Woodward, eds. Cambridge University Press, Cambridge, 1997, 20.
42. Whitmore, T.C., Canopy gaps and the two major groups of forest trees, *Ecology*, 70, 536, 1989.
43. Slik, J.W.F., Assessing tropical lowland forest disturbance using plant morphological and ecological attributes, *Forest Ecology and Management*, 205, 241, 2005.
44. Condit, R., Hubbell, S.P., and Foster, R.B., Assessing the response of plant functional types to climatic change in tropical forests, *Journal of Vegetation Science*, 7, 405, 1996.
45. Denslow, J., Functional group diversity and responses to disturbances, in *Biodiversity and ecosystem processes in tropical forests*, G.H. Orians, R. Dirzo, and J.H. Cushman, eds. Springer, Berlin, 1996, 127.
46. Swaine, M.D. and Whitmore, T.C., On the definition of ecological species groups in tropical rain forests, *Vegetatio*, 75, 81, 1998.
47. Host, G.E. and Pregitzer, K.S., Ecological species groups for upland forest ecosystems of northwestern Lower Michigan, *Forest Ecology and Management*, 43, 87, 1991.
48. Kohler, P., Ditzer, T., and Huth, A., Concepts for the aggregation of tropical tree species into functional types and the application to Sabah's lowland rain forests, *Journal of Tropical Ecology*, 16, 591, 2000.
49. Vanclay, J., Aggregating tree species to develop diameter increment equations for tropical rain forest, *Forest Ecology and Management*, 42, 143, 1991.
50. Lieberman, D., Lieberman, M., Hartshorn, G., et al., Growth rates and age-size relationships of tropical wet forest trees in Costa Rica, *Journal of Tropical Ecology*, 1, 97, 1985.
51. Riddoch, I., Grace, J., Fasehun, F.E., et al., Photosynthesis and successional status of seedlings in a tropical semideciduous rain forest in Nigeria, *The Journal of Ecology*, 97, 39, 1991.
52. Bonal, D., Sabatier, D., Montpied, P., et al., Interspecific variability of delta C-13 among trees in rainforests of French Guiana: Functional groups and canopy integration, *Oecologia*, 124, 454, 2000.
53. Guehl, J.M., Domenach, A.M., Bereau, M., et al., Functional diversity in an Amazonian rainforest of French Guyana: A dual isotope approach (delta N-15 and delta C-13), *Oecologia*, 116, 316, 1998.
54. Metzger, J.P., Tree functional group richness and landscape structure in a Brazilian tropical fragmented landscape, *Ecological Applications*, 21, 1147, 2000.
55. Chazdon, R.L., Careaga, S., Webb, C., et al., Community and phylogenetic structure of reproductive traits of woody species in wet tropical forests, *Ecological Monographs*, 73, 331, 2003.
56. Horvitz, C.C., Pascarella, J.B., McMann, S., et al., Functional roles of invasive non indigenous plants in hurricane-affected subtropical hardwood forests, *Ecological Applications*, 8, 947, 1998.
57. Kalacska, M., Sanchez-Azofeifa, G.A., Calvo-Alvarado, J., et al., Species composition, similarity and diversity in three successional stages of a seasonally dry tropical forest, *Forest Ecology and Management*, 200, 227, 2004.
58. Sanchez-Azofeifa, G.A., Castro, K., Rivard, B., et al., Remote sensing research priorities in tropical dry forest environments, *Biotropica*, 35, 134, 2003.
59. Lambers, H. and Poorter, H., Inherent variation in growth rate between higher plants: A search for physiological causes and ecological consequences, *Advances in Ecological Research*, 23, 187, 1992.

60. Gentry, A.H., Changes in plant community diversity and floristic composition on environmental and geographical gradients, *Annals of the Missouri Botanical Garden*, 75, 1, 1988.
61. Murphy, P.G. and Lugo, A. E., Ecology of tropical dry forest, *Annual Review of Ecology and Systematics*, 17, 67, 1986.
62. Gliessman, S.R., Engles, E., and Krieger, R., *Agroecology: Ecological processes in sustainable agriculture*. Ann Arbor Press, Chelsea, MI, 1998.
63. Nobel, P.S., *Physicochemical and environmental plant physiology*, 3rd ed. Academic Press/Elsevier, New York, 2005.
64. Molina, L.T. and Molina, M.J., Absolute absorption cross sections of ozone in the 185- to 350-nm wavelength range, *Journal of Geophysical Research-Atmospheres*, 91, 14501, 1986.
65. Field, C.B., Ball, J.T., and Berry, J.A., Photosynthesis: Principles and field techniques, in *Plant physiological ecology: Field methods and instrumentation*, R.W. Pearcy, J. Ehleringer, H.A. Mooney, et al., eds. Chapman & Hall, London, 1989, 209.
66. Condon, E.U., Molecular optics, in *Handbook of physics*, E.U. Condon and H. Odishaw, eds. McGraw Hill, NY, 1958.
67. Gates, D.M., Keegan, H.J., Schleter, J.C., et al., Spectral properties of plants, *Applied Optics*, 4, 11, 1965.
68. Castro-Esau, K.L., Sanchez-Azofeifa, G.A., and Caelli, T., Discrimination of lianas and trees with leaf-level hyperspectral data, *Remote Sensing of Environment*, 90, 353, 2004.
69. Nobel, P.S. and De la Barrera, E., Nitrogen relations for net CO_2 uptake by the cultivated hemiepiphytic cactus, *Hylocereus undatus*, Scientia Horticulturae, 96, 281, 2002.
70. Taiz, L. and Zeiger, E., *Plant physiology*, 3rd ed. Sinauer Associates, Sunderland, MA, 2002.
71. Hunt, J.E.R., Rock, B.N., and Nobel, P.S., Measurement of leaf relative water content by infrared reflectance, *Remote Sensing of Environment*, 22, 429, 1987.
72. Castro-Esau, K.L., Sanchez-Azofeifa, G.A., Rivard, B., et al., Variability in leaf optical properties of Mesoamerican trees and the potential for species classification, *American Journal of Botany*, 93, 517, 2006.
73. Short, N. Hyperspectral imaging, in *Remote sensing tutorial*. 2003 [cited; available from http://rst.gsfc.nasa.gov/Front/tofc.html.NASA].
74. Blackburn, G.A., Relationships between spectral reflectance and pigment concentrations in stacks of deciduous broadleaves, *Remote Sensing of Environment*, 70, 224, 1999.
75. Elvidge, C.D. and Chen, Z.K., Comparison of broadband and narrowband red and near-infrared vegetation indices, *Remote Sensing of Environment*, 54, 38, 1995.
76. Thenkabail, P.S., Enclona, E.A., Ashton, M.S., et al., Hyperion, IKONOS, ALI, and ETM+ sensors in the study of African rainforests, *Remote Sensing of Environment*, 90, 23, 2004.
77. Thenkabail, P.S., Smith, R.B., and De Pauw, E., Hyperspectral vegetation indices and their relationships with agricultural crop characteristics, *Remote Sensing of Environment*, 71, 158, 2000.
78. Lee, K.S., Cohen, W.B., Kennedy, R.E., et al., Hyperspectral versus multispectral data for estimating leaf area index in four different biomes, *Remote Sensing of Environment*, 91, 508, 2004.
79. Kalacska, M., Sanchez-Azofeifa, G.A., Rivard, B., et al., Ecological fingerprinting of ecosystem succession: Estimating secondary tropical dry forest structure and diversity using imaging spectroscopy, *Remote Sensing of Environment*, 108, 82, 2007.
80. Food and Agricultural Organization (FAO), *Global forest resources assessment*. FAO of the United Nations, Rome, 2001.
81. Steininger, M., Satellite estimation of tropical secondary forest aboveground biomass: Data from Brazil and Bolivia, *International Journal of Remote Sensing*, 21, 1139, 2000.

82. Thenkabail, P.S., Smith, R.B., and De Pauw, E., Evaluation of narrowband and broadband vegetation indices for determining optimal hyperspectral wavebands for agricultural crop characterization, *Photogrammetric Engineering and Remote Sensing*, 68, 607, 2002.
83. Held, A., Ticehurst, C., Lymburner, L., et al., High resolution mapping of tropical mangrove ecosystems using hyperspectral and radar remote sensing, *International Journal of Remote Sensing*, 24, 2739, 2003.
84. Schnitzer, S.A. and Bongers, F., The ecology of lianas and their role in forests, *Trends in Ecology and Evolution*, 17, 223, 2002.
85. Korner, C., Slow in, rapid out-carbon flux studies and Kyoto targets, *Science*, 300, 1242, 2003.
86. Blackburn, G.A. and Milton, E.J., Seasonal variations in the spectral reflectance of deciduous tree canopies, *International Journal of Remote Sensing*, 16, 709, 1995.
87. Salovaara, K.J., Thessler, S., Malik, R.N., et al., Classification of Amazonian primary rain forest vegetation using Landsat ETM+ satellite imagery, *Remote Sensing of Environment*, 97, 39, 2005.
88. Achard, F., Eva, H., and Mayaux, P., Tropical forest mapping from coarse spatial resolution satellite data: Production and accuracy assessment issues, *International Journal of Remote Sensing*, 22, 2741, 2001.
89. Clark, D.B., Clark, D.A., and Read, J.M., Edaphic variation and the mesoscale distribution of tree species in a neotropical rain forest, *Journal of Ecology*, 86, 101, 1998.
90. Hess, L.L., Melack, J.M., Novo, E.M.L.M., et al., Dual-season mapping of wetland inundation and vegetation for the central Amazon basin, *Remote Sensing of Environment*, 87, 404, 2003.
91. Foody, G.M., Remote sensing of tropical forest environments: Towards the monitoring of environmental resources for sustainable development, *International Journal of Remote Sensing*, 24, 4035, 2003.
92. Foody, G.M., Palubinskas, G., Lucas, R.M., et al., Identifying terrestrial carbon sinks: Classification of successional stages in regenerating tropical forest from Landsat TM data, *Remote Sensing of Environment*, 55, 205, 1996.
93. Alves, D., Meira Filho, L., d'Alge, J., et al., The Amazonia information system, in ISPRS 17 in *ISPRS*, Washington, D.C., 1992.
94. Brown, S. and Lugo, A.E., Tropical secondary forests, *Journal of Tropical Ecology*, 6, 1, 1990.
95. Uhl, C., Buschbacher, R., and Serrao, E.A.S., Abandoned pastures in Eastern Amazonia. I. Patterns of plant succession, *Journal of Ecology*, 76, 663, 1988.
96. Castro-Esau, K., Sanchez-Azofeifa, G.A., and Rivard, B., Monitoring secondary tropical forests using space-borne data: Implications for Central America, *International Journal of Remote Sensing*, 24, 1853, 2003.
97. Lucas, R.M., Honzak, M., Curran, P.J., et al., Mapping the regional extent of tropical forest regeneration stages in the Brazilian legal Amazon using NOAA AVHRR data, *International Journal of Remote Sensing*, 21, 2855, 2000.
98. Schmidtlein, S. and Sassin, J., Mapping of continuous floristic gradients in grasslands using hyperspectral imagery, *Remote Sensing of Environment*, 92, 126, 2004.
99. Koltunov, A., Ustin, S.L., Ashner, G.P., et al., Selective logging changes forest phenology in the Brazilian Amazon, *Proceedings of the National Academy of Sciences*, in press.
100. Asner, G.P., Keller, M., Pereira, R., et al., Canopy damage and recovery after selective logging in Amazonia: Field and satellite studies, *Ecological Applications*, 14, S280, 2004.

3 Hyperspectral Data for Assessing Carbon Dynamics and Biodiversity of Forests

Richard Lucas, Anthea Mitchell, and Peter Bunting

CONTENTS

3.1 INTRODUCTION

The forests of the tropics and subtropics represent a diversity of habitats (e.g., rainforests, mangroves, and wooded savannas) that vary both spatially and temporally. Spatial variables include species diversity, structural attributes (e.g., height, cover, stem density, and vertical stratification), and biomass and are influenced by factors such as soils, geology, climate, topography, and the past biogeographic distributions of species. Temporal variables relate to seasonal phenology (e.g., period of leaf cover ranging from deciduous to evergreen) and growth stage and are influenced primarily by climate (e.g., drought) and hydrology (e.g., flood or tidal inundation). These forests are also changing over time as a consequence of disturbance (e.g., fires, logging) or enhancement of growth (e.g., woody thickening [1]). Such changes are attributable to both natural and direct (e.g., deforestation) or indirect (e.g., climate change) anthropogenic causes. A large proportion of forests are also secondary [2] and exist at varying stages of degradation or regeneration [3,4].

For remote sensing scientists, this spatial and temporal variation represents both an opportunity and a challenge for the use of hyperspectral data. In terms of opportunities, hyperspectral data have provided new options for assessing biological diversity [5] and contributed to assessments of dead and live carbon [6], measures of forest health [7,8], and understanding of ecosystem processes (e.g., through retrieval of foliar biochemicals; [9]). However, the utility of these data in the tropics and subtropics has been limited by environmental conditions (e.g., persistent cloud cover and haze), sensor characteristics (e.g., low dynamic range and signal-to-noise ratios, atmospheric effects, and image artifacts), and the complexity of the forested environment in terms of species diversity, multilayering, and shadowing effects.

This chapter provides on overview of the use of hyperspectral data in tropical and subtropical forests and focuses primarily on forest types that are prevalent in these regions—namely, rainforests (evergreen and semi-evergreen), mangroves, and wooded savannas. The chapter conveys key features of hyperspectral data that allow different levels of information on forests to be extracted compared to multispectral counterparts. Using previously published research and case studies from Brazil and Australia, the use of hyperspectral data for assessing forest biodiversity, carbon dynamics, and health is demonstrated. Finally, the chapter provides some indication of the future directions for hyperspectral remote sensing of tropical and subtropical forests and outlines how existing and future sensors might be integrated to provide options for conserving, restoring, and sustainably utilizing forests.

3.2 HYPERSPECTRAL REMOTE SENSING: RELEVANCE TO FORESTS

3.2.1 Key Benefits of Hyperspectral Remote Sensing Data

The acquisition of hyperspectral data is advantageous in forest studies for several reasons. The provision of data in contiguous narrow bands across the 400- to 2500-nm wavelength range [10] allows differences in the magnitude and shape of spectral reflectance curves for different surfaces to be identified. Such differences can be exploited for discriminating tree species and are often most evident (although not exclusively so) in the near infrared (NIR) wavelength regions. The large number of bands available also allows minor absorption features to be located and described (in terms of wavelength position, magnitude, width, and depth). These features can be related to, for example, the amount of foliar biochemicals (e.g., nitrogen, chlorophyll, and lignin) within plants and used to better understand ecosystem processes (e.g., photosynthesis). Hyperspectral data also allow the scene to be decomposed such that the relative contributions of, for example, soil and water to the overall reflectance can be determined. Such information can be exploited to better quantify the structure of forests or understand ecological functioning and biogeochemical cycling [6].

The observing characteristics of hyperspectral sensors also impact on the type and amount of information that can be retrieved. Most hyperspectral sensors are mounted on aircraft and typically observe at very fine (herein referred to as ≤1 m) and fine (1 to <5 m) spatial resolutions, although other airborne (e.g., AVIRIS) and space-borne (e.g., Hyperion) sensors operate at moderate (5–30 m) spatial resolutions [11]. Sensors observing at finer spatial resolution allow individual trees to be resolved but as the spatial resolution decreases, contributions to the overall spectral reflectance by different surface materials increase, which leads to greater mixing within pixels and the loss of visual detail. Data acquired by fine to very fine spatial resolution sensors are generally used for localized small area studies, while those operating at coarser spatial resolutions provide a wider view of the landscape. As most hyperspectral observations are from aircraft, opportunities for multitemporal analyses have generally been restricted.

3.2.2 PIXEL- VS. OBJECT-BASED ANALYSIS

Historically, most remote sensing studies have focused on a pixel-based analysis where the only statistic available is the intensity (for each wavelength) unless kernels that consider neighboring pixels are used (e.g., for deriving texture measures or detecting edges). However, these are not sympathetic to the boundaries of the image features. For this reason, object-oriented methods are increasingly being investigated, particularly with improvements in computing power and the development of image processing suites such as the Definiens software [12] and SPRING [13]. Object-oriented methods have introduced a preprocessing step that generates segments (or objects) within an image that can be manipulated, described, or classified subsequently to represent real-world features of the landscape. These features might include, for example, tree crowns or discrete forest stands [14,15]. By grouping pixels into objects, attributes other than the pixel intensity (e.g., standard deviation, object shape, area and perimeter, proximity and enclosure measures, and texture measures generated from the constituent pixels) can also be used to assist classification of the scene. As a consequence, accuracies of classification are often increased beyond what is normally achievable using a per-pixel–based approach. Objects can also be merged to form larger objects (e.g., individual tree crowns can be combined to generate maps of the forested area) or can be located in hierarchical layers within which common object boundaries are maintained but the size and number of objects can vary.

3.2.3 SPECTRAL UNMIXING

Unmixing the proportional contribution of materials to the overall pixel or object value is a procedure commonly used with hyperspectral remote sensing. Developed procedures include linear spectral unmixing [16,17], which typically uses, as input, pure spectra for different materials that are referred to as endmembers. Endmember spectra are either extracted from known features within the scene itself or from field or laboratory spectroradiometer measurements. The result of the process is generally a set of images, each of which represents the proportional contribution of the endmember to each pixel within the original data. These proportions can then be used to provide information on the scene, which differs from that obtained from the original reflectance data in that it can often be related to the distribution of biophysical attributes (e.g., the amount of dead woody material) and interpreted accordingly. For example, using multispectral Landsat sensor data of tropical forests in Brazil, Adams et al. [18] illustrated how linear spectral unmixing could be used to retrieve the proportions of photosynthetic vegetation (i.e., green leaves), soil background, and shade endmembers within the scene. Using the endmember fraction images, regenerating forests of different ages and pathway could be discriminated because of the increased shadow in the crown associated with the structural development as the forests aged and differences in the leaf and canopy structure associated with the pioneer genera dominating [19]. Asner and Heidebrecht [6] also highlighted the importance of the nonphotosynthetic vegetation (NPV; e.g., wood) endmember as a common contributor to the reflectance of semi-arid vegetation.

3.2.4 Forest/Nonforest Discrimination

The detection of the forest boundary is a fundamental requirement for mapping and monitoring forests but approaches vary depending upon the forest biome being considered. In particular, forests vary in terms of their canopy closure and also their spectral contrast with adjoining environments. For example, the relatively closed canopy of tropical rainforests provides a distinct contrast with adjoining nonforest environments (e.g., pastures or savannas) but often not with mangroves [20]. In other environments (e.g., woodlands and open forests; [21]), definition of the forest extent is complicated by the different density and crown size distributions of trees and the area mapped often differs depending upon the spatial resolution of the observing sensor.

From broadband optical sensors, discrimination of forest and nonforest vegetation has often been achieved using vegetation indices such as the normalized difference vegetation index (NDVI) or endmember fractions and best results are obtained where the amount of vegetation outside the forested area is minimal. Some studies (e.g., Gougeon and Leckie [22]) have suggested that textual differences between the forested and nonforested areas might also assist discrimination. These same approaches can be applied to hyperspectral data, but additional indices are also available for forest boundary detection. For example, Bunting and Lucas [23] described a forest discrimination index (FDI) that is applicable to hyperspectral data acquired at very fine (e.g., compact airborne spectrographic imager [CASI]), fine (e.g., HyMap), and moderate (e.g., Hyperion) spatial resolutions where

$$FDI = \rho_{838} - (\rho_{714} + \rho_{446}) \tag{3.1}$$

and ρ represents reflectance at particular wavelength centers (e.g., 838 nm). The FDI was developed for subtropical woodlands and open forests and identifies only those pixels where the red edge [24] is reasonably well defined in the spectral profile. A threshold of zero separates forest from nonforest areas and the index is best applied where semi-arid, dry conditions persist and the amount of photosynthetic material within the understory (herbaceous and shrub) layer is minimal.

3.2.5   TREE LOCATIONS

Within the forested area and using very fine spatial resolution hyperspectral data, the location of trees can be approximated by identifying pixels associated with the brightest sunlit component of the conically-shaped crowns. Typically, these pixels represent the high points (tops) of trees, particularly where conically shaped crowns occur (e.g., coniferous species or trees in the earlier phases of growth [25]). However, in some forests (e.g., those dominated by decurrent growth forms with large expansive crowns), the high (bright) points can also occur on the outer edges of crowns.

Hyperspectral data provide a number of reflectance channels that can be used to better identify high points within crowns, with many of these based on the red edge and NIR bands. Such indices are often more successful in more open forests, where crowns are relatively expansive and isolated and typical examples include the C1 and C2 ratios [26]:

$$C1 = \frac{\rho_{740}}{\rho_{680}} \tag{3.2}$$

and

$$C2 = \frac{\rho_{740}}{\rho_{714}} \tag{3.3}$$

However, comparisons with light detection and ranging (LiDAR) data [27–29] suggest that the bright points identified do not always represent the highest points in the crown and may be inconsistent within and between the bands or indices used.

3.2.6   CROWN DELINEATION

The delineation of individual or clusters of tree crowns is enhanced using hyperspectral data, largely because of the inclusion of red edge bands. The fundamental assumption of all crown delineation algorithms is the formation of a "hill and valley" structure within the image whereby crowns form the hills (with the bright point typically associated with the upper portions) and shadowing creates radiometric valleys in-between. Three philosophies (valley following, bright point expansion, and template matching) that all revolve around this basic assumption have been proposed and demonstrated for the extraction of crowns.

Valley-following techniques identify "valley" contours within the radiometric surface to separate crowns. This technique was first developed on Canadian forests [30] but has particular application to closed forests in the tropics and subtropics.

FIG. 3.1   (a) Compact Airborne Spectrographic Imager (CASI) image of a rainforest canopy in Australia and (b) crowns delineated using the algorithm of Gougeon [30]. (Reprinted with permission of Taylor & Francis from Ticehurst, C. et al., in *Third International Conference on Geospatial Information in Agriculture and Forestry*, Denver, CO, 2001.) See CD for color image.

The algorithm first thresholds the image to identify low points (i.e., valley seeds) which are then expanded to fill the gaps between crowns. A rule-base is then used to "jump" between valleys such that crowns within the "hillscape" are split. A recognized limitation of the technique was that gaps of one pixel are retained between crowns and this restricted use in multilayered forests where crowns often overlap and/or adjoin. However, the algorithm has been advanced subsequently and applied successfully to a range of forest structures using data from several hyperspectral and multispectral sensors (fig. 3.1) [22,30,31].

Crown expansion techniques identify the bright points within an image (i.e., the hill tops) and then expand these down to the valleys (equivalent to the crown edge), splitting the surface into crowns while doing so. The valleys within the image are identified in a similar way to the valley-following methods and are used to limit the expansion of the crown centroids. An early example was presented by Culvenor [32], who focused on Eucalyptus-dominated forests in southeast Australia, which supported a relatively closed and homogeneous canopy. While applicable to closed forests, an alternative object-orientated technique was developed by Bunting and Lucas [23] to delineate crowns in subtropical forests in Australia where canopies were more open and consisted of crowns of varying dimensions. In particular, the crowns of many of the taller (>25 m high) *Eucalyptus* and *Angophora* species were expansive and contained a number of locally bright points that were similar in appearance to many of the smaller, more compact crowns of *Callitris* and *Lysocarpus* species. An iterative approach was therefore introduced whereby large differences in the hill and valley surfaces were first identified to split delineated objects along an identified valley. These objects were then classified as crowns using a rule base of shape features and, once identified, were removed from subsequent processing steps. Remaining objects were then split further using other rules that were based around smaller changes (valleys) in the radiometric surface with the aim of identifying smaller crowns. The processes of splitting and classification were repeated for a user-defined number of iterations or until all areas of forest had been assigned as a crown. A process that identified and corrected errors in the delineation process was then implemented.

Template matching techniques use a series of templates that describe and specify the appearance of crowns associated with different species when observed under different viewing and illumination directions. Template matching was first introduced by Pollock [33], who worked on coniferous forests in Canada where the relatively homogeneous nature of the crowns corresponded closely to the hill and valley assumption and was an ideal system for this approach. The technique involves the initial creation of templates representing the appearance of crowns of selected species under different viewing conditions and ideally encompasses all variation within the forest. These templates are then matched to areas of the image and, where a match occurs, the crown is extracted. An advantage of this technique is that the species associated with each crown can be identified if the template is specific to those present.

Each of these three methods has typically generated accuracies of ~70% when applied to the forest on which they were developed, although many evaluated accuracies only on those trees with diameters at breast height (DBH) ≥10 cm. However, accuracies generally decrease with the complexity of the forest. For example, Bunting and Lucas [23] reported accuracies of over 70% for open forests but these decreased to 48% for forests where the stem density was greater, although this included trees with diameters ≥5 cm. This reduction in accuracy was attributed also to intercrown shadowing and overlapping of neighboring crowns in multiple layers. Methods are also often not transferable between forest types. For example, Leckie et al. [31] applied the approach of Gougeon [30] to more complex old-growth coniferous forest stands but only achieved accuracies of 50–60%. Within all methods, divergence has occurred and hence a large number of papers have been published in the field. Many of these approaches are first generation and, to increase the accuracy of crown identification and the extraction of associated tree and crown parameters, are likely to be expanded in the future by incorporating textural measures [34], spectral endmember proportions [15], expert system rule bases (e.g., canopy gap masking; [35]), and other computer vision techniques [36].

3.2.7 Extraction of Spectra for Species Discrimination

Differences in the spectral reflectance of tree crowns are variable between wavelength regions but are greatest within the NIR and, to a certain extent, the shortwave infrared (SWIR) wavebands. Differences in the visible wavelengths are attributable largely to variations in the amount and type of photosynthetic pigments. Within the NIR region, reflectance is largely a function of the type, density, and arrangement of leaves as these influence photon scattering [37]. The SWIR reflectance is related more to differences in crown water concentration and, in particular, the expression of water absorption features at 1400, 1900, and 2700 nm [38] and also the lignin-cellulose absorption features associated with high reflectance contributions from NPV.

The classification of tree species from hyperspectral data has been based on pixels or objects and has utilized a wide range of algorithms, including supervised (e.g., maximum likelihood) classifications (often with a selection of bands determined beforehand through, for example, forward stepwise selection based on linear or multiple discriminant analysis (LDA/MDA) [37,39,40] and spectral angle mapping (SAM [37,41])). In many cases (e.g., Lucas et al. [5] and Clark, Roberts, and Clark [37]), the use of LDA has proved optimal for discrimination. At the tree level, greater accuracies of discrimination can generally be retrieved by (a) extracting spectra from the meanlit area [5,37] rather than the maximum (brightest) point within the crown;

(b) defining this area using specific bands (e.g., the 800-nm NIR [30,37]) or ratios and particularly those involving the red edge (e.g., C2); (c) integrating data across spectral regions (i.e., including the SWIR); and (d) focusing on a fewer number or a broader range of species (e.g., based on different genera). Some studies (e.g., Meyer, Staenz, and Itten [14]) suggested that classes relating to species could also be assigned based on the majority allocated to pixels contained within a delineated crown.

Vegetation discrimination has been highlighted as one of the benefits of using hyperspectral data and the establishment of spectral libraries for different species has been suggested [43]. The ability to discriminate trees to species can be limited, however, by reflectance contributions from other canopy components (e.g., NPV) and the soil background, the greater spectral variability within trees of the same species than between trees of different species, the occurrence of many in the subcanopy layers such that they are obscured from the view of the observing sensor or are shadowed, the variations associated with health (e.g., insect or fungal damage) and regrowth stage, and also the spatial resolution of the observing sensor [5,37]. A limitation also is knowledge of the species occurring within the scene as often rare or less common species occur but are not included in the training or validation data sets.

3.2.8 SPECTRAL INDICES

With the provision of narrowband hyperspectral data, a number of indices have been developed to assist the detection of green (live), non-photosynthetic, and stressed vegetation and the retrieval of foliar chemicals including leaf pigments (e.g., chlorophyll) and structural elements (e.g., lignin and cellulose). As indicated previously, some indices have utility for differentiating forests from adjacent land covers and as input to tree top identification and crown delineation algorithms. Table 3.1 lists key indices that are considered to be useful for hyperspectral remote sensing of tropical and subtropical forests.

3.3 BIODIVERSITY

Hyperspectral data can provide new information on forest diversity, primarily because of the greater amount of spectral information that is available for tree species and forest community discrimination. The following sections outline how hyperspectral remote sensing data can be used to augment our knowledge of tree species distributions across a range of tropical and subtropical forest biomes and thereby contribute to our understanding of biodiversity in general.

3.3.1 TROPICAL FORESTS

Tropical forests maintain over 50% of the world's plant species [44–46] and represent a major repository of global biodiversity. However, detailed spatial information on the composition and distribution of tree species and associated floral and faunal diversity is limited to only a few areas [37]. Furthermore, many forests are now secondary [2] and knowledge of the composition and turnover of species during the succession and their capacity to recovery tree species diversity to predisturbance levels is needed but is limited at the landscape scale.

TABLE 3.1
Known Sensitivities and Key Vegetation Indices Used in the Assessment of Carbon Dynamics and Forest Health/Condition

Acronym	Parameter	Wavelength sensitivity/vegetation index	Ref.
		Green (live) vegetation	
NDVI	Normalized NDVI[a]	$(NIR - R)/(NIR + R)(\rho750 - \rho680)/(\rho750 + \rho680)$	130
Green NDVI	Green NDVI[b]	$(NIR - green)/(NIR + green)(\rho NIR - \rho540 - 570)/(\rho NIR + \rho540 - 570)$	71, 131
SR	Simple ratio index	$\rho800/\rho680$	10
RE NDVI	Red edge NDVI	$(\rho750 - \rho705)/(\rho750 + \rho705)$	132
mSR$_{705}$	Modified red edge simple ratio index	$(\rho750 - \rho445)/(\rho705 - \rho445)$	133
mNDVI$_{705}$	Modified red edge NDVI	$(\rho750 - \rho705)/(\rho750 + \rho705 - 2\rho445)$	133
		Photosynthesis	
	Total chlorophyll concentration[e]	$\rho672/(\rho550 \times \rho708)(\rho850 - \rho710)/(\rho850 - \rho680)$	71
PRI	Photochemical reflectance index[d]	$(\rho531 - \rho570)/(\rho531 + \rho570)$	134
RG ratio	Red green ratio index[e]	$\sum_{i=600}^{699} R_i \Big/ \sum_{i=500}^{599} R_i$	135
		Canopy biochemicals—leaf pigments	
CRI1	Carotenoid reflectance index 1	$(1/\rho510) - (1/\rho550)$	136
CRI2	Carotenoid reflectance index 2	$(1/\rho510) - (1/\rho700)$	136
ARI2	Anthocyanin reflectance index 2	$\rho800\,[(1/\rho550) - (1/\rho700)]$	137
NDNI	Normalized difference nitrogen index[f]	$[\log(1/\rho1510) - \log(1/\rho1680)]/[\log(1/\rho1510) + \log(1/\rho1680)]$	138
		Canopy water	
WBI	Water band index[g]	$\rho900/\rho970$	139
MSI	Moisture stress index	$\rho1599/\rho819$	140
NDWI	Normalized difference water index	$(\rho860 - \rho1240)/(\rho860 + \rho1240)$	141
NDII	Normalized difference infrared index	$(\rho819 - \rho1649)/(\rho819 + \rho1649)$	142
SWAM	Spectroscopic water absorption metric	$[\int_{\rho=930}^{1040} \max\ (\rho930 - 1040) - \rho] \times \rho = 930$ $[\int_{\rho=1100}^{1230} \max\ (\rho1100 - 1230) - \rho]\ \rho = 1100$	10

continued

TABLE 3.1 (continued)
Known Sensitivities and Key Vegetation Indices Used in the Assessment of Carbon Dynamics and Forest Health/Condition

Acronym	Parameter	Wavelength sensitivity/vegetation index	Ref.
		Nonphotosynthetic vegetation	
CAI	Cellulose absorption index[h]	$0.5 [(\rho 2000 - \rho 2200)/\rho 2100]$	143
NDLI	Normalized difference lignin index[i]	$[\log (1/\rho 1754) - \log (1/\rho 1680)]/[\log (1/\rho 1754) + \log (1/\rho 1680)]$	138
PRSRI	Plant senescence reflectance index[j]	$(\rho 680 - \rho 500)/\rho 750$	144
		Health	
RVSI	Red-edge vegetation stress index[k]	$[(\rho 714 + \rho 752)/2] - \rho 733$	145
	Wavelength position of red edge[l]	Red edge inflection point near 733 nm	145
REP	Red edge position index[m]	Wavelength of maximum derivative of reflectance from 690 to 740 nm	73

[a] Estimate of green vegetation cover.
[b] Five times more sensitive than NDVI to chlorophyll concentration.
[c] Chlorophyll-a + chlorophyll-b.
[d] Measures the down-regulation of photosynthesis during stress.
[e] Indicative of leaf redness caused by anthocyanins.
[f] Relative canopy nitrogen content.
[g] Relative water content.
[h] Relative amount of dry plant material.
[i] Relative canopy lignin content.
[j] Plant litter (dry and decomposing leaves).
[k] Identifies inter- and intracommunity multitemporal stress trends based on spectral changes in upper red-edge geometry.
[l] Main inflection point of the slope between red and NIR. A shift toward shorter (longer) wavelengths indicates increasing stress (vigor).
[m] Sensitive to changes in chlorophyll concentration.

The main difficulty of using optical remote sensing data to discriminate tree species is that the diversity within tropical rainforests is high and species are typically rare. Therefore, extensive areas have to be searched to obtain a representative sample for training classifications [37]. The constant growing season also results in few systematic and predictable patterns in flowering, fruiting, and leaf fall [37] and hence confuses selection of an optimal time period for observation. Within-canopy variation, even within the same tree, is commonplace as leaves may be of varying age (with associated differences in maturation of the mesophyll and concentrations of pigments and water) and suffer from different levels of necrosis, herbivory, and epiphyll cover [37,47] (fig. 3.2). At the crown level, contributions from nonfoliar surfaces (e.g., bark) and variations in shading and anisotropic multiple scattering

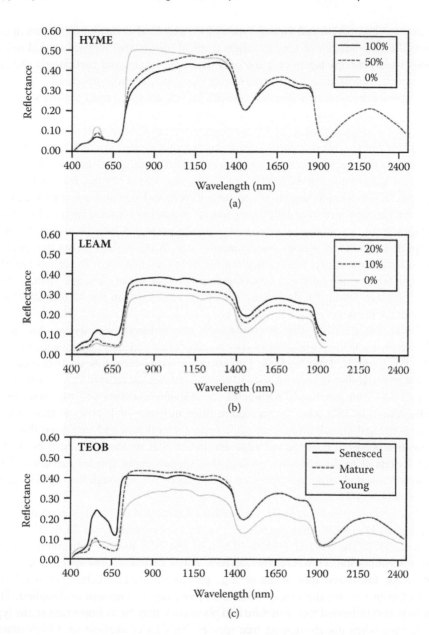

FIG. 3.2 The impact of changing (a) cover of a single species of epiphyll, (b) area of leaf herbivory, and (c) leaf age on the spectral reflectance of the tropical rainforest species *Hymenolobium mesoamericanum* (HYME), *Lecythis ampla* (LEAM), and *Terminalia oblonga* (TEOB). All spectra were taken from upper canopy leaves. The senescent leaf of *Terminalia* was collected from the ground. (Reprinted from Clark, M. et al., *Remote Sensing of Environment*, 96, 375–398. With permission from Elsevier.)

relative to illumination and view geometry [37] may lead to further confusion in the spectral discrimination of species. Many tropical forests also support several (often three or four) canopy layers [44] and only the upper layers, and particularly the tall emergent trees with their large expansive crowns, are generally observed.

Using moderate spatial resolution multispectral sensors, broad classifications of forest type (e.g., Nelson [48]) and also the composition and stage of forests following different pathways of regeneration have been obtained. For example, Lucas et al. [19] noted that forests following different pathways of regeneration and dominated by the pioneer species *Vismia* and *Cecropia* exhibited spectral reflectance trajectories that were most distinct in the SWIR and, in the early stages of regeneration, the NIR regions. This was attributed largely to variations in leaf size and orientation as well as overall canopy structure [19]. Finer spatial resolution sensors, such as IKONOS and Quickbird, are providing data over tropical forests that have been used for tree crown delineation and species discrimination (e.g., Wang et al. [49]), but the spectral resolution and the existence of only a few bands in broad wavelength regions are still not optimal. For this purpose, hyperspectral data that are of very fine spatial resolution and can detect discriminatory spectral features within the 400- to 2500-nm range are required.

Some success with the discrimination of rainforest tree species has been achieved using field and laboratory spectroradiometer and also hyperspectral image data. For example, Fung and Siu [39] used simulated branch-scale hyperspectral visible and NIR data to discriminate 12 subtropical tree species with an overall accuracy of 84% and producer's accuracies for individual species ranging from 56 to 100%. Cochrane [50] achieved reasonable discrimination of 11 tropical tree species using simulated branch and crown-scale hyperspectral data and suggested that best separation occurred near the red edge and in the NIR regions. Ticehurst et al. [51] also attempted to discriminate tropical tree species using spectra extracted from crowns delineated using the algorithm of Gougeon [30], although the high diversity of species limited success.

For seven targeted species in Costa Rican forests and using LDA, meanlit spectra, and 60 HYDICE bands, producer's accuracies ranging from 74 to 95% were obtained [37]. The SWIR1 (~1420–1920 nm) and NIR bands contributed most to this discrimination. However, the analysis noted that over 300 species occurred within the forest and that different (e.g., probability-based) techniques would be needed to ensure that the species other than those targeted remain unclassified. The analysis also indicated that structure and phenology may be as important as the leaf reflectance when discriminating tree species. This latter analysis also highlighted the benefits of hyperspectral over multispectral data as higher classification accuracies were obtained and measurements of the shape and position of key spectral features, such as the liquid water absorption features in the NIR, could be utilized. Purely analytical hyperspectral techniques, including spectral shape filters [50] and first- and second-order derivatives, have been shown to increase the accuracy of tree species discrimination over reflectance spectra [42,52], although many conclusions have been based on field or laboratory spectroscopy rather than image data. In most cases, the accuracies obtained were greater than could be achieved through visual analysis (e.g., of aerial photography) alone. Computer-based classifications also permit the automation and removal of subjectivity from the process [37].

3.3.2 WOODED SAVANNAS

Wooded savannas are distributed throughout the tropics and subtropics and are well represented on either side of the tropical forest belt but particularly in southern Africa, Brazil, Central America, and Australia. Compared to tropical rainforests, the diversity of tree species within wooded savannas is typically less and the wider spacing of crowns facilitates better delineation and their discrimination to species. Using hyperspectral data, many studies have focused on more open forests (including woody savannas) in Australia [5,53,54].

Commonly, stereo aerial photography has been used to outline the extent of different forest structural formations and the species contained are described primarily through visual interpretation. As an example, Tickle et al. [55] assessed the species composition of forests near Injune in central Queensland, Australia, by interpreting 150 true color stereo aerial photographs acquired over 150- × 500-m areas in a 10 × 15 grid, with each separated by 4 km in the east–west and north–south directions. As with wooded savannas worldwide, many of the forests were dominated by a few genera—namely, *Eucalyptus, Angophora, Callitris,* and *Acacia*—and many were occurring in association with and as a function of terrain, climate, and soils. Hyperspectral CASI (1-m spatial resolution) and HyMap (2.6-m spatial resolution) were acquired over the same grid and time period data to establish whether refinements to the species mapping could be achieved using hyperspectral data. The CASI and HyMap data sets were each calibrated to surface reflectance using an empirical line technique and HyCorr atmospheric correction software, respectively, such that the reflectance of surfaces within both data sets was similar. These data sets were then registered to each other using automated tie-point extraction and registration procedures [5]. To generate maps of tree species distributions, crowns of varying size and shape were first delineated within the CASI data using a crown expansion algorithm [23] and spectra were then extracted using a threshold of the C2 ratio (equation 3.2) to define the meanlit area. LDA was then applied to spectra extracted from delineated crowns of known species, as identified by referencing forest inventory data and stereo aerial photography. For the LDA classification, the pooled within-class covariance matrix and predictor reflectance band variables from the training samples were used to construct discriminant functions for each class. These functions were then used to associated delineated crowns with a species type.

Using the CASI data alone, 81.7% of the crowns used to train the classification were correctly classified and the classification accuracy for any one species exceeded 72% (table 3.2). Accuracies in the classification of the testing data set were lower (75.9%) but those for the individual species were still above 66%. The functions were then used subsequently to classify all unknown delineated crowns, thereby allowing tree species maps to be generated from the CASI data (fig. 3.3). By integrating spectra extracted from HyMap data, accuracies in the overall classification increased for both the training and testing data sets (87.4 and 86.1%, respectively) and for the individual species concerned. Within the CASI data, most confusion was between

TABLE 3.2
Indicative Accuracies (%) in the Classification of Delineated Tree Crowns to Species Using Mean Lit Spectra Extracted from CASI and HyMap Data

Total: 242		ANE	BGL	CP-	PBX	SBA	SLI	SWB	Total
CASI	Training (%)	88.9	85.7	72.7	81.3	75	77.3	100	81.7
	Testing (%)	66.7	100	66.7	71.4	66.7	82.1	80	75.9
HyMap with SWIR	Training (%)	88.9	100	94.1	96.7	66.7	82.6	100	87.4
	Testing (%)	85.7	100	90.9	82.6	71.4	92	90	86.1

Notes: ANE = *A. neriifolia*; BGL = *A. harpophylla*; CP- = *C. glaucophylla*; PBX = *E. populnea*; SBA = *A. leiocarpa*; SLI = *E. melanaphloia*; SWB = *E. mitchelli*.

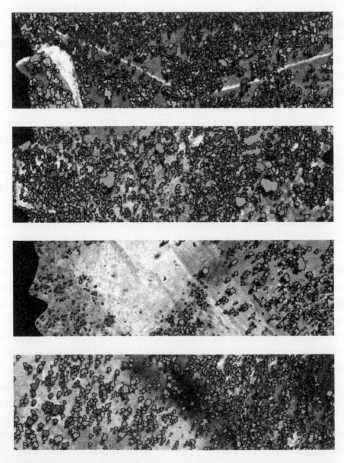

FIG. 3.3 Classifications of tree species delineated for four 500- × 150-m areas and overlain onto the CASI imagery: *C. glaucophylla* (dark green), *E. melanaphloia* (cyan), *E. populnea* (orange), *A. leiocarpa* (orange; B only), *E. chlorochlada* (pink), *A. harpophylla* (magenta), *E. mitchelli* (mid-green), and understory (light green). See CD for color image.

TABLE 3.3

Indicative Accuracies (%) in the Classification of Delineated Tree Crowns to Species Using Mean Lit Spectra Extracted from CASI and HyMap Data

Total: 242		ANE	BGL	CP-	PBX	SBA	SLI	SWB	Total
HyMap without SWIR	Training (%)	85.2	66.7	76.5	90	61.9	87	100	81.9
	Testing (%)	67.9	100		81.8	82.6	57.1	72	100	75.7

Notes: ANE = *A. neriifolia*; BGL = *A. harpophylla*; CP- = *C. glaucophylla*; PBX = *E. populnea*; SBA = *A. leiocarpa*; SLI = *E. melanaphloia*; SWB = *E. mitchelli*.

silver-leaved ironbark (*E. melanaphloia*) and white cypress pine (*C. glaucophylla*) and also *Acacia* species (e.g., *A. neriifolia*) and both poplar box (*E. populnea*) and sandalwood box (*E. mitchelli*), but this was reduced when the HyMap SWIR data were included.

To evaluate whether the increase in classification accuracy was attributable to the inclusion of the HyMap SWIR wavebands, the classification procedure was repeated using only the HyMap wavebands equating to the spectral range of the CASI data (i.e., 445–837 nm). Here, the overall classification accuracies were reduced (table 3.3) compared to when the SWIR wavebands were included (table 3.2) and were similar to those obtained using the CASI data. Classification accuracies for some species (e.g., *E. mitchelli*, *E. populnea*) were increased but for others (e.g., *A. leiocarpa*) a decrease was observed. The analysis therefore suggested that the SWIR wavelengths allowed species to be better classified but spectral mixing within the coarser spatial resolution (2.6 m) of the HyMap led to variations (increases or decreases) in the classification accuracy of individual species, largely because of differences in the size and architecture (e.g., leaf/branch size and density) of their crowns.

To expand the classification of tree crowns beyond the 150- × 500-m area of the CASI scenes, a segmentation technique developed within Definiens Professional [12] was applied to the HyMap data. This segmentation was designed to separate large individual crowns (e.g., those belonging to individuals of *A. leiocarpa*) that were visible within the 2.6-m resolution from clusters of trees with similar spectral values. The technique involved a series of spectral difference segmentations and classifications. Initially, a forest mask was generated using the FDI (equation 3.1), which was split subsequently where large differences (based on automatically identified thresholds) in spectral values between objects occurred. When objects reduced to within a certain size threshold, these were no longer split. All remaining objects were subjected to further spectral difference segmentation in an iterative process whereby the thresholds applied were reduced. The meanlit spectra for the HyMap data were then extracted (based on the C2 ratio) from the resulting segments (of known species) and used to train the classification of the remaining segments. An example of the classifications generated independently using the CASI and HyMap segmented objects is shown in figure 3.4.

The analysis highlighted the benefits of very fine spatial resolution hyperspectral data for tree species identification in wooded savannas but also indicated that high accuracies in the discrimination of species could be achieved where crowns

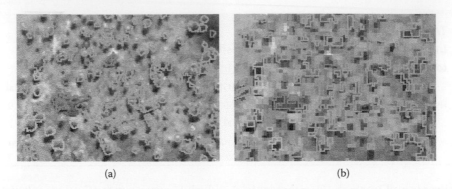

(a)   (b)

FIG. 3.4 Classifications of crown/crown clusters delineated within (a) CASI and (b) HyMap data to species: *Eucalyptus populnea* (orange), *Acacia harpophylla* (pink), *Eucalyptus melanaphloia* (purple), and *Eremophila mitchelli* (green). See CD for color image.

were relatively widely spaced and were of a size at which a reasonable (i.e., mean-lit) spectra could be extracted. At progressively coarser spatial resolutions, the likelihood of identifying individual tree crowns reduces but is dependent upon species. For example, many of the larger crowns associated with *Eucalyptus* and *Angophora* species were observed as discrete objects, even within 20-m spatial resolution Hyperion data.

3.3.3 MANGROVES

Mangrove communities differ from both tropical rainforests and subtropical wooded savannas in that they are comparatively species poor and are often dominated by one or a few species. Worldwide, there are 9 orders, 20 families, and 27 genera and between 60 and 70 species of mangroves (trees and shrubs [56,57]). The Atlantic East Pacific region is comparatively species poor (3 taxa) compared to the Indo West Pacific (17 taxa [58]).

To some extent, the relatively low diversity of mangrove species justifies the establishment of spectral libraries, but the leaf spectral reflectance properties alone often do not permit discrimination. In particular, spectral confusion occurs because of tidal (inundation) effects on the soil, differences in the physiological status of plants and their morphological properties, and the dynamics of the land/ocean interface over time (diurnally and seasonally [56,59]). The delineation of tree crowns and their discrimination to species is also complicated because of the high density of individuals and the occurrence of many in the subcanopy. Nevertheless, many mangrove species can be distinguished at the stand level as they often occur in contiguous zones that parallel the water's edge and are dominated by one or several species. When zones are mapped, a general indication of the species composition and variable descriptions of structure (e.g., high or low, dense or open) are often provided. In many cases, however, the remoteness and inaccessibility of many mangroves limits the acquisition of field and finer spatial resolution data to support these interpretations. Identification is also more difficult when mangroves occur in narrow fringes or are fragmented [60].

The greater number of wavebands associated with hyperspectral data and the inclusion of those along the red edge do increase the capacity to differentiate mangrove communities. In Kakadu National Park (KNP) in northern Australia, true color stereo aerial photographs of the West Alligator River (acquired in 1991) provided an overview of mangrove extent, and while zones could be differentiated, confusion between these occurred because of similarities in the reflectance of the dominant species in the visible wavelengths. However, when these same mangroves were observed using CASI data (acquired in 2002; fig. 3.5), the spectral differences between the seaward zone dominated by *Sonneratia alba* and those inland dominated by *S. alba, Avicennia marina, Rhizophora stylosa, R. stylosa,* and *A. marina* (on the landward side) were enhanced. Within this imagery, the crowns of larger individuals could be discerned but because most trees were densely packed (with several occurring within the area of one 1-m^2 pixel), algorithms for individual tree crown delineation and species discrimination were less applicable. Nevertheless, broad species classifications based on standard supervised (e.g., maximum likelihood) algorithms were generally sufficient and negated the need for algorithms tailored for hyperspectral analysis. As with all approaches to mapping species, rare or less common species (in this case, *Camptostemon schultzii* and *Ceriops tagal*) were unable to be discerned largely because of their presence within the subcanopy.

The ability to discriminate tree species or communities in tropical forests, mangroves, and/or wooded savannas, which is a prerequisite to biodiversity assessment, is dependent upon the spatial resolution of the observing sensor, the spectral differences and variation between and within species [37], and also the timing and frequency of observation [6]. To discriminate vegetation species at the crown or crown cluster level, for example, very fine spatial resolution data are required but their use is compromised by contributions from different components of the crown (leaves and NPV), the understory, soil, and also shadow. At fine spatial resolutions, the location of larger trees can often be resolved but as the resolution becomes coarser, species discrimination is more difficult unless species occur in relatively homogeneous stands across large areas (e.g., mangroves). In these cases, communities alone are more likely to be discriminated although some inference as to the composition of the communities in terms of, for example, dominant species can be made [61,62]. In some environments (e.g., wooded savannas), individual crowns may be too small for detection or may occur in the subcanopy. Within the tropical forests, the high diversity of species occurring within multiple layers of canopy also complicates retrieval. Therefore, the capacity to retrieve information on the diversity of species from hyperspectral data is also dependent upon the forest environment being considered and approaches and results are likely to be inconsistent within and between forest types and growth stages.

3.4   CARBON DYNAMICS

Increasingly, attention is focusing on using hyperspectral data to better understand the carbon budget of forests in the tropics and subtropics, with emphasis placed on refining estimates of photosynthetic activity, assessing forest growth stage, and identifying the amount and proportions of non-photosynthetic material within the

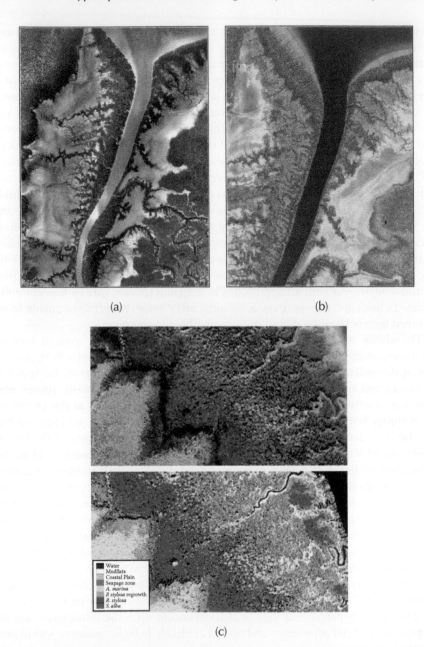

(a)   (b)

Water
Mudflats
Coastal Plain
Seapage zone
A. marina
R stylosa regrowth
R. stylosa
S. alba

(c)

FIG. 3.5   **(See color insert following page 134)** Mangroves along the West Alligator River of KNP as observed from (a) true color stereo photography and (b) CASI data. (c) A classification of the main communities on the west bank of the West Alligator River.

vegetation or on the ground surface (e.g., for fuel load assessment). The following sections outline how hyperspectral data can be used to increase understanding of the distribution and dynamics of carbon within forested landscapes.

3.4.1 PHOTOSYNTHESIS

Photosynthesis is the process by which plants convert light energy to reduce carbon dioxide (CO_2) to sugars and starch, which are then reconstituted to organic carbon (cellulose and lignin) and used to increase and maintain biomass [63]. As such, this process relates directly to the amount of carbon taken up by vegetation, with net primary production (NPP) describing the increase in carbon over a period of time (typically a year). For photosynthesis, around 70–90% of light energy (primarily in the 400- to 700-nm region) is absorbed by the leaf pigments chlorophyll a and b and also carotenoids (alpha-carotene, beta-carotene, and xanthophylls [64]). Chlorophyll a is the most abundant pigment and absorbs strongly in the blue (430 nm) and red (660 nm) wavelengths. Chlorophyll b is absorbed more in the 450- to 460-nm and 640- to 650-nm wavelength regions. Carotenoids also absorb energy and further protect chlorophyll against photodestruction by excess light [63]. Chlorophyll is highly correlated with foliar nitrogen (N) as this is contained within the chlorophyll molecules in the form of the carbon-fixing enzyme RUBG-carboxylase [65,66] and the amount relates closely to rates of net photosynthesis and also respiration. At broad temporal and spatial scales, canopy N is indicative of the maximum photo-synthetic rate [67] and has been related to NPP, litter fall N, and N mineralization. Remotely sensed estimates of N can therefore provide insight into terrestrial carbon and N cycles and indicate ecosystem productivity [68].

As photosynthetic pigments are responsible for capturing solar energy, amounts within vegetation can be quantified using reflectance data and linked to photosynthetic production. A particular advantage of hyperspectral remote sensing data is that the increased spectral resolution facilitates more specific retrieval of foliar chemicals, either through relationships established with particular wavelength regions or indices, by considering the shape of the spectral reflectance curve, or through model inversion techniques using hyperspectral remote sensing data as input (e.g., Jacquemoud et al. [69] and Demarez and Gastellu-Etchegorry [70]). As examples, estimates of chlorophyll have been generated through empirical relationships with reflectance data and band ratios between sensitive (i.e., visible) and insensitive (i.e., NIR) regions [71] or through consideration of red edge measures [72–74]. Foliar chlorophyll concentration has also been associated with variations in the shape of spectra and, in particular, a minor deepening and major broadening of the chlorophyll absorption feature [75] and changes in the red edge position (REP) [24,74]. However, relationships can be masked by differences in species, growth stage, leaf layering, and leaf water content.

Foliage N concentration has also been retrieved from hyperspectral data acquired by airborne sensors, including airborne AVIRIS and HyMap (e.g., references 67, and 76 through 81) and space-borne Hyperion data [82]. These retrievals are based largely on known N absorption features [83,84], although Kokaly and Clark [81] implied that an accuracy of ~0.5% retrieval from hyperspectral data was necessary to distinguish between ecosystems with differences in N sufficiently large to affect photosynthesis. The slopes of both fresh and dry leaf absorption spectra at locations near known protein absorption features have also been correlated with total N, particularly in the region of 2150 and 2170 nm [85]. Retrieval has further been explained by relationships existing with chlorophyll b at 460 nm and starch and proteins at 2250 and 2300 nm [82].

The potential to retrieve information on the photosynthetic process and hence the cycling of carbon by forests is therefore increased with the provision of hyperspectral data, largely because these data can be used to enhance input to forest productivity models. However, a current limitation is the lack of temporal coverage by hyperspectral sensors and other influencing factors such as the signal-to-noise ratio, viewing and illumination conditions, the state of the atmosphere, and the physical nature of the vegetation canopy in terms of leaf area, orientation and clumping, and the soil background [78].

3.4.2 DISTRIBUTION OF LEAF AND WOOD

Within all forests, leaves represent a small fraction (typically less than 5–10%) of the total aboveground biomass and the majority is stored as carbon in the trunks, branches, and roots (e.g., in the case of mangroves). Within the relatively closed-canopy tropical rainforests and mangroves, the woody material (or NPV) is generally not visible to optical sensors because of the predominance of leaves in the upper canopy. However, within wooded savannas, the amount of NPV observed is often quite large and typically comprises live woody material within the crowns (branches) or dead material, either standing (as dead or senescent trees) or lying on the ground. Furthermore, NPV can be in the form of leaf litter or dead grass, shrubs, or other herbaceous vegetation. The amounts of dead NPV depend upon the intensity, extent, and timing of processes such as drought and anthropogenic disturbances (e.g., poisoning, ring barking, or inundation) [86,87] as well as upon rates of decomposition.

Extracting the proportions of leaf and wood (live and dead) material separately within forested areas has been achieved, particularly within more open forested environments. Common measures of leaf cover include the leaf area index (LAI, m^2 m^{-2}) and foliage projected cover (FPC, %). LAI is defined as the total one-sided area of leaves in the canopy within a defined region (typically 1 m^2; [88]) and is a key indicator of evapotranspiration and photosynthesis and hence stand productivity [89]. Canopy light interception, NPP, stemwood production, and volume growth are all directly proportional to LAI [90]. LAI has typically been estimated by applying regressions between field-based measures and remotely sensed data and derived indices including the NDVI [91] and the simple ratio (SR [6]), although these generally represent a combination of leaf area and both canopy cover and architecture [10]. Furthermore, vegetation indices tend to saturate at an LAI of between 3 and 4 [10] because visible and NIR radiation reach a maximum as canopy closure and the density of foliage increase. LAI is also more readily determined where canopy closure approaches 90% [92] and is therefore a more appropriate measure in tropical rainforests and mangroves.

LAI is a less useful descriptor of wooded savannas because of the clumped and more open nature of the forest canopy and, in Australia, FPC has been adopted as an alternative. FPC provides a better estimate of the photosynthetic potential of a landscape under low foliage conditions and is defined as the horizontal percentage cover of photosynthetic foliage in all strata [93]. From remote sensing data, FPC has been estimated from a multiple regression between field-based estimates (themselves derived from basal area) and both reflectance data (from Landsat sensor data) and vapor pressure deficit, with the latter included because of a known correlation between the evaporative potential and FPC [93,94]. However, one of the difficulties in estimating both LAI and FPC in more open forest environments and using broadband sensors has been the contributions from the soil background and understory [88], the nonrandom distribution of foliage, and the influence of NPV within the canopy. To compensate, many studies have utilized broadband spectral indices such as the soil adjusted vegetation index (SAVI) [95] and modifications of these (e.g., the transformed SAVI [TSAVI]).

However, hyperspectral data can be used to generate a greater range of narrowband spectral indices, which can better compensate for background effects, particularly where vegetation cover is low. For example, the presence of bare soil cover is often indicated by a well-defined OH absorption feature at 2200 nm. Others (e.g., Wessman et al. [96]) have used spectral unmixing to estimate the subpixel proportions of green vegetation from hyperspectral data. From this information, better insights into the fraction of intercepted photosynthetically active radiation (fAPAR) can be obtained and exploited to further understand and quantify the photosynthetic process (e.g., by using fAPAR as input to forest productivity models).

Woody vegetation or NPV is frequently present within the canopies of subtropical woodlands or on the ground surface (e.g., as leaf litter or fallen logs) and comprises largely cellulose, lignin, and starch. Lignin is a structural element that accounts for about 25–33% of the dry mass of wood and forms vessels (e.g., tracheids, xylem fibers) that transport water through the plant. Cellulose is the primary structural component of green vegetation and forms the walls of cells in the leaves. Dry carbon spectral absorption features near 2100 and 2400 nm are associated with the presence of cellulose, starch, and soils [83], as quoted by Asner and Heidebrecht [6]. Lignin concentration is highly correlated with site N availability (through N mineralization) and has been retrieved successfully using AVIRIS data [77]. However, the absorption features for these elements are often obscured by water absorption that occurs within the SWIR region [97]. The number of absorption features of NPV is less compared to green vegetation but these are more distinct and relate primarily to stretching, bending, and overtones of C–H and O–H bonds common to organic carbon compounds [83]. Compared to green vegetation, increases in the amount of NPV within the canopy decrease the strength of the 680-nm absorption feature, the magnitude of the NIR plateau, and the difference in the magnitudes of the NIR plateau (at 1100 nm) and local SWIR maximum (1680 and 2200 nm; [97]: fig. 3.6). Reflectance across the SWIR region is also increased [97].

The contribution of NPV to the overall reflectance tends to be greater within more open canopies (e.g., LAI < 5.0) and where leaves are more vertically oriented (as is typical of many *Eucalyptus* canopies), as this exposes a greater proportion of between and below-canopy (ground) NPV and also within-canopy NPV. Retrieval of ground NPV is more likely where the dimensions of gaps between crowns are

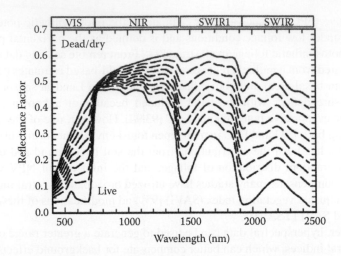

FIG. 3.6 Typical spectral reflectance curves for (a) photosynthetic and (b) vegetation NPV. (Modified after and reprinted from Asner, G. P., *Remote Sensing of Environment*, 64, 234–253, 1998. With permission from Elsevier.)

greater than the spatial resolution of the observing sensor. The presence of NPV within the landscape is illustrated in figure 3.7a, which highlights areas of dead standing timber, coarse woody debris (dead trunk), and senescent grasses and herbaceous shrubs. To quantify ground surface and within-canopy NPV, spectral unmixing is often applied using endmembers for NPV and also shade/moisture, soil, and/or photosynthetic vegetation (e.g., Asner et al. [10]). Typically, endmember spectra are extracted either from the image data themselves or through reference to field spectroradiometer data. An example of spectral unmixing of the scene presented in figure 3.7b is given in figure 3.7c. Here, the NPV in the area outside the delineated crowns is identified while within the crowns the proportion of NPV varies as a function of canopy openness and relative reflectance contributions from branches. While spectral unmixing of the endmembers using the full spectral range can provide reliable estimates of NPV, the estimates of other endmembers (e.g., shade and photosynthetic vegetation) are often less well estimated. For this reason, Asner and Heidebrecht [6] advocated the use of endmembers extracted entirely from the SWIR (2000–2120 nm) region [97,98].

3.4.3 Fire Fuel Loads

Leaf litter, bark, and elevated plant material are typical components of the forest fuel load that contribute to fire behavior and can be used to predict fire hazard. Traditionally, assessments of fuel hazard have been undertaken using ground measurements and on a scale that is limited in coverage and does not adequately quantify the variation across the landscape. A number of studies have focused on retrieving broad forest structural attributes from empirical/statistical (e.g., regression [99]) or inverse (e.g., radiative transfer) modeling [100,101]. For example, Chafer, Noonan, and Macnought [102] demonstrated a correlation between the NDVI derived from

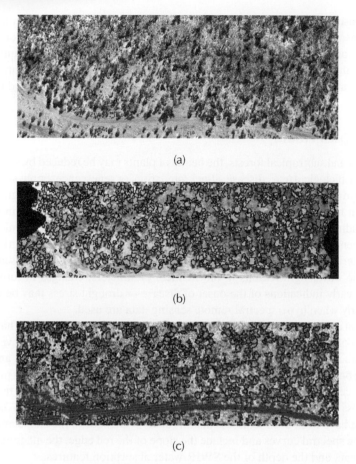

FIG. 3.7 (a) Stereo aerial photography and (b) CASI image of wooded savannas in Queensland, Australia, dominated by *E. populnea* and *E. melanaphloia*. (c) A classification of CASI data showing the distribution of NPV within delineated tree crowns (orange represents more NPV) and in open areas (yellow represents more NPV) in the form of dead grass, standing dead tree trunks, and coarse woody debris. See CD for color image.

SPOT-4 high resolution visible infrared (HRVIR) data and overall fuel loads, but saturation of the NDVI occurred with increasing fractional vegetation cover (reaching a maximum at 60–90% cover; [59,103]). Asner et al. [10] suggested the potential of hyperspectral indices for predicting fire fuel loads in tropical rainforests in Amazonia. However, retrieval of understory attributes, which contribute significantly to the fuel hazard, has been limited.

The integration of estimates of canopy condition (e.g., based on forest biochemistry or the relative proportions of photosynthetic vegetation and NPV) derived from hyperspectral data with information on the vertical and horizontal structure of forests (i.e., fuel arrangement) estimated from canopy penetrating LIDAR may offer the best opportunity to characterize fuel hazard at the landscape level. A novel approach by Roff et al. [104] combined airborne HyMap and LiDAR data and ground-based measurements of forest attributes to quantify fuel loads in the Jilliby State Conservation area in New South Wales, Australia. In certain dry forest types, different combinations of narrowband vegetation indices, including the anthocyanin reflective index (ARI) and cellulose absorption index (CAI), and structural parameters (e.g., maximum canopy height, as estimated from the LiDAR) yielded improved correlations between observed and predicted fuel scores.

3.4.4 FOREST HEALTH

In tropical and subtropical forests, the health of plants may be reduced by stress agents that can be biological (e.g., disease, attack/predation, or senescence) or physiochemical (e.g., dehydration, growth inhibitors, or environmental conditions) in origin. Typical vegetation responses to stress include structural deformity (mostly within canopies and including wilting and stunted growth), changes in internal biochemistry (e.g., leaf pigment concentration and canopy water), and partial or complete degradation of plant material (e.g., reductions in crown leaf area). By exploiting known sensitivities in specific wavelength regions to biochemical parameters, changes in forest health and even early indications of the onset of disease or drought stress may be detected, particularly when hyperspectral remote sensing data are used.

Recent advances in remote sensing and detection of vegetation health have arisen as a result of increased understanding of the reflectance characteristics of vegetation. In general, a healthy canopy will absorb the majority of incident visible and SWIR radiation but a drought-stressed or dying senescent canopy, with a greater proportion of lignin, cellulose, and starch in the canopy components, will scatter the majority of light across the full spectrum but particularly in the SWIR bands (see fig. 3.6). Attributes that have been used to assess health relate to changes in the shape and magnitude of the spectral curves and include the slope of the red edge, the magnitude of the NIR plateau, and the depth of the SWIR water absorption features.

Differences in reflectance between health and stressed vegetation tend to be most apparent in the red edge, green peak, and SWIR while less so in the chlorophyll well (red) and the NIR plateau. Hence, band ratios between these sensitive and insensitive reflectance bands have been used to detect stress and differentiate healthy vegetation [26,71,105]. Shifts in the red edge inflection point (REIP) have also proved useful for indicating health with a shift to the shorter wavelengths associated with stressed vegetation [106–108] and the typical cause being a decline in chlorophyll content. Long-term stress in forests may also be detected through changes in leaf lignin concentration. In many environments, species often collectively exhibit the same spectral response transitions but the detection of stressed vegetation is, nevertheless, complicated, as the chemical content (e.g., chlorophyll and leaf water) may also vary as a result of seasonal leaf flush and senescence.

The health of forests is indicated by the level of photosynthetic activity but also by the proportions of leaf and wood material within the canopy. Normal growth and functioning induce a characteristic spectral response in vegetation, as related to species and growth stage, as well as seasonal effects. Factors considered beyond normal functioning that intermittently induce stress (e.g., the onset of drought or disease) result in changes in spectral shape and magnitude. Hyperspectral remote sensing offers many opportunities for the detection and mapping of these spectral shifts and there is potential to provide early warning of changing forest health and condition.

3.5 DISCUSSION

Worldwide, forested ecosystems have been severely depleted as a consequence of deforestation (largely for agricultural expansion) and degradation (e.g., through selective logging). Indirect anthropogenic disturbance has occurred because of increased fire activity (e.g., in selectively logged areas [109]) and changes in climate [46], including more intense drought and flooding. On the coastal fringe, significant areas of mangrove forest have been destroyed for various uses (e.g., aquaculture), but also because of natural events, including cyclones and tsunamis. Mangroves are also responding to fluctuating sea levels as a consequence of climatic change. The extent and condition of forests have therefore been reduced substantially (fig. 3.8) and this trend is continuing, particularly in the tropical and subtropical regions. While remote sensing observations have and continue to play a role in mapping and

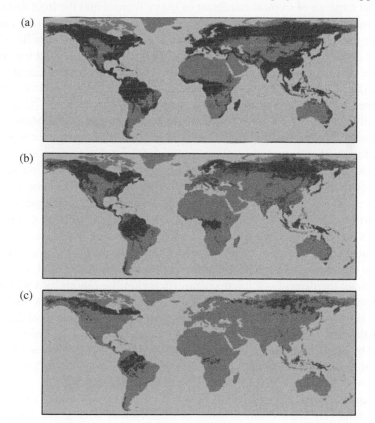

FIG. 3.8 The extent of forests (a) 8000 years ago and (b) the present; (c) represents the extent of remnant forest. (Modified after Byrant, D. et al., *The Last Frontier Forests: Ecosystems and Economics on the Edge*, World Resources Institute, Washington, D.C., 1997.) See CD for color insert.

monitoring deforestation, this section focuses specifically on the potential benefits of integrating hyperspectral remote sensing data for this purpose and also in relation to forest restoration.

3.5.1 TROPICAL FORESTS

Tropical rainforests have experienced substantial losses in extent and condition, with these largely fueled by expanding populations and the associated demand for agricultural land and also increases in logging to support the timber trade. The loss of biodiversity resulting from tropical deforestation and degradation is substantial as many species are rare and populations are generally confined to relatively small areas of forest. The removal or alteration of tree communities, either through wholesale or selective clearing, also impacts on the complex interactions among tree species and associated organisms including seed dispersers, herbivores, and symbiotic fungi.

In an ideal situation, information on the distribution of individual tree species is needed to better evaluate and maintain biodiversity, but is difficult to achieve because of the vast expanse of tropical rainforest in many regions and the high diversity of species, many of which occur in the subcanopy. The spatial resolution of currently operating space-borne hyperspectral sensors (e.g., Hyperion) is also limited and excessive cloud cover and haze in many tropical regions further restrict the utility of these data. For these reasons, most studies revert to airborne hyperspectral data but their use is often prohibited by the high costs of flying and the availability of the sensors. Nevertheless, the use of these sensors can be targeted such that (a) areas with high biodiversity or vulnerability to change (e.g., occurring on deforestation fronts) are preferentially observed and/or (b) tree species that are of particular importance are discriminated and mapped.

For example, trees of the genus *Dipteryx* provide a major seed resource and nesting cavities for the great green macaw (chapter 6) but many have been removed from protected areas, which has led to further declines in this endangered species. Using airborne hyperspectral data, Clark et al. [37] classified this genus with a producer's and user's accuracy exceeding 90% and suggested that the resulting maps could contribute to the conservation of the macaw in terms of mapping habitat and migration corridors. An alternative approach proposed by Asner et al. (chapter 12) is to estimate the diversity of foliar chemicals within the canopy as a whole using hyperspectral data and to relate this to faunal and floral distributions. Other studies have similarly linked biodiversity to the type and heterogeneity of land cover [45]. Further improvements in diversity assessment could occur by integrating data acquired by sensors that provide structural information on the dimensions and distributions of trees (e.g., emergents) within the vertical profile, including LiDAR [110]) or polarimetic/interferometric synthetic aperture radar (SAR) [111]).

For quantifying carbon within primary tropical rainforests, hyperspectral data are relatively limited unless used in combination with active sensors (e.g., LiDAR). However, Asner et al. [10] noted that hyperspectral data can be used to detect changes in the physiological state of rainforest canopies. Notably, the study established that metrics of canopy moisture content (the spectroscopic water absorption feature [SWAM]) and light use efficiency (the photochemical reflectance index

[PRI], which relates to the amount of atmospheric CO_2 update by vegetation per unit of energy absorbed (and generated from Hyperion data), expressed sensitivity to seasonal drought conditions and could be used to constrain models of NPP. The study also highlighted sensitivity of the ARI to drought conditions, where anthocyanin is a pigment generally indicative of newly formed foliage prior to the full development of chlorophyll pigments. Hyperspectral remote sensing is arguably best used to assess species recovery associated with regrowth on land that has experienced different levels of disturbance in the form of agricultural land use, fires, and logging. Although few studies of tropical forest regeneration have been undertaken using hyperspectral data, Lucas et al. [19] noted that regenerating forests as old as 15–20 years and dominated by the pioneer genera *Cecropia* and *Vismia* were spectrally distinguishable in Landsat SWIR and, in the earlier stages of succession, the NIR wavebands. The dominance of these pioneer genera was attributed to differences in the fire history, periods of active land use, and frequency of vegetation reclearance, as mapped using time-series of Landsat sensor data; those dominated by *Cecropia* species were more diverse in terms of tree species. Furthermore, Shuttle Imaging Radar (SIR-C) L-band SAR backscatter from forests of the same age but occurring on land with different histories was lower for those dominated by *Vismia* species, suggesting a change in structure and/or biomass as the regeneration proceeded and as a function of prior land use [112]. Further discrimination of species within forests following different pathways of regeneration might be achieved using hyperspectral remote sensing data. These studies suggest that by integrating maps of species distributions with land cover change data sets and estimates of biomass retrieved from active sensors, information on the dynamics of regeneration following clearance might be obtained across the landscape. Such knowledge would assist predictions of how regenerating forests might recover carbon and tree species diversity to predisturbance levels and identify areas where restoration of forest cover might best be targeted.

3.5.2 MANGROVES

Mangroves are prolific along coastal and river margins of the tropics and subtropics and, as such, are responsive to the natural change, including that which is direct (e.g., tsunami damage, cyclones, storm surges) or indirect (e.g., sea level rise). Despite recognition of their ecological and economic importance, mangroves are also being exploited unsustainably and at an extraordinary rate in many regions. In 1980, the IUCN *World Mangrove Atlas* [113] estimated that the global area of mangrove was 19.8 million ha, but had declined to ~18 million ha at the end of the twentieth century. However, the International Tropical Timber Organization (ITTO) estimated that 100,000 ha have been destroyed annually and active areas of expansion were few. The Food and Agricultural Organization (FAO), in 2003, estimated that, by the end of 2000, only 15 million ha of mangroves remained. The greatest change has occurred in the tropical regions of Asia, where mangrove ecosystems are disappearing at an alarming rate largely because of demographical pressures and agricultural expansion. Similar pressures, albeit of lower intensity, have also impacted on the mangroves along the coasts of the African and American continents [114].

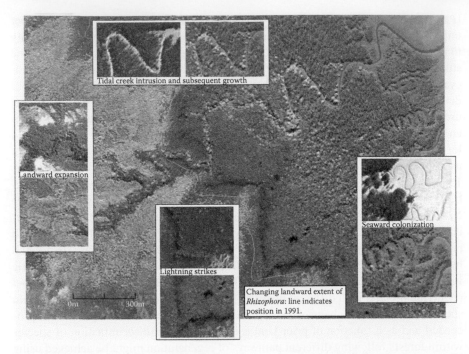

FIG. 3.9   Changes in mangrove forest structure and composition as observed using time-series aerial photography (1991) and hyperspectral CASI data (2002). See CD for color image.

At a regional level and for mapping and monitoring changes in mangrove extent associated with deforestation or regeneration, multispectral remote sensing data acquired by space-borne sensors are generally adequate. However, in many areas, mangrove extent might remain similar, but changes in species composition, structure, and biomass can occur, with these reflecting diversification, encroachment or loss of species, and growth and dieback of individuals. Therefore, more detailed base-line data sets of species distributions as well as structure and biomass are required in addition to extent, with the level of detail required depending upon the nature and extent of change [20]. As an illustration, changes in the extent and condition of mangroves in KNP have been reported over the past few decades [115,116] and many have been observed to extend inland [117–119]. Concurrent with the spread of man-groves has been the expansion of the tidal creek network, ongoing channel cutting, and subsequent saltwater intrusion into adjacent freshwater wetlands—processes that are also prevalent throughout the Alligator Rivers region in Australia's Northern Territory [119]. In contrast to many mangroves in the Indo-Pacific region, those in KNP have remained relatively undisturbed and are therefore responding to the more subtle impacts of coastal environmental change, including those associated with sea level and climatic variation. Within the mangroves of the West Alligator River, a comparison of stereo aerial photographs and CASI data acquired in 1991 and 2002, respectively, revealed both landward (along tidal creeks) and seaward expansion of mangroves, regrowth within and expansion of *Rhizophora stylosa* forest, reductions in stature through cyclone damage, the formation of canopy gaps following lightning strikes, establishment of *Avicennia marina* along intruding tidal creeks, and expansion of the crown area of existing trees [119,120] (fig. 3.9).

The changes have, in general, all occurred on a spatial scale ranging from 1 to 200 m and within a period of 12 years. However, comparison with 1950 aerial photography suggested that the rate of change might be accelerating. The changing extent of these mangroves was quantified by comparing classifications of both data sets, although, as indicated earlier, mapping of species was only possible using the CASI data. This case study therefore highlights the importance of monitoring changes in species composition and the benefits of integrating fine spatial resolution hyperspectral data, particularly where change is more subtle. By monitoring similar sites throughout the tropics and subtropics, insights into the longer-term impacts of changing climate and sea levels can be obtained, although disaggregating these from changes that might occur as part of natural processes is fundamental.

3.5.3   WOODED SAVANNAS

Tropical and subtropical savannas, which contain varying proportions of tree and grass cover, occupy ~11.5% (~16 million km^2) of the Earth's land surface (between ~30° N and 35° S; [121]). This ecosystem is an important repository of biodiversity [122] and plays a major role in the carbon cycle, accounting for ~30% of net terrestrial primary production [123] (typically 5–12 Mg C ha^{-1}.yr^{-1}). Savannas are also particularly sensitive to disturbance and climatic alteration and associated changes in their extent, condition, and dynamics can influence the carbon cycle in either a positive (e.g., through regrowth and woody thickening) or negative (e.g., through deforestation, herbivory, fires, and degradation) way [124]. As an illustration, woody thickening has been reported globally [1] and, in Australia, the associated carbon sequestered was estimated to be equivalent to that released through vegetation clearance [125]. In many countries, this ecosystem remains largely unprotected and clearing is ongoing. Many communities targeted are often considered to be of low commercial value and are relatively easy to clear; their biodiversity and carbon values are, however, often overlooked.

At the tree level, hyperspectral data have proved useful for species discrimination but difficulties arise in the classification of rare or uncommon species because of the lack of a priori knowledge. Such information is necessarily obtained through reference to aerial photography (through experienced interpretation), existing vegetation mapping, site surveys, or expert knowledge but greater automation of this process is desirable. Community descriptions also need to be generated from the tree-level mapping and this can be achieved through, for example, the application of area-based standard diversity measures (e.g., Shannon–Wiener diversity) and canonical analysis. An alternative is to link mapped distributions of trees by applying thresholds to graph-based minimum spanning trees that can be weighted by considering the physical environment (e.g., topography, soils, and local climate), biogeographical distributions of species, and interactions and associations within and between species. Polygon boundaries can then be placed around trees of the same species where these occur within a certain distance of each other and then intersected subsequently with those generated for other species to generate

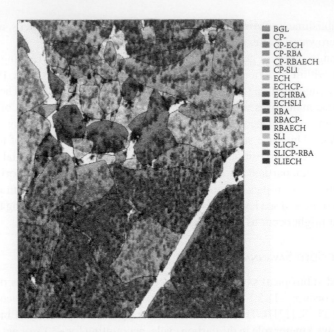

BGL
CP-
CP-ECH
CP-RBA
CP-RBAECH
CP-SLI
ECH
ECHCP-
ECHRBA
ECHSLI
RBA
RBACP-
RBAECH
SLI
SLICP-
SLICP-RBA
SLIECH

FIG. 3.10 Forest communities mapped by applying a graphs-based minimum spanning tree to individual tree crowns/clusters delineated within 1-m spatial resolution CASI data (background) and discriminated to species using stepwise discriminant analysis (RBA = *Angophora floribunda*; refer to table 3.1 for other species codes). See CD for color image.

a community-based classification (fig. 3.10). Through this approach, an important step in understanding the information content of remote sensing data acquired by both active (e.g., radar) and passive sensors at coarser spatial resolution can be better established and the overall contribution of vegetation and other materials at these resolutions better ascertained.

Hyperspectral data can also contribute to an understanding of carbon dynamics within wooded savannas by providing information on foliar chemistry and the distribution of NPV, particularly that associated with dead standing or fallen trees. Both measures are important as input to parameterize and constrain carbon cycle models. Spatial estimates of the distribution of aboveground and component (leaf, branch, and trunk) biomass can also be generated by applying species-specific allometric equations to tree crowns identified to species from hyperspectral data and using height (and, by inference, diameter) retrieved from co-registered LiDAR data (fig. 3.11; [126,127]). Such an approach is most applicable to forests where tree crowns are relatively large and evenly spaced but less so in multilayered forests with overlapping crowns and/or a high density of small trees [126]. Changes in the amount of dead and live carbon and also the overall health of the forests might also be determined by comparing crown delineations and associated measures (species, biomass, and height) based on data acquired over a time-series [87]. As with mangroves, the detection of change within the forest area can lead to a better understanding of impacts of longer-term variability and trends associated with, for example, climatic change.

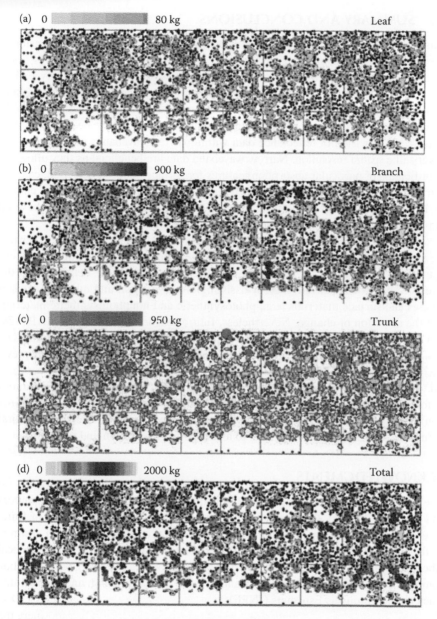

FIG. 3.11 Estimates of (a) leaf, (b) branch, (c) trunk, and (d) total aboveground biomass generated by applying species-specific allometric equations to delineated crowns discriminated to species and associated with a height estimate from LiDAR. (Reprinted from Lucas, R.M. et al., *International Journal of Remote Sensing,* 2007, in press. With permission from Taylor & Francis.) See CD for color image.

3.6 SUMMARY AND CONCLUSIONS

This chapter has provided an overview of the role of hyperspectral remote sensing in forest characterization, mapping, and monitoring, focusing particularly on key issues in the analysis of hyperspectral data and in relation to assessments of biodiversity, carbon cycle science, and vegetation health. The chapter reviews a wide range of studies from the tropics and subtropics and also draws on selected case studies from our own work. The key advantages of using hyperspectral remote sensing data have been highlighted: namely, the provision of reflectance data within narrow wavebands and often at fine to very fine spatial resolution. Narrow waveband data are beneficial as they allow the identification of particular absorption features for surface material detection, calculation of indices that facilitate more specific retrieval of foliar chemicals (e.g., chlorophylls, carotenoids, lignins), and retrieval of endmember spectra that can be used subsequently in classification procedures (e.g., linear spectral unmixing).

Fine to very fine spatial resolution data are provided largely because most sensors are supported by airborne platforms and space-borne sensors with equivalent configurations that are not, as yet, available. These data allow the detail within the forests to be better resolved, thereby assisting discrimination and mapping of plant species and surface materials (e.g., photosynthetic and nonphotosynthetic material) and the detection of change. Nevertheless, future sensors, including ENMAP [128] and FLORA [129], are in the process of planning and are anticipated to provide data of a far greater spatial, spectral, temporal, and radiometric quality than has previously been available. These data, particularly if integrated with other sources (e.g., LiDAR, SAR), are anticipated to provide new insights into the state and dynamics of forests in the tropics and subtropics. As such, a better understanding of the functioning of these ecosystems, particularly in response to anthropogenic and natural influences, can be obtained and used to assist their conservation and restoration.

ACKNOWLEDGMENTS

The authors would like to thank the Queensland Department of Natural Resources and Water (QDNRW), the Queensland Herbarium, the Australian Bureau of Rural Sciences, the Australian Research Council, the Environmental Research Institute of the Supervising Scientist (ERISS), Parks Australia North (PAN) and the local Aboriginal landowners, the University of New South Wales (UNSW), the Australian National University (ANU), CSIRO Canberra, Definiens Imaging, and the Instituto Nacional de Pesquisas Especiais (INPE).

REFERENCES

1. Scholes, R.J. and Archer, S.R., Tree-grass interaction in savannas, *Annual Review of Ecology and Systematics*, 28, 517, 1997.
2. Brown, S.A. and Lugo, A.E., Tropical secondary forests, *Journal of Tropical Ecology*, 6, 1, 1990.
3. Lucas, R.M., Honzak, M., Curran, P.J., et al., Tropical forest regeneration in Cameroon: An analysis of NOAA AVHRR data, *International Journal of Remote Sensing*, 21, 2831, 2000.

4. Lucas, R.M., Honzak, M., Curran, P.J., et al., The regeneration of tropical forests within the legal Amazon, *International Journal of Remote Sensing*, 21, 2855, 2000.
5. Lucas, R.M., Bunting, P., Paterson, M., et al., Classification of Australian forest communities using aerial photography, CASI and HyMap data, *Remote Sensing of Environment*, in press.
6. Asner, G.P. and Heidebrecht, K.B., Spectral unmixing of vegetation, soil and dry carbon cover in arid regions; comparing multispectral and hyperspectral observations, *International Journal of Remote Sensing*, 23, 3939, 2002.
7. Stone, C., Chisholm, L., and Coops, N., Spectral reflectance characteristics of Eucalypt foliage damaged by insects, *Australian Journal of Botany*, 49, 687, 2001.
8. Stone, C., Chisholm, L., and McDonald, S., Effects of leaf age and psyllid damage on the spectral reflectance properties of *Eucalyptus saligna* foliage, *Australian Journal of Botany*, 53, 45, 2005.
9. Ollinger, S.V., Smith, M.L., Martin, M.E., et al., Regional variation in foliar chemistry and N cycling among forests of diverse history and composition, *Ecology*, 83, 339, 2002.
10. Asner, G., Keller, M., Pereira, R., et al., Canopy damage and recovery after selective logging in Amazonia: Field and satellite studies, *Ecological Applications*, 14, S280, 2004.
11. Lucas, R.M., Rowlands, A.R., Niemann, O., et al., Hyperspectral sensors: Past, present and future, in *Advanced image processing techniques for remotely sensed hyperspectral data*, P.K. Varshney and M.K. Arora, eds. Springer, New York, 2004, 11.
12. Definiens, A.G. 2005 [cited; available from: www.definines.com].
13. INPE (Instituto Nacional de Pesquisas Espaciais) and Georeferenciadas and SPRING (Sistema de Prcessamento de Imagens Georeferenciadas), *Manual De Usuario, Version 2.0, Release 2.0.4*. 1997, Net GIS. Geoprocessamento e Informaica Ltda, Sao Jose dos Campos, SP, Brazil.
14. Meyer, P., Staenz, K., and Itten, K.I., Semi-automated procedures for tree species identification in high spatial resolution data from digitized color infrared-aerial photography, *ISPRS Journal of Photogrammetry and Remote Sensing*, 51, 5, 1996.
15. Leckie, D., Gougeon, F., Hill, D., et al., Combined high density LiDAR and multispectral imagery for individual tree crown analysis, *Canadian Journal of Remote Sensing*, 29, 633, 2003.
16. Settle, J.J. and Drake, N.A., Linear mixing and the estimation of ground cover proportions, *International Journal of Remote Sensing*, 14, 1159, 1993.
17. Shimabukuro, Y.E., Batista, G.T., Mello, E.M.K., et al., Using shade fraction image segmentation to evaluate deforestation in Landsat thematic mapper images of the Amazon region, *International Journal of Remote Sensing*, 19, 535, 1998.
18. Adams, J.B., Sabol, D., Kapos, V., et al., Classification of multispectral images based on fractions of endmembers; application to land-cover change in the Brazilian Amazon, *Remote Sensing of Environment*, 52, 137, 1995.
19. Lucas, R.M., Held, A., Phinn, S., et al., Remote sensing of tropical forests, in *Manual of remote sensing: Natural resources and environment*. 2004, 239.
20. Lucas, R.M., Mitchell, L., Rosenqvist, A., et al., The potential of L-band SAR for quantifying mangrove characteristics and change: Case studies from the tropics, *Aquatic Conservation: Marine and Freshwater Ecosystems*, 17, 245, 2007.
21. Dowling, R. and Accad, A., Vegetation classification of the riparian zone along the Brisbane River, Queensland, Australia using light detection and ranging (LiDAR) data and forward looking digital video, *Canadian Journal of Remote Sensing*, 29, 556, 2003.
22. Gougeon, F.A. and Leckie, D.G., Forest information extraction from high spatial resolution images using an individual tree crown approach. Information Report Bc-X-396, Natural Resources Canada, Canadian Forest Service, Victoria, BC, 2003.

23. Bunting, P. and Lucas, R.M., The delineation of tree crowns in Australian mixed species forests using hyperspectral compact airborne spectrographic imager (CASI) data, *Remote Sensing of Environment*, 101, 230, 2006.
24. Horler, D.N.H., Dockray, M., and Barber, J., The red edge of plant leaf reflectance, *International Journal of Remote Sensing*, 4, 273, 1983.
25. Gougeon, F.A. and Moore, T., Classification individuelle des arbres a partir d'images a haute resolution spatiale, in M. Bernier, F. Bonn, and P. Gagnon, eds. *Teledetection et gestion des resouces,* vol. 6e. Congres De L'association Quebecoise de Teledetection, Sherbrooke, QC, 1988.
26. Carter, G. and Young, D.R., Responses of leaf spectral reflectance to plant stress, *American Journal of Botany*, 154, 239, 1993.
27. Holmgren, J. and Persson, A., Identifying species of individual trees using airborne laser scanner, *Remote Sensing of Environment*, 90, 415, 2004.
28. Hyyppa, J., Hyyppa, H., Litkey, P., et al., Algorithms and methods of airborne laser scanning for forest measurements, *International Archives of Photogrammetry, Remote Sensing and Spatial Information Sciences*, 36-8/W2, 82, 2004.
29. Hyyppa, J., Kelle, O., Lehikoinen, M., et al., A segmentation-based method to retrieve stem volume estimates from 3-D tree height models produced by laser scanner, *IEEE Transactions on Geoscience and Remote Sensing*, 39, 969, 2001.
30. Gougeon, F.A., A crown-following approach to the automatic delineation of individual tree crown in high spatial resolution aerial images, *Canadian Journal of Remote Sensing*, 21, 274, 1995.
31. Leckie, D.G., Gougeon, F.A., Tinis, S., et al., Automated tree recognition in old-growth conifer with high-resolution digital imagery, *Remote Sensing of Environment*, 94, 311, 2005.
32. Culvenor, D., Tida: An algorithm for the delineation of tree crown in high spatial resolution remotely sensed imagery, *Computers and Geosciences*, 28, 33, 2000.
33. Pollock, R.J., The automatic recognition of individual trees in aerial images of forests based on a synthetic tree crown image model, in *Computer science.* University of British Columbia, Vancouver, BC, Canada, 1996, 177.
34. Warner, T.A., Lee, J.Y., and McGraw, J.B., *Delineation and identification of individual trees in the Eastern deciduous forest.* Natural Resources Canada, Canadian Forest Service, Victoria, BC, Canada, 1998.
35. Blackburn, G.A., Remote sensing of forest pigments using airborne imaging spectrometer and LiDAR imagery, *Remote Sensing of Environment*, 82, 311, 2002.
36. Pinz, A., Zaremba, M.B., Bischof, H., et al., Neuromorphic methods for recognition of compact image objects, *Machine Graphics and Vision*, 2, 209, 1993.
37. Clark, D.A., Roberts, D.A., and Clark, D.A., Hyperspectral discrimination of tropical rain forest tree species at leaf to crown scales, *Remote Sensing of Environment*, 96, 375, 2005.
38. Gausman, H.W., Plant leaf optical properties in visible and near-infrared light, in *Graduate studies no. 2,* Lubbock, TX, 1985.
39. Fung, T. and Siu, W.L., Hyperspectral data analysis for subtropical tree species recognition, in *IGARSS '98 Sensing and Managing the Environment, IEEE International Geoscience and Remote Sensing*, Seattle, WA, 1998.
40. Galvao, L.S., Formaggio, A.R., and Tisot, D.A., Discrimination of sugarcane varieties in southeastern Brazil with Eo-1 Hyperion data, *Remote Sensing of Environment*, 94, 523, 2005.
41. Demuro, M. and Chisholm, L., Assessment of Hyperion for characterizing mangrove communities, in *Proceedings of the International Conference AVIRIS 2003 Workshop*, NASA JPL, Pasadena, CA, 2003.
42. Gong, P., Ru, R., and Yu, B., Conifer species recognition: An exploratory analysis of in situ hyperspectral data, *Remote Sensing of Environment*, 62, 189, 1997.

43. Roberts, D.A., Batista, G., Pereira, J., et al., Change identification using multitemporal spectral mixture analysis: Applications in Eastern Amazonia, in *Remote sensing change detection: Environmental monitoring applications and methods*, C. Elvidge and R. Luneta, eds. Ann Arbor Press, Ann Arbor, MI, 1998, 137.

44. Whitmore, T.C., *An introduction of tropical rain forests*. Clarendon, Oxford, England, 1990.

45. Foody, G., Remote sensing of tropical forest environments: Towards the monitoring of environmental resources for sustainable development, *International Journal of Remote Sensing*, 24, 4035, 2003.

46. Thomas, C.D., Cameron, A., Green, R.E., et al., Extinction risk from climate change, *Nature*, 427, 145, 2004.

47. Roberts, D.A., Gardner, M., Church, R., et al., Mapping chaparral in the Santa Monica Mountains using multiple endmember spectral mixture models, *Remote Sensing of Environment*, 65, 267, 1998.

48. Nelson, B.W., Natural forest disturbance and change in the Brazilian Amazon, *Remote Sensing Reviews*, 10, 105, 1994.

49. Wang, L., Sousa, W.P., Gong, P., et al., Comparison of IKONOS and Quickbird images for mapping mangrove species on the Caribbean Coast of Panama, *Remote Sensing of Environment*, 91, 432, 2004.

50. Cochrane, M.A., Using vegetation reflectance variability for species level classification of hyperspectral data, *International Journal of Remote Sensing*, 21, 2075, 2000.

51. Ticehurst, C., Lymburner, M., Held, A., et al., Mapping tree crowns using hyperspectral and high spatial resolution imagery, in *Third International Conference on Geospatial Information in Agriculture and Forestry*, Denver, CO, 2001.

52. van Aardt, J.A. and Wynne, R.II., Spectral separability among six southern tree species, *Photogrammetric Engineering and Remote Sensing*, 67, 1367, 2001.

53. Coops, N., Dury, S., Smith, M.L., et al., Comparison of green leaf Eucalypt spectra using spectral decomposition, *Australian Journal of Botany*, 50, 567, 2002.

54. Goodwin, N., Turner, R., and Merton, R., Classifying Eucalyptus forests with high spatial and spectral resolution imagery: An investigation of individual species and vegetation communities, *Australian Journal of Botany*, 53, 337, 2005.

55. Tickle, P.K., Lee, A., Lucas, R.M., et al., Quantifying Australian forest floristics and structure using small footprint LiDAR and large scale aerial photography, *Forest Ecology and Management*, 223, 379, 2006.

56. Blasco, F., Gauquelin, T., Rasolofoharinoro, M., et al., Recent advanced in mangrove studies using remote sensing data, *Marine and Freshwater Research*, 49, 287, 1998.

57. Alongi, D.M., Present state and future of the world's mangrove forests, *Environmental Conservation*, 29, 331, 2002.

58. Duke, N.C., Lo, E.Y.Y., and Sun, M., Global distribution and genetic discontinuities and genetic discontinuities of mangroves emerging patterns in the evolution of *Rhizophora*, *Trees*, 16, 65, 2002.

59. Meza Diaz, B. and Blackburn, G.A., Remote sensing of mangrove biophysical properties: Evidence from a laboratory simulation of the possible effects of background variation on spectral vegetation indices, *International Journal of Remote Sensing*, 24, 53, 2003.

60. Manson, F.J., Loneragan, N.R., McLeod, I.M., et al., Assessing techniques for estimating the extent of mangroves: Topographic maps, aerial photographs and Landsat TM images, *Marine and Freshwater Research*, 52, 787, 2001.

61. Gao, J., A comparative study on spatial and spectral resolutions of satellite data in mapping mangrove forests, *International Journal of Remote Sensing*, 20, 2823, 1999.

62. Lucas, R.M., Honzak, M., do Amaral, I., et al., Forest regeneration on abandoned clearances in Central Amazonia, *International Journal of Remote Sensing*, 23, 965, 2002.

63. Hopkins, W.G., *Introduction to plant physiology*. Wiley, New York, 1999.

64. Gausman, H.W., Visible light reflectance, transmittance and absorptance of differently pigmented cotton leaves, *Remote Sensing of Environment*, 13, 233, 1982.
65. Kruijt, B., Ongeri, S., and Jarvis, P.G., Scaling par absorption, photosynthesis and transpiration from leaves to canopy, in *Scaling-up from cell to landscape*, Society for Experimental Biology Seminar Series, P.V. Gardingen, G. Foody, and P. Curran, eds. Cambridge University Press, Cambridge, U.K., 1997, 79.
66. Daughtry, C.S.T., Walthall, C.L., Kim, M.S., et al., Estimating corn leaf chlorophyll concentration from leaf and canopy reflectance, *Remote Sensing of Environment*, 74, 229, 2000.
67. Martin, M.E. and Aber, J.D., High spectral resolution remote sensing of forest canopy lignin, nitrogen and ecosystem processes, *Ecological Applications*, 7, 431, 1997.
68. Coops, N.C., Stone, C., Merton, R., et al., Assessing Eucalypt foliar health with field-based spectra and high spatial resolution hyperspectral imagery, in *Proceedings of the IEEE 2001 International Geoscience and Remote Sensing Symposium*, University of New South Wales, Sydney, Australia, 9–13 July 2001.
69. Jacquemoud, S., Bacour, C., Poilive, H., et al., Comparison of four radiative transfer models to simulate plant canopies reflectance: Direct and inverse mode, *Remote Sensing of Environment*, 74, 471, 2000.
70. Demarez, V.G. and Gastellu-Etchegorry, J.P., A modeling approach for studying forest chlorophyll content, *Remote Sensing of Environment*, 71, 226, 2000.
71. Gitelson, A. and Merzlyak, M.N., Remote estimation of chlorophyll content in higher plant leaves, *International Journal of Remote Sensing*, 18, 2691, 1997.
72. Clevers, J.G.P.W., Imaging spectrometry in agriculture plant vitality and yield indicators, in *Imaging spectrometry: A tool for environmental observations*. J. Hill and J. Megier, eds. Kluwer Academic, Dordrecht, 1994, 193.
73. Curran, P.J., Windham, W.R., and Gholz, H.L., Exploring the relationship between reflectance red edge and chlorophyll concentration in slash pine leaves, *Tree Physiology*, 15, 203, 1995.
74. Jago, R.A., Cutler, M.E., and Curran, P.J., Estimating canopy chlorophyll concentration from field and airborne spectra, *Remote Sensing of Environment*, 68, 217, 1999.
75. Banninger, C., Phenological changes in the red edge shift of Norway spruce needles and their relationship to needle chlorophyll content, in *Proceedings of the 5th International Colloquium—Physical Measurements and Signatures in Remote Sensing*, Noordwijk, 1991.
76. Matson, P., Johnson, L., Billow, C., et al., Seasonal patterns and remote spectral estimation of canopy chemistry across the Oregon transect, *Ecological Applications*, 4, 280, 1994.
77. Johnson, L.F., Hlavka, C.A., and Peterson, D.L., Multivariate analysis of AVIRIS data for canopy biochemical estimation along the Oregon transect, *Remote Sensing of Environment*, 47, 216, 1994.
78. Gastellu-Etchegorry, J.P., Zagolski, F., Mougin, E., et al., An assessment of canopy chemistry with AVIRIS—A case study in the Landes Forest, South-West France, *International Journal of Remote Sensing*, 16, 487, 1995.
79. LaCapra, V.C., Melack, J.M., Gastil, M., et al., Remote sensing of foliar chemistry of inundated rice with imaging spectrometry, *Remote Sensing of Environment*, 55, 50, 1996.
80. Dury, S.J., Xiuping, J., and Turner, B., From leaf to canopy: Determination of nitrogen concentration of Eucalypt tree foliage using HyMap image data, in *10th Australasian Remote Sensing and Photogrammetry Conference*, Adelaide, Australia, 2000.
81. Kokaly, R.F. and Clark, R.N., Spectroscopic determination of leaf biochemistry using band-depth analysis of absorption features and stepwise multiple linear regression, *Remote Sensing of Environment*, 67, 267, 1999.
82. Coops, N.C., Stanford, M., Old, K., et al., Assessment of Dothistroma needle blight of *Pinus radiata* using airborne hyperspectral imagery, *Phytopathology*, 33, 1524, 2003.

83. Curran, P.J., Remote sensing of foliar chemistry, *Remote Sensing of Environment*, 30, 271, 1989.

84. Murray, A.L. and Williams, P.M., *Chemical principles of near-infrared technology*. American Association of Cereal Chemists, 1987, St. Paul, MN.

85. Johnson, L.F. and Billow, C.R., Spectrometric estimation of total nitrogen concentration in Douglas-fir foliage, *International Journal of Remote Sensing*, 17, 489, 1996.

86. Randall, L., Lucas, R.M., Austin, J., et al., Understanding the dynamics of land cover change: Coupling remote sensing data with farmer interviews, in *Proceedings 13th Australasian Remote Sensing and Photogrammetry Conference*, Canberra, Australia, 2006.

87. Lucas, R.M., Accad, A., Randall, L., et al., Assessing human impacts on Australian forests through integration of airborne/spaceborne remote sensing data, in *Patterns and processes in forest landscapes: Multiple uses and sustainable management*, R. Lafortezza, J. Chen, G. Sanesi, et al., eds. In press.

88. Gong, P., Pu, R., Biging, G.S., et al., Estimation of forest leaf area index using vegetation indices derived from Hyperion hyperspectral data, *IEEE Transactions on Geoscience and Remote Sensing*, 41, 1355, 2003.

89. Chason, J.W., Baldocchi, D.D., and Huston, M.A., Comparison of direct and indirect methods for estimating forest canopy leaf-area, *Agriculture Forestry Meteorology*, 57, 107, 1991.

90. Jensen, R.R. and Binford, M., Measurement and comparison of LAI estimators derived from satellite remote sensing techniques, *International Journal of Remote Sensing*, 25, 4251, 2004.

91. Woodcock, C.E., Collins, J.B., Jakabhazy, V.D., et al., Inversion of the Li–Strahler canopy reflectance model for mapping forest structure, *IEEE Transactions on Geoscience and Remote Sensing*, 35, 405, 1997.

92. Franklin, S.E., Lavigne, M.B., and Dueling, M.J., Estimation of forest leaf area index using remote sensing and GIS data for modeling net primary production, *International Journal of Remote Sensing*, 18, 3459, 1997.

93. Specht, R.L. and Specht, A., *Australian plant communities. Dynamics of structure, growth and biodiversity*. Oxford University Press, Melbourne, Australia, 1999.

94. Lucas, R.M., Cronin, N., Moghaddam, M., et al., Integration of radar and Landsat-derived foliage projected cover for woody regrowth mapping, Queensland, Australia, *Remote Sensing of Environment*, 100, 407, 2006.

95. Huete, A.R. and Liu, H.Q., An error and sensitivity analysis of the atmospheric and soil correcting variants of the NDVI for the MODIS-EOS, *IEEE Transactions on Geoscience and Remote Sensing*, 32, 897, 1994.

96. Wessman, C.A., Nel, E.M., Bateson, C.A., et al., A method to access absolute fPAR of vegetation in spatially complex ecosystems, in *Proceedings of the 7th Annual Airborne Earth Science Workshop, NASA JPL*, Pasadena, CA, 1998.

97. Asner, G.P., Biophysical and biochemical sources of variability in canopy reflectance, *Remote Sensing of Environment*, 64, 234, 1998.

98. Ustin, S.L., Zarco-Tejada, P.J., and Asner, G.P., The role of hyperspectral data in understanding the global carbon cycle, in *Proceedings of the Tenth JPL Airborne Earth Science Workshop*, Pasadena, CA, 2001.

99. Gobron, N., Pinty, B., and Verstraete, M.M., Theoretical limits to the estimation of the leaf area index on the basis of visible and near-infrared remote sensing data, *IEEE Transactions on Geoscience and Remote Sensing*, 35, 1438, 1997.

100. Kimes, D., Knjazikhin, Y., Privette, J.L., et al., Inversion methods for physically based models, *Remote Sensing Reviews*, 18, 381, 2000.

101. Atzberger, C., Object-based retrieval of biophysical canopy variables using artificial neural nets and radiative transfer models, *Remote Sensing of Environment*, 93, 53, 2004.

102. Chafer, C.J., Noonan, M., and Macnought, E., The post-fire measurement of fire severity and intensity in the Christmas 2001 Sydney wildfires, *International Journal of Wildland Fire*, 13, 227, 2004.

103. Jiang, Z., Huete, A., Chen, J., et al., Analysis of NDVI scaled difference vegetation index retrievals of vegetation fraction, *Remote Sensing of Environment*, 101, 366, 2006.

104. Roff, A.M., Taylor, G.R., Turner, R., et al., Hyperspectral and LiDAR remote sensing of forest fuel loads in Jilliby Sate Conservation Area, in *Proceedings of the 13th Australasian and Remote Sensing Conference*, Canberra, 2006.

105. Treitz, P.M. and Howarth, P.J., Hyperspectral remote sensing for estimating biophysical parameters of forest ecosystems, *Progress in Physical Geography*, 23, 359, 1999.

106. Rock, B.N., Hoshizake, T., and Miller, J.R., Comparison of in situ and airborne spectral measurements of the blue shift associated with forest decline, *Remote Sensing of Environment*, 24, 109, 1988.

107. Hoque, E. and Hutzler, P.J.S., Spectral blue-shift of red edge monitors damage class of beech trees, *Remote Sensing of Environment*, 39, 81, 1992.

108. Vogelmann, J.E., Rock, B.N., and Moss, D.M., Red edge spectral measurements from sugar maple leaves, *International Journal of Remote Sensing*, 14, 1563, 1993.

109. Kuntz, S. and Siegert, S., Monitoring of deforestation and land use in Indonesia with multitemporal ERS data, *International Journal of Remote Sensing*, 14, 2835, 1999.

110. Drake, J.B., Dubayaha, R.O., Clark, D.B., et al., Estimation of tropical forest structural characteristics using large-footprint LiDAR, *Remote Sensing of Environment*, 79, 305, 2002.

111. Hajnsek, I., Kugler, F., Papathanassiou, K., et al., INDREX II—Indonesian airborne radar experiment campaign over tropical forest in L- and P-band: First results, in *Proceedings IGARSS*, 6, 2005.

112. Prates-Clark, C., Lucas, R.M., and dos Santos, J.R., Dynamics of tropical forest regeneration in the Brazilian Amazon—Histories of land use, *Remote Sensing of Environment*, in press.

113. Spalding, M., Blasco, F., and Field, C., *World mangrove atlas*. International Society for Mangrove Ecosystems, Okinawa, Japan, 1997, 180.

114. Blasco, F., Aizpuru, M., and Gers, C., Depletion of the mangroves of continental Asia, *Wetlands Ecology and Management*, 9, 245, 2001.

115. Finlayson, C.M., Davidson, N.C., Spiers, A.G., et al., Global wetland inventory current status and future priorities, *Marine and Freshwater Research*, 50, 717, 1999.

116. Ellison, A.M. and Farnsworth, E.J., Simulated sea level change alters anatomy, physiology, growth and reproduction of red mangrove (*Rhizophora mangle* L), *Oecologia*, 112, 435, 1997.

117. Knighton, A.D., Woodroffe, C.D., and Mills, K., The evolution of tidal creek networks, Mary River, Northern Australia, *Earth Surface Processes and Landforms*, 17, 167, 1992.

118. Cobb, S.M., Saynor, M.J., Eliot, I., et al., Saltwater intrusion in the Alligator Rivers region, Northern Australia, in *Assessment and monitoring of coastal change in the Alligator Rivers region, Northern Australia*, I. Eliot, M. Saynor, M. Eliot, and M. Finlayson, eds. *Supervising scientist report*, Canberra, Australia, 2000, 157.

119. Mitchell, A.L., Lucas, R.M., Donnelly, B.E., et al., A new map of mangroves for Kakadu National Park, Northern Australia, based on stereo aerial photography, *Journal of Aquatic Conservation*, 17, 446, 2007.

120. Lucas, R.M., Ellison, J.C., Michell, A., et al., Use of stereo aerial photography for quantifying changes in the extent and height of mangroves in tropical Australia, *Wetlands Ecology and Management*, 10, 161, 2002.

121. Olson, J.S., Watts, J.A., and Allison, L.J., *Carbon in live vegetation in major world ecosystems*. Oak Ridge National Laboratory, Oak Ridge, TN, 1983.

122. Furley, P.A., Nature and diversity of neotropical savanna vegetation with particular reference to the Brazilian cerrados, *Global Ecological Biogeographic Letters*, 8, 223, 1999.

123. Field, C.B., Behrenfeld, M.J., Randerson, J.T., et al., Primary production of the biosphere: Integrating terrestrial and oceanic components, *Science*, 281, 237, 1998.
124. Frost, P.G.H., The ecology of Miombo Woodlands, in *The Miombo in transition: Woodlands and welfare in Africa*, B. Campbell, ed. Center for International Forestry Research, Bogor, West Java, 1996, 11.
125. Burrows, W.H., Henry, B.K., Back, P.V., et al., Growth and carbon stock change in Eucalypt woodlands in North-East Australia; ecological and greenhouse sink implications, *Global Change Biology*, 8, 1, 2002.
126. Lucas, R.M., Lee, A., and Bunting, P., Retrieving forest biomass through integration of CASI and LiDAR data, *International Journal of Remote Sensing*, in press.
127. Lee, A.C. and Lucas, R.M., A LiDAR-derived canopy density model for tree stem and crown mapping in Australian forests, *Remote Sensing of Environment*, in press.
128. Stuffler, T., Kaufmann, C., Hofer, S., et al., The enmap hyperspectral imager—An advanced optical payload for future applications in Earth observation programs, in *Proceedings of the 57th IAC (International Astronautical Congress)*, Valencia, Spain, 2006.
129. Asner, G.P., Ungar, S., Green, R., et al., Leaping from AVIRIS to high fidelity spaceborne imaging spectroscopy, in *14th Annual JPL Airborne Earth Science Workshop*, Pasadena, CA, 2005.
130. Rouse, J.W., Haas, R.H., Schell, J.A., et al., Monitoring vegetation systems in the Great Plains with ERTS, in *Proceedings of the 3rd Earth Resources Technology Satellite-1 Symposium*, NASA SP-351, U.S. Government Printing Office, Washington, D.C., 1973, 309.
131. Gitelson, A.A., Kaufman, Y.J., and Merzlyak, M.N., Use of a green channel in remote sensing of global vegetation from EOS-MODIS, *Remote Sensing of Environment*, 58, 289, 1996.
132. Gitelson, A. and Merzlyak, M.N., Quantitative estimation of chlorophyll a using reflectance spectra: Experiments with autumn chestnut and maple leaves, *Journal of Photochemistry, Photobiology B: Biology*, 22, 247, 1994.
133. Datt, B., Visible/near infrared reflectance and chlorophyll content in *Eucalyptus* leaves, *International Journal of Remote Sensing*, 20, 2741, 1999.
134. Gamon, J.A., Lee, L.F., Qiu, H.L., et al., A multi-scale sampling strategy for detecting physiologically significant signals in AVIRIS imagery, in *Summaries of the 7th Airborne Earth Science Workshop, January 12–16, NASA JPL*, California, 1998.
135. Gamon, J.A. and Surfus, J.S., Assessing leaf pigment content and activity with a reflectometer, *New Phytologist*, 143, 105, 1999.
136. Gitelson, A.A., Zur, Y., Chivkunova, O.B., et al., Assessing carotenoid content in plant leaves with reflectance spectroscopy, *Photochemistry and Photobiology*, 75, 272, 2002.
137. Gitelson, A.A., Merzlyak, M.N., and Chivkunova, O.B., Optical properties and non-destructive estimation of anthocyanin content in plant leaves, *Photochemistry and Photobiology*, 74, 38, 2001.
138. Serrano, L., Penuelas, J., and Ustin, S., Remote sensing of nitrogen and lignin in Mediterranean vegetation from AVIRIS data: Decomposing biochemical from structural signals, *Remote Sensing of Environment*, 81, 355, 2002.
139. Penuelas, J., Pinol, J., Ogaya, R., et al., Estimation of plant water concentration by the reflectance water index (R900/R970), *International Journal of Remote Sensing*, 18, 2869, 1997.
140. Hunt, E.R. and Rock, B.N., Detection of changes in leaf water content using near- and middle-infrared reflectances, *Remote Sensing of Environment*, 30, 43, 1989.
141. Gao, B.C., NDWI: A normalized difference water index for remote sensing of vegetation liquid water from space, *Remote Sensing of Environment*, 58, 257, 1996.
142. Hardisky, M.A., Klemas, V., and Daiber, F.D., Remote sensing salt marsh biomass and stress detection, *Advances in Space Research*, 2, 219, 1983.

143. Daughtry, C.S.T., Discriminating crop residues from soil by shortwave infrared reflectance, *Agronomy Journal*, 93, 125, 2001.
144. Merzlyak, M.N., Gitelson, A., Chivkunova, O.B., et al., Non-destructive optical detection of pigment changes during leaf senescence and fruit ripening, *Physiologica Plantarum*, 106, 135, 1999.
145. Merton, R.N., *Multi-temporal analysis of community scale vegetation stress with imaging spectrometry*. University of Auckland, Auckland, NZ, 1998.
146. Byrant, D., Nielson, D., and Tangley, L., *The last frontier forests: Ecosystems and economics on the edge*. World Resources Institute, Washington, D.C., 1997.

4 Effect of Soil Type on Plant Growth, Leaf Nutrient/Chlorophyll Concentration, and Leaf Reflectance of Tropical Tree and Grass Species

Julio C. Calvo-Alvarado, Margaret Kalacska,
G. Arturo Sanchez-Azofeifa, and Lynne S. Bell

CONTENTS

4.1   INTRODUCTION

An understanding of how leaf spectral reflectance changes as a function of soil type is key to interpreting intra- and interspecies spectral reflectance differences. Because leaf traits are influenced by leaf age, canopy position, and soil fertility, a comprehensive model of leaf spectral reflectance must include a clear description of soil/nutrient/plant interactions. Soil properties play an intrinsic and important role in affecting plant health that, when combined with soil nutrients, contributes to a given spectral reflectance at the leaf and canopy levels. In general, soil effects on plant/leaf spectral reflectance have not been studied in depth, and in some cases, the plant/soil/ nutrient interactions are decoupled from overall spectral observations.

Pioneer work such as that conducted by references 1–10, among many others, has focused on describing and linking spectral reflectance properties to chlorophyll/ carotenoid concentrations and other photoprotection pigments. In general very few studies are from tropical regions and none has specifically considered the effect of soil type and nutrient availability on spectral reflectance. The closest studies considering plant/soil interactions have been those exploring the spectral expression of plant stress (e.g., references 11 through 13). In addition, a common denominator of several studies mentioned is a focus on spectral response across different wavelengths (e.g., Clark, Roberts, and Clark [2]) without consideration for soil spatial heterogeneity and nutrient availability. Therefore, there is a strong need to link the effects of soil properties and nutrients to the spectral response of plants in order to better understand leaf and canopy spectral reflectance.

In this first of three chapters (see also chapters 5 and 9) that examine the application of machine learning to hyperspectral data, we begin with a brief background of hyperspectral data characteristics and focus on the nearest neighbor (k-NN) classification process as a form of nonparametric learning that can produce favorable results in classifying challenging data sets. Second, we apply the concepts to an investigation of the effect of soil type/treatment on reflectance for two tropical trees and one grass species in an example of the application of the classification methodology.

The overall objectives of the application section of this chapter are to (1) evaluate the effects of five soil types on plant growth and foliar nutrient concentration in seedlings of two tree species, (2) investigate the effect of plant genetic variation on plant growth as a response to soil type, (3) investigate the utility of the indirect leaf chlorophyll measurements from the SPAD as a tool to evaluate plant vigor, and (4) assess the effect of soil type on the spectral response of tropical tree seedlings (from seeds and clones) and a tropical grass species.

In this study we hypothesize that, given the distinct physical and chemical properties of the selected soil types, plant growth, leaf nutrients, and chlorophyll concentration will correlate with soil fertility, and soil type differences will be expressed in the leaf reflectance of the various species.

4.2 HYPERSPECTRAL DATA PROPERTIES

In this section we briefly examine some properties of hyperspectral data that differentiate it from more conventional two- and three-dimensional forms. One of the most common reasons for using hyperspectral data is to enhance the available information, especially for complex research questions. However, the larger number of dimensions poses complications: an exponential increase in computational effort and problems with parameter, density, or state estimations [14]. In parameter estimation, the objective is to derive a parametric description of an object or class. Density estimation, on the other hand, involves fitting a function to describe the distribution of the data, while state estimation is the assignment of a classification label or the derivation of a parametric description that varies in time or space [14,15].

Figures 4.1a and 4.1b illustrate two common representations of hyperspectral imagery: (a) as a three-dimensional image cube where the z axis illustrates the spectral information in n-dimensions, and (b) as an n-dimensional scatter plot where individual pixels are represented as points in the selected n-dimensions (five dimensions in this example; each dimension is a band). For statistical purposes, hyperspectral data can be described in terms of hyperspheres, hypercubes, hyperellipsoids, or in more intuitive density functions (e.g., Gaussian or uniform distributions). A hypersphere is the generalization of a circle (two dimensions) or a sphere (three dimensions). Its volume, where r is the radius, d is the number of dimensions, and Γ the gamma function, is [14]

$$V_s(r) = \frac{2r^d}{d} \times \frac{\pi^{\frac{d}{2}}}{\Gamma\left(\frac{d}{2}\right)} \tag{4.1}$$

A hypercube is the generalization of a square (two dimensions) or cube (three dimensions) in n dimensions (figs. 4.1c–4.1f). Its volume in $[-r\ r]^d$ is [14]

$$V_c(r) = (2r)^d \tag{4.2}$$

Jimenez and Landgrebe [14] illustrate the following five unusual geometric, statistical, and asymptotic characteristics of hyperdimensionality that affect analysis techniques:

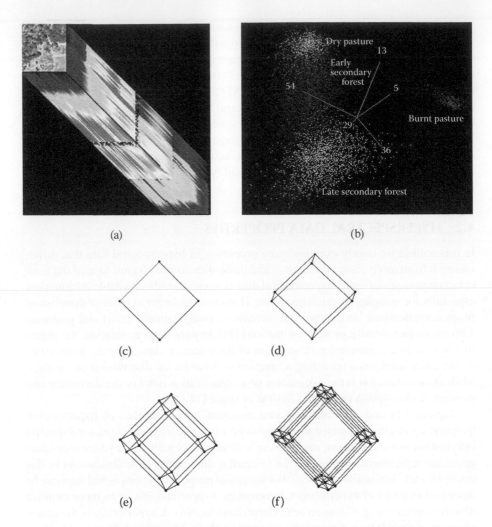

FIG. 4.1 (a) A three-dimensional representation of a hyperspectral image where the z axis is the reflectance in every pixel along the top and right edges for each of the 125 bands; (b) a five-dimensional scatter plot representing reflectance from bands 5 (518 nm), 13 (636 nm), 29 (865 nm), 36 (963 nm), and 54 (1226 nm) for four classes from the image illustrated in (a): (e): burnt pasture (red), dry pasture (green), early secondary forest (blue), and late secondary forest (yellow); (c) representation of a hypercube in two dimensions—the analogue of a square; three dimensions (d)—the analogue of a cube; and higher dimensions: four dimensions (e); and five dimensions (f). See CD for color image.

1. As the number of dimensions of a hypercube increases, the volume (data) becomes concentrated in the corners ($\lim_{d\to\infty} f_{d1} = 0$, where f_{d1} is the fractional volume), whereas for a hypersphere or hyperellipsoide, the volume becomes concentrated in an outside shell (fig. 4.2). The importance of this characteristic is that the majority of the hyperdimensional data can be projected into a lower dimensional subspace without losing class separability. It also means that Gaussian data will be concentrated in the tails of the

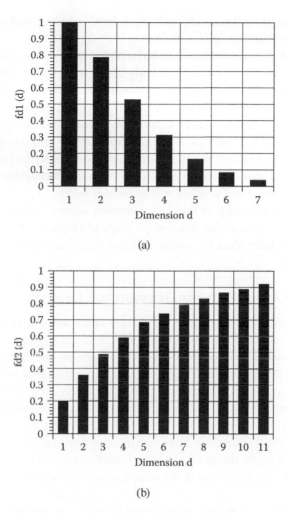

FIG. 4.2 (a) The fractional volume of a hypersphere (fd1) inscribed in a hypercube as dimensionality increases; (b) fraction of the volume of a hypersphere contained in the outside shell (defined by a sphere with radius r − ε) as a function of dimensionality where ε = r/5. (Reprinted from Jimenez, L.O. and Landgrebe, D.A., *IEEE Transactions on Systems, Man, and Cybernetics—Part C: Applications and Reviews,* 28(1), 39–54, 1998. With permission from IEEE.)

distribution and uniformly distributed data will be concentrated in the corners, complicating density estimations.

2. The angle between the diagonal and a Euclidean coordinate approaches orthogonality with increased dimensions. This implies that if a data cluster is projected onto a diagonal, such as by averaging features, information about its location in the *n*-dimensional space is lost.

3. As dimensionality increases, lower dimensional linear projections of high-dimensional data have a tendency to be normal or be a mixture of normal densities. This characteristic implies that once hyperdimensional data are

reprojected into a linear subspace any feature reduction technique can be used, even those that assume normality.

4. The size of the training data (labeled samples) must increase with increasing dimensionality in order to properly estimate the multivariate densities of the classes. This is one of the reasons why it is important to reduce the dimensions of the data—through feature selection, for example—rather than applying computational techniques at full dimensionality such as PCA, discriminant analysis, etc., unless the number of training samples is very large. In addition, the Hughes phenomenon states that increasing the number of dimensions improves the performance of a classifier only up to a point after which performance will degrade [16–18].

5. As the dimensionality of the data increases, second-order statistics (i.e., variance, covariance) that describe the shape of a class are more relevant for class separability than first-order statistics (i.e., class mean) that describe its location.

4.2.1   k-NN Classification with Hyperspectral Data

Both the distribution of the data in the feature space and the number of desired classes directly influence separability. Figure 4.3a illustrates in two dimensions a two-class Gaussian data set; each of the three illustrated classifiers (linear, neural network, k-NN) adequately separates the classes. In some cases the distribution of the data is such that parametric learning cannot adequately separate the classes. With nonparametric instance-based learning, the hypotheses are constructed directly from the training data sets. Figures 4.3b–4.3e illustrate an eight-class data set where the classes have varying distributions. The linear discriminant classifier (LDC) (figs. 4.3b, 4.3d) cannot adequately define the decision boundaries for these eight classes with varying distributions, regardless of the samples size (i.e., 100 vs. 1000). In addition to assuming normal distributions for the eight classes, the LDC assumes that the covariance matrix for each is the same—minimizing the Mahalanobis distance [15]. If $P(w_k)$ is the probability of a given class and N_k is the number of samples within a class, w_k can be described by a multinomial distribution where the prior probability of class w_k is N_k/N_s where N_s is the total number of samples. The variance of the prior probability is [15]

$$Var[P(w_k)] = \frac{P(w_k)(1 - P(w_k))}{N_s} \qquad (4.3)$$

In figure 4.3b, if the probability of class "a" is 0.13, with a permitted variability of 10%, the required number of samples would be 669. If the permitted variability decreases to 5%, the total number of samples increases to 2677 following:

$$N_s = \frac{1 - P(w_k)}{P(w_k) \times Var[P(w_k)]} \qquad (4.4)$$

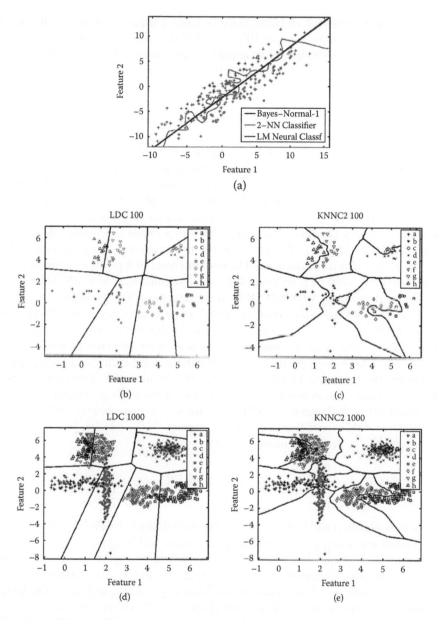

FIG. 4.3 Effect of increased number of classes. (a) Two classes, normal distribution 200 data points; (b) eight classes of varying distributions, linear discriminant classifier (LDC) 100 points; (c) 2-NN classifier (KNNC2) 100 points; (d) eight classes LDC 1000 points; (e) eight classes KNNC2 1000 points. See CD for color image.

The nearest neighbor classifier (k-NN) more adequately separates the eight classes with both sample sizes of 100 and 1000 (figs. 4.3c and 4.3e). The nearest neighbor classifier (k-NN) is a simple nonparametric instance-based learning method [19]. The premise of the k-NN approach is that for any given data point, its properties will be similar to those in its feature space neighborhood. The value of k denotes the number of neighbors taken into consideration in the classification problem; the optimal value for k will vary with each classification problem [19]. However, with hyperdimensional data it is essential to perform a feature selection or other dimension reduction technique prior to the implementation of the k-NN because of the reasons noted previously and also because, in high-dimensional spaces, the calculation for the nearest neighbors becomes computationally intensive. The neighborhood that encompasses each data point can become very large (almost spanning the entire input feature space), negating the purpose of the nearest neighbor approach [19]. For example, assuming the data are a hypercube with a radius (r), dimensions (d), and size N (number of points), the volume of the hypercube following equation 4.2 is $(2r)^d$ or b^d, where b is the diameter. The neighborhood of k points surrounding any given data point must be a fraction of the total volume: $V_c = b^d = k/N$. Solving for b, the formula becomes $b = (k/N)^{1/d}$ [19]. If the data have 100 dimensions ($d = 100$), $k = 10$, and $N = 1000$, b is equal to 0.95, or 95% of the entire feature space. With only 10 dimensions and fewer neighbors ($k = 3$), the neighborhood encompasses 56% of the total feature space. In two dimensions, the neighborhood is less than 1%.

Another way to conceptualize the neighborhood surrounding each data point is to consider the input data as a hypersphere with a volume of V_s (equation 4.1). If a data point of interest is denoted by x, the objective of the k-NN classification process is to select a spherical region around x, such that it encompasses only k other points. The volume of this second sphere (V_x) will depend on the location of x within the feature space. The density estimation is [15]

$$p(x \mid w_k) = \frac{k}{NV_x} \qquad (4.5)$$

where w_k is a given class. Therefore, where the probability $(x \mid w_k)$ is large, the volume (V_x) is small and where $p(x \mid w_k)$ is small, the volume will be large.

For remotely sensed data where classes are difficult to separate due to their similarity, two or three dimensions may not be sufficient. Figure 4.4 illustrates a simulated data set with two classes, each of which has a spherical distribution. The linear classifier (46% error) and neural network classifier (32% error) cannot separate the data in two dimensions (fig. 4.4a); even the performance of the k-NN classifier is not adequate (31% error). Adding one more dimension (fig. 4.4b) does not improve separability (minimum 21% error). With 30 dimensions, the error decreases to 7% for the k-NN classifier (two neighbors) and does not improve with additional features (i.e., 7% error remains even with 100 dimensions). With more than 100 dimensions the error increases.

In the following section we examine the application of pattern classifiers such as the k-NN, the LDC, a neural network classifier, a quadratic classifier, and a decision tree to separate the reflectance of two species (and one clone) of tropical

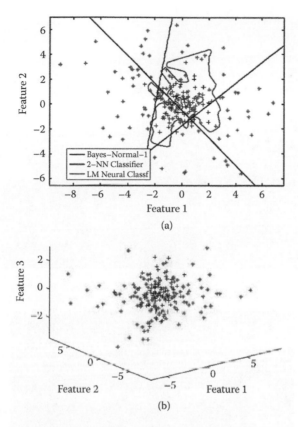

FIG. 4.4 Separating spherically distributed data (two classes) in two (a) and three (b) dimensions. See text for additional dimensions. See CD for color image.

tree seedlings and one grass species grown on different soil types and with different treatments.

4.3 APPLICATION—EFFECT OF SOIL TYPE ON SPECTRAL RESPONSE

4.3.1 EXPERIMENTAL DESIGN

An experiment with seedlings of introduced and native tree species, commonly used in reforestation or agroforestry systems, was established in January 2004 in the Forest Nursery at the San Carlos Campus of the Instituto Tecnológico de Costa Rica (ITCR). A total of four experimental blocks was established using a randomized complete block design. The size of the experimental plots was 2 × 2 trees planted in black plastic pots, of 30 cm diameter by 35 cm high. All seedlings were transferred to the pots from the nursery when they were at least 10 cm tall. The experimental blocks were established directly in the open; hence, the seedlings were exposed directly to all meteorological events (fig. 4.5a). Because the site is humid with rain year round and our intention was to replicate field conditions during reforestation,

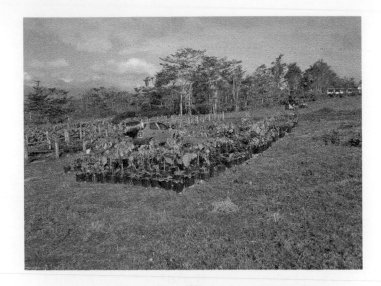

(a)

(b)

FIG. 4.5   (a) Experimental design with the tree seedlings at the ITCR campus in San Carlos, Costa Rica; (b) pasture grass *P. conjugatum* measured in situ in northwestern Costa Rica. See CD for color image.

irrigation of these blocks was not required. Weeding was conducted manually once during the entire experiment. A complete evaluation of height, steam diameter, survival, aerial, and root biomass of the seedlings was carried out at the end of the experiment in April 2004.

4.3.2   SELECTION OF TREE SPECIES

Two species were included in this nursery experiment: *Tectona grandis* (Linn), as the best exotic species used in reforestation projects in Costa Rica, and *Swietenia macrophylla* King (mahogany), as a precious wood tree species in danger of extinction [20]. Seedlings of these species were purchased at a commercial nursery and came from standard seed propagation methods. We also included in the experiment selected clones for *T. grandis* reproduced at the experimental ITCR nursery in San Carlos. All clones were reproduced from at least five different genetic sources that were selected for best growth, phenotypic characteristics, and adaptability. The grass species examined in this experiment—*Paspalum conjugatum* (Nees)—is a common invasive pasture grass. It was measured on-site in a pasture in the northwestern part of the country (fig. 4.5b).

4.3.3 SELECTION OF SOIL TYPES

Three basic soil types were used in the nursery experiment representing contrasting soil conditions for reforestation projects in Costa Rica: (a) a loamy textured Andisol (Volcanic) from the western slope of the Irazú Volcano, (b) a clayey textured acid Ultisol (Residual) from the southern region, and (c) a clayey, loamy textured Inceptisol (Alluvial) from the flood plains of the Térraba River. We collected the soils from the A horizon in each site and transported them to the nursery, where each soil type was homogenized by constant mixing before filling the pots for the seedlings. For each homogenized soil we took a 0.5-kg soil sample for soil chemical analysis to the soil laboratory at the University of Costa Rica. The soil elemental constituents were estimated for the samples by the methods listed in table 4.1. Based on Sanchez [21], we calculated the following parameters: (a) effective cation exchange capacity (CEC) as the sum of acidity and bases, (b) base saturation (in percentage) from the sum of exchangeable base cations,

$$Bases\%_{Total} = \frac{Ca + Mg + K}{ECEC} \times 100 \tag{4.6}$$

and (c) exchangeable acid saturation (in percentage) as

$$AS\% = \frac{Acidity}{ECEC} \times 100 \tag{4.7}$$

TABLE 4.1
Estimation Methods for Soil Elements, pH, and Acidity

Soil characteristic/element	Estimation method
pH	Water 1:2
Al (exchangeable acidity)	Titration
Ca, Mg	KCl 1 M 1:10 extraction
P, K, Zn, Fe, Mn, Cu	Modified Olsen pH 8.5 1:10 extraction
S, B, N, OM	Atomic absorption spectrophotometry

Total organic carbon was calculated by dividing organic matter in percentage by a constant of 1.724 [22]. A summary of results from the soil elemental analysis is illustrated in table 4.2.

In addition to the three basic soils we prepared two additional soil substrata by liming with $CaCO_3$ (calcium carbonate) or by fertilization with NPK (nitrogen, phosphorous, and potassium). One substratum was prepared by fertilization with 75 g/pot of NPK (10-30-10) and by correcting the soil acidity of the Ultisol by liming to reduce the soil exchangeable acid saturation from 55 to 10%. This was achieved by adding $CaCO_3$ at an equivalent rate of 1.7 tons of $CaCO_3$ per hectare according to the equation suggested by Cochrane, Salinas, and Sanchez [23]:

$$\frac{CaCO_3}{ha}(Ton) = \frac{1.5 \times (AS_{existing} - AS_{desired}) \times ECEC}{100} \qquad (4.8)$$

The second substratum was prepared by adding only 75 g/pot of NPK (10-30-10) to the volcanic soil. In theory, liming and fertilization with NPK of these soils will correct most of the soil fertility deficiencies [21]. In summary, the nursery experiment included five soil types ranked from high to low fertility (tables 4.2 and 4.3).

In order to provide a comparison with pasture we included one additional soil type: an Inceptisol from the northwestern region of Guanacaste (northwestern sector of the country). The soil chemistry analysis for these pasture experiment samples was conducted by the same soil lab at the University of Costa Rica following the procedures described earlier and in table 4.1. The difference between the untreated Inceptisol and the treated Inceptisol (Inceptisol$_{Trt}$) is that the former has been altered following the subsurface decomposition of 800 lb of organic matter from buried cattle carcasses from an ongoing forensic experiment [24,25]. The purpose of including this type of soil alteration in the chapter is that cadaveric decomposition is in an integral, though marginally understood, part of ecosystem functioning [26], with localized effects that are both different and greater than those from the decomposition of plant material (e.g., leaf litter). As a global phenomenon and contributor to landscape heterogeneity (at various scales), we explore the potential effect of this microsere on the spectral properties of a tropical grass species.

Sixteen months prior to sampling, both the untreated and treated Inceptisol were excavated to a depth of 1.5 m. During excavation and refilling with a motorized backhoe, the horizons were mixed. The fertility and properties of the Inceptisol are similar to the volcanic soil (Andisol) since this soil has *andic* properties inherited from old volcanic ash depositions (tables 4.2 and 4.3). The approximate breakdown of the elemental constituents of the increased organic matter (table 4.3) in the Inceptisol$_{Trt}$ is adapted from Emsley [27]. However, it is important to note that the release of the majority of the elements during decomposition will be in the form of various compounds rather than in pure elemental form.

4.3.4 FOLIAR ANALYSIS

A composite foliar sample (approximately 100 g dried leaves) was collected at the end of the nursery experiment from each treatment. Each sample was dried at 70°C

TABLE 4.2

Soil Types Used in the Experiment and Their Chemical Analysis

Soil type	Site	pH in H$_2$O	Acidity	Ca	Mg	K	CEC	Total bases	% AS
					cmol(+)/L				
Andisol	Cartago, Irazú Volcano	5.6 M	0.30 L	9.00 M	1.57 L	0.69 H	11.56 M	11.26 M	3 L
Inceptisol	Terraba River flood plain, Palmar Norte	5.5 M	0.28 L	21.40 M	3.50 L	0.99 H	26.17 H	25.89 H	1 L
Ultisol	Buenos Aires de Osa	5.0 L	1.40 H	0.83 L	0.18 L	0.13 L	2.54 L	1.14 L	55 H
Inceptisol[a]	Los Inocentes, Guanacaste	6.4 (0) M	0.15 (0.03) L	8.24 (0.22) M	2.61 (0.07) L	0.13 (0.01) L	11.12 (0.33) M	10.98 (0.30) L	1 (0.26) L
Inceptisol$_{Trt}$[a,b]	Los Inocentes, Guanacaste	6.3 (0) M	0.16 (0.01) L	5.51 (0.12) M	2.42 (0.05) L	0.61 (0.03) H	8.70 (0.15) M	8.54 (0.15) L	2 (0.08) L

continued

TABLE 4.2 (continued)
Soil Types Used in the Experiment and Their Chemical Analysis

Soil type	Site	mg/L							%				Fertility classification
		P	Cu	Fe	Mn	Zn	B	S	N	OM	C	C/N	
Andisol	Cartago, Irazú Volcano	12.40 M	27.3 H	22.5 M	10.1 M	5.7 M	1.04	35.50	—	—	—	—	Medium to high
Inceptisol	Terraba River flood plain, Palmar Norte	12.50 M	118.0 H	104.0 M	7.90 M	6.30 M	0.64	66.90	—	—	—	—	High
Ultisol	Buenos Aires de Osa	3.10 L	10.30 M	501.0 H	3.10 L	2.40 M	1.58	3.50	—	—	—	—	Low
Inceptisol[a]	Los Inocentes, Guanacaste	3.00 (0) L	6.00 (0) L	37.00 (1.73) M	143.67 (3.51) H	1.97 (0.06) L	—	13.67 (0.58)	0.13 (0.01)	1.23 (0.06)	0.72 (0.03)	5.65 (0.26)	Medium
Inceptisol_fn[a,b]	Los Inocentes, Guanacaste	3.00 (0) L	7.00 (1.0) L	30.67 (4.51) L	197.33 (21.08) H	2.03 (0.31) M	—	23.67 (0.58)	0.15 (0.01)	1.63 (0.06)	0.95 (0.03)	6.46 (0.24)	Medium to high

Notes: CEC = effective cation exchange capacity, OM = total organic matter. Elemental concentrations classified as H for high, M for medium, and L for low. AS = acid saturation (aluminum), Ca = calcium, Mg = magnesium, K = potassium, P = phosphorus, Cu = copper, Fe = iron, Mn = manganese, Zn = zinc, B = boron, S = sulfur, N = nitrogen, C = carbon, C/N = carbon/nitrogen ratio.

a Results represent mean and (standard deviation) from three samples.

b Results from 16 months following beginning of decomposition.

TABLE 4.3

General Characterization of Soil Types Used in the Experiment

Soil type	Soil characteristics
Inceptisol: Fluventic Eutropept	Loamy texture with high fertility. This soil has optimal soil physical and chemical characteristics for plant growth. In dry climate the low water-holding capacity of this soil could be a limitation for plant growth.
Andisol: Typic Haplundands (volcanic)	Loamy texture. Medium to high fertility. Low available P, N, and K because of the presence of allophane that fixes P, K, and organic matter. Even though the texture is loamy the water-holding capacity is high because of the high organic matter content of this soil. The addition of fertilizer, such as NPK, is a common solution to correct the nutrient limitations of this soil.
Ultisol: Ustic Kandihumult	Clayey texture. Very acid soil with severe chemical limitation for optimal plan growth. Soils are low in bases (Ca, Mg, and K), P, and N, and are very acid with low pH and high concentration of Al and Fe. Soil color is red because of iron oxides. The application of $CaCO_3$ to reduce soil acidity from 55 to 10% will favor plant growth by balancing available cations and microelements and by fixing Al. The use of dolomite will increase available Ca and Mg. The application of NKP fertilizer will increase these nutrients and correct for NPK deficiencies of this soil.
Inceptisol: Andic Ustic Humitropet	Clayey texture. Moderate to high fertility with low concentration of P and Mg. These soils have some allophane because of the influence of past volcanic ash depositions that had rejuvenated the fertility of the A horizon, given the extra classification of Andic in the taxonomic name.
Inceptisol: Andic Ustic Humitropet + subsurface decomposed cadavers	Same as above. The 800 lb of subsurface cadavers is composed roughly of the following: 223 kg (oxygen); 83 kg (carbon); 36 kg (hydrogen); 9 kg (nitrogen); 5 kg (calcium); 4 kg (phosphorus); 726 g (potassium); 726 g (sulfur); 518 g (sodium); 492 g (chlorine); 98 g (magnesium); 22 g (iron); 13 g (fluorine); 12 g (zinc); 5 g (silicon); 0.3 g (aluminum); 0.4 g (copper); 0.06 g (manganese); other trace elements.

for 72 hours for foliar chemical analysis at the soil laboratory at the University of Costa Rica. Total Kjeldahl nitrogen (TKN) was determined by using the macro-Kjeldahl method, and concentrations of P, Ca, Mg, K, S, Fe, Cu, Zn, Mn, and B were determined using wet ash extraction (nitric/perchloric acid) and atomic absorption. Nutrient contents are reported in percentage of dry matter for N, P, Ca, Mg, K, and S and in milligrams per kilogram for Fe, Cu, Zn, Mn, and B (table 4.4). The macro-nutrients C, H, N, O, P, and S; macrominerals K, Ca, Fe, and Mg; and microminerals Cu, Zn, B, and Mn examined in the foliar analysis are important both for plant growth and health and as the building blocks of compounds directly affecting leaf reflectance (i.e., chlorophyll, carotenoids, xanthophylls).

4.3.5 HANDHELD RELATIVE CHLOROPHYLL MEASUREMENTS

We used the SPAD 502 (Minolta—Osaka, Japan) chlorophyll hand meter as an indirect way to estimate leaf chlorophyll concentration [28]. The SPAD-502 has a

TABLE 4.4

Foliar Nutrient Content for Tree Seedlings in the Five Soil Treatments

| Soil type | % | | | | K | S | mg/kg | | | | | Ca/K | Mg/K | (Ca + Mg)/K | Chlorophyll (SPAD) |
	N	P	Ca	Mg			Fe	Cu	Zn	Mn	B				
T. grandis seedlings																
Andisol + NPK	2.63	0.41	1.26	0.19	2.07	0.18	255	18	57	90	703	0.61	0.09	0.70	27.9
Andisol	3.00	0.41	1.25	0.27	2.41	0.21	567	19	44	134	853	0.52	0.11	0.63	31.0
Ultisol + CaCO$_3$ + NPK	2.81	0.60	0.82	0.23	2.55	0.21	120	20	40	79	901	0.32	0.09	0.41	29.0
Ultisol	3.20	0.41	0.84	0.18	1.94	0.22	575	23	39	195	1121	0.43	0.09	0.53	31.3
Inceptisol	3.20	0.43	1.27	0.36	2.22	0.22	190	19	41	228	684	0.57	0.16	0.73	31.8
T. grandis clones																
Andisol + NPK	2.45	0.60	0.66	0.29	2.58	0.17	130	22	42	67	646	0.26	0.11	0.37	25.5
Andisol	1.93	0.30	1.05	0.22	2.19	0.32	107	12	37	41	793	0.48	0.10	0.58	31.6
Ultisol + CaCO$_3$ + NPK	2.00	0.37	0.73	0.29	1.60	0.14	139	15	32	51	857	0.46	0.18	0.64	25.0
Ultisol	2.18	0.13	0.72	0.37	0.89	0.17	127	12	25	168	932	0.81	0.42	1.22	24.4
Inceptisol	2.81	0.73	0.91	0.45	2.47	0.24	220	30	62	110	1375	0.37	0.18	0.55	24.5
S. macrophylla																
Andisol + NPK	3.49	0.24	1.33	0.18	1.70	0.32	153	11	29	81	523	0.78	0.11	0.89	25.0
Andisol	2.63	0.17	1.17	0.19	1.55	0.34	102	9	20	45	711	0.75	0.12	0.88	22.4
Ultisol + CaCO$_3$ + NPK	2.34	0.20	1.22	0.21	1.79	0.36	82	9	23	56	708	0.68	0.12	0.80	26.5
Ultisol	2.23	0.10	0.78	0.21	0.98	0.30	168	9	20	37	713	0.80	0.21	1.01	32.8
Inceptisol	2.57	0.25	1.20	0.22	1.83	0.35	86	11	23	45	714	0.66	0.12	0.78	17.6

0.06-cm^2 measurement area and calculates a unitless index based on absorbance at 650 and 940 nm. The reported accuracy of the SPAD-502 is ±1 SPAD unit. For each seedling, just before final harvesting, we measured four leaves at random and used the arithmetic mean of these measurements as the mean value for the plant.

The SPAD values are indirect measures of chlorophyll content [28], where higher values have been shown to be indicative of a greater concentration of photosynthetic pigments and foliar nitrogen. However, the wavelengths assessed by the SPAD (650 and 940 nm) may also be affected by plant water stress; hence, results may not be indicative of a straightforward relationship with foliar nitrogen or chlorophyll.

4.3.6 LEAF REFLECTANCE MEASUREMENTS

Reflectance was measured for 10–13 leaves from each seedling that survived until the end of the experiment (4 months) with a Unispec Spectral Analysis System (PP Systems, Amesbury, Massachusetts) [29,30]. This spectrometer has a spectral range from 350 to 1100 nm with a spectral resolution (full width at half maximum [FWHM]) <10 nm and a sampling interval of 3.3 nm. The field of view of the bare foreoptic is 40° and contains a bifurcated fiber optic with an internal 7-W halogen light source and an attached leaf clip.

For the grass samples, 30 blades each were sampled from the Inceptisol and the Inceptisol$_{Trt}$. Reflectance was measured with an ASD Fieldspec handheld spectrometer (Analytical Spectral Devices, Boulder, Colorado) [29,31]. This spectrometer has a spectral range of 325–1075 nm, a spectral resolution (FWHM) of 3.5 nm, and a sampling interval of 1.6 nm. The spectrometer was fitted with a plant probe (with an internal light source) and leaf clip that holds the samples in place, excludes ambient light, and ensures a constant geometry for the light source and foreoptic.

The integration time for both instruments was automatically set using a 99% reflective Spectralon™ white reference panel in the leaf clip. A dark current correction was performed to eliminate instrument noise from spectral measurements. Reflectance of the samples was computed as a ratio of each sample spectrum to the white reference spectrum. The areas of high instrument noise, below 450 nm and above 950 nm, were removed from all the spectra.

4.3.7 STATISTICAL ANALYSIS

We used an analysis of variance (ANOVA) and least significant difference (Fisher LSD) to test for differences in seedling growth and SPAD values among soil treatments at the $P < .05$ level. To evaluate if these differences had a relationship with foliar and soil nutrient concentrations we used a Pearson correlation analysis using all collected variables with $P < .05$ and $P < .1$. STADISTICA v. 6 [32] was used for the statistical analysis. Each significant correlation was examined graphically by plotting the two correlated variables to determine the validity of the relationship. If the plot revealed that the significance of the correlation was due to the influence of outliers, it was discarded.

4.3.8   CLASSIFICATION OF LEAF SPECTRA

Potential effects of soil type on leaf reflectance were examined through a selection of the optimal wavelengths (i.e., to reduce the dimensions—number of wavelengths of the data) followed by classification of the spectra with a number of pattern classifiers using PRTools 2004 in Matlab [15]. A forward feature selection procedure with a "1-nearest neighbor leave-one-out" classification criterion was used to reduce the number of bands. The first band chosen is the one with the greatest separability of the data. Bands are added in the order in which they improve class separability based on the criterion. We chose the optimal number of features for each classifier based the training and validation errors of the classifiers using varying numbers of features (wavelengths). The pattern classifiers applied to the spectra are k-NN, feed-forward neural network, decision tree, linear discriminant, and quadratic [15].

Initially, for the seedlings, a classification matrix was constructed examining each of the 10 possible combinations from the five soil types (i.e., soil 1 vs. soil 2, soil 1 vs. soil 3 ..., soil 4 vs. soil 5). Second, we examined the separability of leaves growing on the three natural untreated soils (Inceptisol, Andisol, and Ultisol). Finally, we examined the separability of the reflectance from all five soil types in one classification problem. For the grass, there was only one comparison: Inceptisol vs. Inceptisol$_{TRT}$.

4.4   RESULTS

4.4.1   SOIL CHEMICAL CHARACTERIZATION

Table 4.2 summarizes the soil chemical analysis for each of the basic soil substrata used in this study. Table 4.3 lists the basic characteristics and their classifications (USDA soil taxonomy). Since we improved the fertility levels of Ca, K, and P of the Ultisol and Andisol by the application of NPK and CaCO$_3$, we recalculated the concentration of these nutrients. In general, we estimated that in the case of the Ultisol the untreated levels in table 4.2 changed from Al = 1.4, Ca = 0.83, K = 0.13, and P = 3.1 to treated levels of Al = 0.5, Ca = 2.53, K = 0.58, and P = 20. For the Andisol the untreated levels changed from K = 0.69 and P = 12 to treated levels of K = 1.14 and P = 20. The treatments applied to the Ultisol and Andisol did not affect the other nutrients; hence, they are as indicated in table 4.2 for the untreated Ultisol and Andisol.

4.4.2   PLANT GROWTH AND VIGOR

Table 4.5 summarizes the results from the statistical analysis for each tree species comparing soil type, chlorophyll, mortality, plant height, and plant diameter. A brief discussion of these comparisons follows.

4.4.2.1   *T. grandis* Seedlings

Relative chlorophyll concentration was contradictory to soil fertility (table 4.5); chlorophyll is significantly the lowest in the best substratum (Inceptisol) and highest in the treated Ultisol (Ultisol + CaCO$_3$ + NPK). The other soil treatments resulted

TABLE 4.5
Summary of Statistical Analysis of Effect of Five Soil Types on Leaf Chlorophyll, Plant Height, Plant Diameter, and Mortality Rate

Variable	Ultisol	Ultisol + NPK + CaCO₃	Andisol	Andisol + NPK	Inceptisol
		T. grandis, df = 49			
Chlorophyll (SPAD)	31.3 (ab)	31.8 (a)	29.0 (ab)	31.0 (ab)	27.9 (b)
Height (cm)	12.8 (c)	26.1 (c)	33.8 (a)	15.7 (bc)	23.8 (ab)
Diameter (mm)	6.8 (b)	14.6 (a)	15.1 (a)	9.2 (b)	15.2 (a)
Mortality (%)	39	33	33	39	55
		T. grandis clones, df = 41			
Chlorophyll (SPAD)	24.4 (b)	25.0 (b)	31.6 (a)	25.5 (b)	24.5 (b)
Height (cm)	11.2 (c)	31.6 (a)	13.6 (bc)	18.3 (bc)	22.2 (b)
Diameter (mm)	4.4 (c)	12.3 (a)	8.2 (b)	10.6 (ab)	11.0 (ab)
Mortality (%)	17	8	42	33	17
		S. macrophylla, df = 49			
Chlorophyll (SPAD)	32.8 (a)	26.5 (ab)	22.4 (bc)	25.0 (b)	17.6 (c)
Height (cm)	22.5 (d)	40.8 (ab)	29.1 (dc)	50.1 (a)	38.1 (bc)
Diameter (mm)	10.4 (c)	16.5 (b)	13.0 (c)	20. (a)	16.7 (b)
Mortality (%)	39	33	44	33	39

Note: Letters in parentheses represent the Fisher LSD homogeneous groups with $\alpha = 0.5$.

in similar chlorophyll values. Leaf chlorophyll had a narrow range from 27.9 to 31.8 SPAD values (range of 3.9 SPAD values) and consequently it is difficult to detect any correlation between chlorophyll and soil variables.

Plant growth was significantly lowest in the Ultisol and highest in the Andisol and Ultisol + CaCO₃ + NPK. The Andisol + NPK and Inceptisol resulted in intermediate plant growth. The positive response of *T. grandis* in the Ultisol + CaCO₃ + NPK confirms that this species is very sensitive to Ca and Mg availability, especially when compared to its growth in the untreated Ultisol, a soil that is very low in bases (table 4.2). Overall, the Ultisol + CaCO₃ + NPK treatment is superior to the Inceptisol and Andisol because the soil is clayey and hence it has a better soil moisture storage capacity. Soil moisture storage capacity impacts plant vigor and growth rate; the Ultisol + NPK + CaCO₃ plants can grow better on the days when there is a water deficit and are not as water stressed as in the other soil types.

The only significant correlation is between plant diameter and soil phosphorus (0.87, $P < .1$). This correlation is related to the fact that soil P deficiency affects plant size and reduces plant growth [21]. The high mortality in the Inceptisol is unexpected and cannot be explained except by the effect of strong water stress (Inceptisols have a poor water-holding capacity). The impact of the soil moisture storage capacity can also be observed in the mortality figures for the Andisol and Andisol + NPK—substrata that are also prone to causing plant water stress. The mortality in the Ultisol is more related to the low fertility, high acidity, and aluminum toxicity of this soil (tables 4.2 and 4.5). When this soil is limed and fertilized (Ultisol + $CaCO_3$ + NPK), the mortality is the lowest among all treatments (table 4.5).

4.4.2.2 *T. grandis* Clones

Relative leaf chlorophyll is significantly highest in the Andisol, with insignificant variation among the other substrata (table 4.5) and hence chlorophyll did not correlate with soil fertility or other plant growth variables. These outcomes confirm results found with the *T. grandis* seedlings (section 4.4.2.1).

T. grandis clones' growth was significantly higher in the Ultisol + $CaCO_3$ + NPK, followed by the Inceptisol and the Andisol + NPK (table 4.5). The least growth was observed on the Ultisol and the Andisol. Even though these results are not exactly the same as for the *T. grandis* seedlings, the similar tendency confirms that *T. grandis* is very sensitive to soil acidity and responds favorably to liming and fertilization. Only plant diameter correlated significantly with soil phosphorus (0.9, $P < .05$). Again, this correlation is consistent with results from *T. grandis* seedlings (section 4.4.2.1).

Mortality rate is very high on the two Andisols and very low in the treated Ultisol (table 4.5). Mortality in the Inceptisol is lower (17%) in comparison with the *T. grandis* seedlings (55%). This difference makes it more difficult to explain the mortality response of this species among soil types. Most likely this mortality occurred during the first weeks following transplanting, when the clones were very small and subject to death from the stress caused by changes in condition from the greenhouse to the open nursery.

4.4.2.3 *S. macrophylla*

Relative chlorophyll SPAD values followed an unexpected trend with respect to soil type and plant growth. Relative chlorophyll concentration was significantly highest in the worst substratum (Ultisol) and was significantly lowest in one of the best substrata (Inceptisol) (table 4.5). This implies that high chlorophyll concentration does not necessarily correspond to good soil. Relative chlorophyll was negatively correlated with soil Ca (−0.92), Mg (−0.9), CEC (−0.89), total bases (−0.9), and S (−0.9) at $P < .05$, a clear indication that chlorophyll concentration is negatively correlated to soil fertility levels in this species. Also contrary to what is expected, chlorophyll did not have a significant relationship with any of the plant growth variables. *S. macrophylla* had the broadest range of chlorophyll from 32.8 to 17.6 SPAD values (range of 15.2 SPAD values) in comparison to *T. grandis* clones and seedlings (table 4.5). Such a broad range indicates that the chlorophyll concentration of *S. macrophylla* may be strongly influenced by the soil water-holding capacity. As indicated in section 4.3.5 the wavelengths assessed by the SPAD (650 and 940 nm) could be affected by plant water stress and hence results may not be indicative of a straightforward relationship with foliar nitrogen or chlorophyll. This potential explanation is supported in this case because the treatment with the highest relative chlorophyll concentrations are in the two Ultisols (best water-holding capacity) and the lowest relative chlorophyll concentration was in the Inceptisol (lowest water-holding capacity).

S. macrophylla plant growth was significantly the lowest on the Ultisol and highest on the Andisol + NPK, followed by the Ultisol + CaCO$_3$ + NPK, Inceptisol, and Andisol (table 4.5). Given that this species reacted very well to the application of fertilizer in the Ultisol and the Andisol, it can be concluded that *S. macrophylla* is very sensitive to soil NPK availability. This observation is confirmed by the correlation analysis; plant height is correlated to soil P (0.9, $P < .05$) and K (0.83, $P < .1$) and plant diameter is correlated to soil K (0.88, $P < .05$) and P (0.87, $P < .1$). Mortality values of this species also confirm *S. macrophylla*'s sensitivity to soil NPK availability. Mortality was lowest in fertilized substrata and highest in the unfertilized substrata (table 4.5).

4.4.3 FOLIAR NUTRIENT CONCENTRATION

Table 4.4 summarizes the results of foliar nutrient concentrations for all treatments. It is important to note that this analysis comes from one composite sample per treatment and hence it lacks statistical variability. We therefore only provide generalized observations. The following description will concentrate on the correlation of the data to explain plant growth and leaf chlorophyll concentration.

4.4.3.1 *T. grandis* Seedlings

Relative chlorophyll was correlated with foliar N (0.96) at $P < .05$ and with S (0.87) at $P < .1$. These correlations are related to the effect of N and S on chlorophyll; N is an elemental compound of the chlorophyll while S directly affects photosynthetic rate [33]. Foliar Ca was strongly correlated with soil pH (0.99, $P < .05$) and soil aluminum saturation (acidity) (–1.0, $P < .05$). Foliar Ca (0.82, $P < .1$) and Mg (0.84, $P < .1$) were correlated with the total soil bases. These correlations demonstrate how foliar Ca and Mg are affected not only by the soil Ca and Mg reserves but also by soil pH, which ultimately regulates cation solubility in the soil solution [21].

4.4.3.2 *T. grandis* Clones

Foliar P was correlated with soil K (0.91, $P < 0.05$), foliar K was correlated with soil pH (0.90, $P < .05$), soil aluminum saturation (acidity) (–0.92, $P < .05$), and soil K (0.97, $P < .05$). Total foliar bases were correlated with soil pH (0.91, $P < 0.05$), soil aluminum saturation (acidity) (–0.90, $P < .05$), and soil K (0.91, $P < .05$). Potassium (K) is a very soluble cation that regulates plant water content and osmosis; hence, K plays an important role in water and soil nutrient absorption [34], thus explaining the correlation of foliar P and bases with soil K. Correlations of foliar bases with soil pH and aluminum saturation are associated to the negative effect of soil acidity on reducing soil nutrient solubility for plant absorption [21].

In the case of *T. grandis* clones there were no significant correlations between relative chlorophyll and foliar nutrients. Contrary to soil nutrient correlations, where either *T. grandis* seedlings or clones showed similar correlations, foliar correlations of *T. grandis* seedlings are different from *T. grandis* clones. We are not able to explain this tendency but our conjecture is that the lower genetic variability of clones, in comparison to the seedlings, could sensibly reduce foliar nutrient concentrations in response to soil variability. This observation is supported additionally by the fact that foliar concentrations of N, Mg, Fe, Zn, and Mn are consistently lower in all treatments for the *T. grandis* clones than for the *T. grandis* seedlings (table 4.4).

4.4.3.3   *S. macrophylla* Seedlings

Foliar P (0.94, $P < .05$), Ca (0.9, $P < .05$), and K (0.84, $P < .1$) were correlated to soil K. Foliar Ca (0.92, $P < .05$) and K (0.84, $P < .1$) were correlated with soil P. These correlations are strongly linked to plant growth and are explained by the effect of soil K in increasing the absorbability of soil P and Ca by plant roots (see section 4.4.3.2) and by the effect of soil P in root systems [21]. Plant diameter was correlated with foliar P (0.89), Ca (0.88), Zn (0.94), and Mn (0.89) at $P < .05$. Plant height was correlated with foliar P (0.86, $P < .1$) and Ca (0.87, $P < .1$). Relative leaf chlorophyll was not correlated with any foliar nutrient, not even with foliar N or Mg, important components of the chlorophyll molecules. This lack of correlation can be explained by water stress as discussed in section 4.4.2.3.

As indicated in table 4.4 the best soil treatment for plant growth was Andisol + NPK, while the worst was the Ultisol. This contrast is also reflected in the foliar nutrient concentrations; N-P-Ca-K are lowest in the Ultisol and highest in the Andisol + NPK.

4.4.4   Leaf Reflectance

Results from the classification of the reflectance data are presented in tables 4.6–4.9 as follows:

- *Classifier* indicates which of the five classifiers (section 4.3.8) provided the best results in separating the groups. The groups are composed of reflectance measured from leaves of seedlings/grass growing on the various soil types.
- *No. of features* indicates the optimal number of bands for separating each group.
- *Training error* is a measure of error calculated for the best classifier during the learning process.
- *Testing error* is the validation error, where the classifiers were tested on a new data set not used during the learning process.
- *Overall error* is the average error between the training and testing errors.
- *No. training samples* indicates the number of spectra that were used to train the classifiers. Because half the data was used for training and half for testing (validation), the value in this column also represents the validation data set.

4.4.4.1   *T. grandis* Seedlings

The highest overall classification error (12.8%) was between the Inceptisol and Ultisol + $CaCO_3$ + NPK soil classes (table 4.6). The best wavelengths for separating these two classes were all above 800 nm (fig. 4.6a), indicative of foliar water and leaf internal structure. Plant health, stress, and growth could all influence leaf structure. The Ultisol+ $CaCO_3$ + NPK has a better water-holding capacity, which would

TABLE 4.6

Individual Soil Comparisons of Leaf Reflectance for *T. grandis* from Seeds

Comparison	Classifier	No. features	Training error (%)	Testing error (%)	Overall error (%)	No. training samples
Andisol NPK vs. Ultisol	lmnc5	10	0	15.5	7.8	116
Andisol NPK vs. Andisol	lmnc4	10	0	4.4	2.2	91
Andisol NPK vs. Inceptisol	lmnc5	20	0	14.4	7.2	97
Andisol NPK vs. Ultisol CaCO₃ NPK	loglc, lmnc2,4,5	10	0	1.4	0.7	71
Ultisol vs. Andisol	lmnc5	20	0	15.4	7.7	117
Ultisol vs. Inceptisol	2-NN	10	0	17.9	8.9	123
Ultisol vs. Ultisol CaCO₃ NPK	lmnc5	20	1	10.3	5.7	97
Andisol vs. Inceptisol	2-NN	20	0	14.3	7.1	98
Andisol vs. Ultisol CaCO₃ NPK	lmnc4	10	1.4	1.9	7.6	72
Inceptisol vs. Ultisol CaCO₃ NPK	lmnc3	10	1.3	25.6	12.8	78
Inceptisol vs. Andisol vs. Ultisol	1-NN	20	0	26.6	13.3	169

reduce water stress on dry days. The lowest overall classification error (0.7%) was between the Andisol + NPK and Ultisol + CaCO₃ + NPK soil classes (table 4.6). The wavelengths chosen by the feature selection are in the 450- to 500-nm and 640- to 690-nm ranges (fig. 4.6b). These wavelength regions are primarily influenced by leaf pigment contents as well as leaf nitrogen.

In comparison to the *T. grandis* clones (table 4.7) and *S. macrophylla* (table 4.8), the overall classification errors for *T. grandis* seedlings are considerably higher (5.7–8.9%) (table 4.6), indicating that from a remote sensing point of view, the effects to plant growth and foliar chemistry from most soil types (tables 4.4 and 4.5) are expressed correspondingly in the spectral response of *T. grandis* seedlings. The exceptions are the two improved soils, Andisol + NPK and Ultisol + CaCO₃ + NPK, which had undergone improved fertilization and liming (in the case of the Ultisol)— strongly affecting reflectance (i.e., lowest classification error 0.7%) in the areas of the spectrum controlled by pigment content (fig. 4.6b).

It is important to note that the main compounds affecting reflectance in the visible wavelengths are the pigments that are composed of C, H, and O, and, in the case of chlorophyll, also N and Mg. The main absorption features in the spectral signature will be from chlorophyll, carotenoids, and xanthophylls and the inflection point of the red edge will also be influenced by N content [35,36]. Other foliar elements may be important in the overall health and structure of the leaves potentially affecting the near-infrared region. Nevertheless, in the foliar constituents that affect reflectance, *T. grandis* seedlings are either similarly affected by most soil types or only minimally affected by most soil types, resulting in the relatively high classification errors; the only exceptions are the two improved soils, as discussed previously.

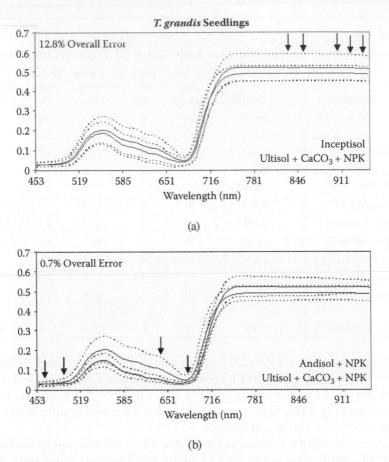

FIG. 4.6 Mean and one standard deviation of reflectance for plants on soils with the highest (a, c, e) and lowest (b, d, f) classification error and mean reflectance of *P. conjugatum* (g) growing on the unmodified Inceptisol and treated Inceptisol following cadaveric decomposition. Arrows indicate wavelength regions highlighted by the feature selection.

4.4.4.2 *T. grandis* Clones

The highest overall classification error was between the Andisol + NPK and Inceptisol soil types (9.5%) (table 4.7). The wavelengths with the greatest separability were 690 nm and the region above 749 nm (fig. 4.6c). Similar to the *T. grandis* seedlings, the worst classification results were from soils where the separability relies on the near-infrared, rather than the visible wavelengths, indicating that pigment concentrations are similar in the seedlings growing on these two soil types. The best classification results (1% overall error) were between the Inceptisol and Ultisol + CaCO₃ + NPK (fig. 4.6d). All the wavelengths with the greatest separability were in the visible wavelengths except for one (fig. 4.6d), indicating that these soil types influence the pigment content of the leaves, and those differences are being translated into the spectral response. In general, the classifications resulted in low overall errors (<5%), except for the worst result described earlier, indicating contrasting effects between

TABLE 4.7

Individual Soil Comparisons of Leaf Reflectance for *T. grandis* Clones

Comparison	Classifier	No. features	Training error (%)	Testing error (%)	Overall error (%)	No. training samples
Andisol NPK vs. Ultisol	lmnc4	10	0	6	3	64
Andisol NPK vs. Andisol	2-NN	10	0	9	4.5	93
Andisol NPK vs. Inceptisol	2-NN	10	0	19	9.5	59
Andisol NPK vs. Ultisol CaCO₃ NPK	2-NN	10	0	6	3	78
Ultisol vs. Andisol	2-NN	10	0	6	3	79
Ultisol vs. Inceptisol	2-NN	10	0	8	4	45
Ultisol vs. Ultisol CaCO₃ NPK	2-NN	10	0	3	1.5	63
Andisol vs. Inceptisol	2-NN	10	0	8	4	74
Volcanic vs. Ultisol CaCO₃ NPK	2-NN	20	0	6	3	93
Inceptisol vs. Ultisol CaCO₃ NPK	2-NN	10	0	2	1	59
Inceptisol vs. Andisol vs. Ultisol	1-NN	20	0	17.2	8.6	99

the soil types on reflectance (table 4.7). For some soil pairs, the potential explanation is clear, such as Ultisol vs. Ultisol + $CaCO_3$ + NPK (1.5% overall error), which have substantial differences in soil nutrient contents that are translated into foliar nutrient and chlorophyll contents (tables 4.2 and 4.4; fig. 4.7).

4.4.4.3   S. macrophylla

This species has the lowest classification errors among the studied species, implying that *S. macrophylla* foliage displays contrasting reflectance properties when leaves from plants growing in different soils are examined. Furthermore, foliar K, N, and chlorophyll concentrations were also found to be contrasting among the soil types in this species (table 4.4; figs. 4.7e and 4.7f), enhancing differences in the spectral response. The highest overall classification error (6.7%) was between the Inceptisol and Andisol + NPK (table 4.8; fig. 4.6e). All the classification wavelengths were from the visible region of the spectrum (fig. 4.6e). Other than the second highest overall classification error (5.7% Ultisol vs. Ultisol + $CaCO_3$ + NPK), the classification errors were very low (<3%; table 4.8). The best classification results (0% overall error) were obtained from a number of soil pairs. An example, Inceptisol vs. Ultisol + $CaCO_3$ + NPK, is illustrated in figure 4.6f. Given the range of variation in chlorophyll and N concentration in this species, foliar K concentration may not be as important in the explanation of the classification errors. While no one foliar element can be responsible for the very low classification results on its own, the most probable is a combined effect of chlorophyll, foliar N, and foliar K (i.e., for plant growth and health) (table 4.8; figs. 4.6e, 4.6f, and 4.7).

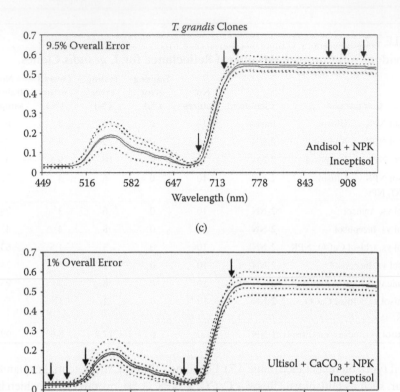

FIG. 4.6 (continued)

For all the studied species, plant and leaf age is also an important consideration. All the plants examined in the nursery experiment were seedlings and as such more readily affected by changes in environmental and soil conditions than mature trees. Castro-Esau et al. [37] found that environmental conditions affected the reflectance of leaves from mature trees; therefore, the effects we found here may translate from seedlings to mature trees.

4.4.4.4 Comparison of Overall Classification Results from All Tree Species

When all soil types are considered in the same problem (table 4.9), all the species (*T. grandis* seedlings and clones and *S. macrophylla*) have a higher overall classification error in comparison to the individual soil matrix results (tables 4.6–4.8), indicating a higher degree of difficulty for separating the leaf reflectance from plants growing in a number of soil types. *S. macrophylla* has the lowest error (0% training error, 18% testing error, 9% overall error) compared to the other species. This may be because *S. macrophylla* preferentially grows in forest openings where the microclimate is regulated as opposed to exposed to the sun and meteorological elements in this

TABLE 4.8
Individual Soil Comparisons of Leaf Reflectance for *S. macrophylla*

Comparison	Classifier	No. features	Training error (%)	Testing error (%)	Overall error (%)	No. training samples
Andisol NPK vs. Ultisol	1-NN	10	0	2.5	1.3	40
Andisol NPK vs. Andisol	lmnc4,5, qdc, 1-NN	10	0	0	0	45
Andisol NPK vs. Inceptisol	lmnc5, 1-NN	10	0	13.3	6.7	45
Andisol NPK vs. Ultisol CaCO₃ NPK	lmnc4	10	0	2.2	1.1	45
Ultisol vs. Andisol	lmnc4	20	0	0	0	35
Ultisol vs. Inceptisol	lmnc2	10	0	0	0	35
Ultisol vs. Ultisol CaCO₃ NPK	lmnc5	10	0	11.4	5.7	35
Andisol vs. Inceptisol	loglc, lmnc4	10	0	0	0	40
Andisol vs. Ultisol CaCO₃ NPK	lmnc3	10	0	5	2.5	40
Inceptisol vs. Ultisol CaCO₃ NPK	loglc, qdc, lmnc2,3,5	10	0	0	0	40
Inceptisol vs. Andisol vs. Ultisol	ldc	10	0	3.6	1.8	55

study. The lower error indicates that the various soil types affected foliar reflectance properties of this species the most, thereby facilitating the classification. Another reason for the increased sensitivity may be that *S. macrophylla* is more affected by water stress in the soils with loamy textures (Inceptisol and Andisol) as well as by soil nutrient content (sections 4.4.2.3 and 4.4.3.3). These two soil types (Inceptisol and Andisol) are also the ones with the greatest overall error from the classification matrix (table 4.8). *T. grandis,* however, is a secondary successional species that has been domesticated for forest plantations and hence this species is more adapted to growing in open sky conditions and will be more resistant to water stress. Overall, leaf reflectance properties were either less affected by varying soil conditions or the majority of the soils had similar effects on reflectance for *T. grandis* (tables 4.6, 4.7, and 4.9; figs. 4.6a–4.6d).

Tables 4.6–4.8 (last row of each table) also summarize classification of species on the three basic unmodified soil types: Ultisol (lowest fertility), Andisol (medium fertility), and Inceptisol (high fertility). The purpose for this comparison was to examine the contrasting effect of soil fertility on foliar nutrient content and reflectance properties. For the *T. grandis* seedlings, the overall error (13.3%) is only slightly higher than the highest error from the matrix (12.8%) (fig. 4.6a). For the *T. grandis* clones the overall error from the three basic soil types (8.6%) is less than the highest error from the matrix (9.5%) (fig. 4.7c). For *S. macrophylla,* the 1.8% error from the three soil types is within the range of most single comparisons in the matrix (table 4.8) and less than the two highest errors from the individual soil matrices (6.7 and 5.7%) (table 4.8; fig. 4.6e). These results indicate that differences in soil nutrient availability affect the spectral response of the studied species

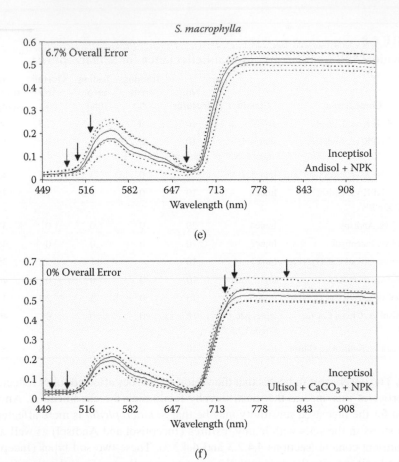

FIG. 4.6 (continued)

TABLE 4.9

Classification Results with All Five Soil Types/Treatments for Each Species

Species	Classifier	No. features	Training error (%)	Testing error (%)	Overall error (%)	No. training samples
T. grandis (clone)	3-NN	30	0	32	16	177
T. grandis (seed)	2-NN	20	0	38	18	240
S. macrophylla	2-NN	10	0	18	9	100

to different degrees. As with the results from the individual classification matrix (tables 4.6–4.8), *S. macrophylla* was affected the most by the different soil types, reinforcing that soil effects are *species specific*. The four-percentage-point difference between the *T. grandis* seedlings and clones may indicate that the lower genetic variation of the *T. grandis* clones results in these plants being more sensitive to the

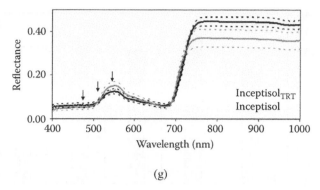

(g)

FIG. 4.6 (continued)

differences in the soil types; in other words, the reflectance of *T. grandis* clones in these soil types was separable with a lower error than the spectral response of *T. grandis* seedlings.

4.4.4.5 Separability of Grass Spectra

The leaf spectra of the grass from the Inceptisol and the treated Inceptisol were perfectly separable (0% training, testing, and overall errors) with all classifiers using 10 wavelengths (fig. 4.6g). The wavelengths with the best discriminatory power are from the 460- 550-nm range, indicative of pigment differences between the blades of grass from the two Inceptisols. The differences in pigment concentration are likely due to differences between mineral and soil nutrient availability (table 4.2). The overall chemical properties of both soils rate them at medium fertility (although the unaltered soil is slightly better in Ca concentration) (table 4.2). Comparing the elemental composition of the two soils (table 4.2), differences exist in Ca, Mg, K, CEC, acidity, Fe, Mn, Zn, S, N, C, and organic matter, and as a result, in the amount of total bases and the C/N ratio (table 4.2). Similar differences in elemental concentrations in grave soil following cadaveric breakdown have also been shown with "cadaver decomposition islands" associated with surface decomposition [38–42].

In contrast to the enhanced nutrients and pH balance obtained by modifying the Ultisol and Andisol, as described in the previous sections through liming and the addition of NPK, the modifications to the Inceptisol through cadaveric decomposition are initially toxic [26]. The decomposition of a body produces over 80 volatile organic compounds [39,43,44], including toluene and benzene, that are toxic to both vegetation and soil microbes similar to the release of hydrocarbons from the rupture of underground fuel tanks. Body decomposition also produces heat, gas, and water vapor emission, also affecting the soil microbes. The water vapor can bring to the surface some of the toxic compounds, which when it reaches the surface, condensates precipitating soluble cations, salts, and toxic compounds. Movement of gas and water vapor is facilitated because of the initial excavations followed by infilling, decreasing the compaction of the soil and increasing the soil porosity within the soil profile [45]. Following dissipation of the initial toxicity, the substratum may become enriched [26] with nutrients and minerals in two ways:

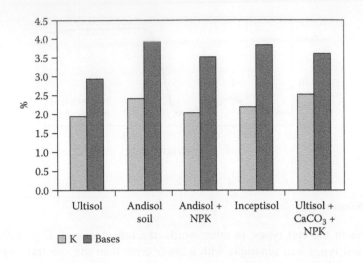

(a)

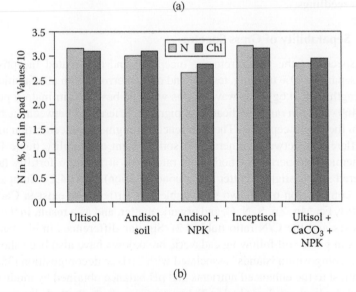

(b)

FIG. 4.7 Foliar K, total foliar bases, foliar nitrogen, and chlorophyll (SPAD) for *T. grandis* seedlings (a, b), *T. grandis clones* (c, d), and *S. macrophylla* (e, f).

1. The gas and water emission could translocate nutrients. For example, S could be deposited on the soil surface following vapor condensation (concentration of S is nearly double on the treated Inceptisol in comparison to the untreated; table 4.2). Other elements may also be translocated in a similar manner.

2. Increased temperature during decomposition and gas/water emission could increase the liberation of nutrients tided in the soil colloids in the interlayers (in this case 2:1 clays and allophane). Potassium is one of the cations tided in these colloids and could be liberated by increased soil temperature that could favor conditions for soil meteorization (soil K is substantially higher in the treated Inceptisol; table 4.2).

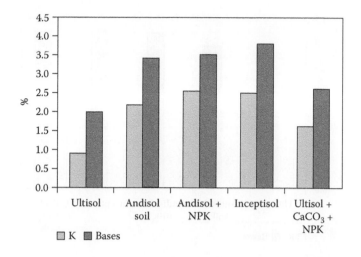

(c)

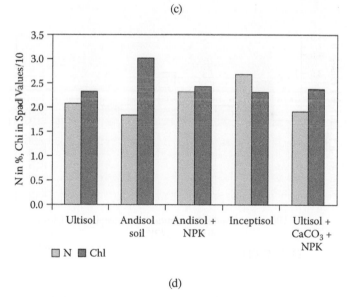

(d)

FIG. 4.7 (continued)

Allophane and 2:1 clays preferentially retain K in the interlayers. At the time the leaf spectra were collected (16 months following burial), the revegetation over the disturbed ground containing the carcasses was incomplete, in comparison to the simple disturbance without chemical alteration—suggesting that effects of the decomposition were still present [24,25]. These findings coincide with observations from Dannell, Berteaux, and Braathen [38] and Towne [42] that indicate surface cadaveric decomposition can alter soil chemistry and the vegetation community for up to 10 years.

Polar or organic compounds will have a preferentially high adsorption in 2:1 clay-based soils (such as the Inceptisol in the test site) whereby the compounds adhere to

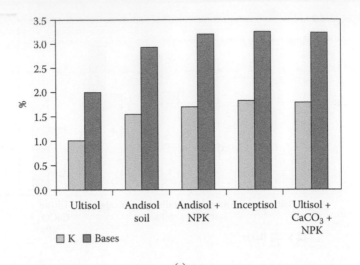

(e)

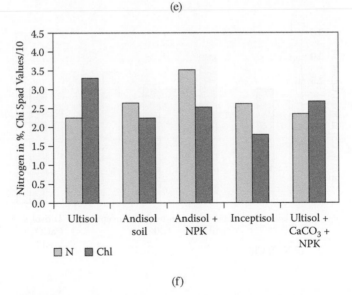

(f)

FIG. 4.7 (continued)

the soil colloids, resisting leeching. Two of the compounds produced in the largest volume are dimethyl disulfide (CH_3-S-S-CH_3) (13.39 nmol/L) and toluene (C_7H_8) (10.11 nmol/L) [44]. Other compounds released in high concentrations are gases such as hydrogen sulfide, carbon dioxide, ammonia, sulfur dioxide, and hydrogen as well as compounds such as cadaverine ($NH_2(CH_2)_5NH_2$), putrescine ($NH_2(CH_2)_4NH_2$), indole (C_8H_7N), and skatole (C_9H_9N) [26,43,44,46]. The release of such compounds is the potential link (and their translocation as described earlier) to the increased elemental concentrations such as N and S. However, based on table 4.3, the maximum additional influx of S is less than 1 kg (approximately 726 g); therefore, a portion of the S translocated to the surface must also be from its liberation from soil minerals in the same mechanism as described for K before. Soil contamination from such compounds (and high temperature through decomposition) can cause severe chemical changes even with low concentrations and can affect the metabolism of soil microbes and plants. Increased soil concentration of Mn can also be explained by the effect of gas emission and high temperature from decomposition that could liberate this toxic element directly by meteorization of soil minerals rich in Mn. The maximum influx of Mn from the decomposition is less than 1 g (table 4.3), whereas the difference in Mn concentration in the altered and unaltered Inceptisol is 53.66 mg/L, indicating that there must be a liberation of this element from the soil minerals.

There is also an increase in C and total organic matter, which could be the result of the influx of organic compounds from the decomposition aided by the increased temperature (also seen with surface decomposition grave soils [26]). Even though Ca and P are present in the body in relatively large amounts (table 4.3), these elements are preferentially found in the skeleton. The pH of the soil is not acidic enough to demineralize the skeleton (initial demineralization may be observed in bony apatite at pH 4 and below); thus, the contribution of the skeleton is considered to be low, while soft tissues, which decompose more readily due to a hydrated and heated system exposed to a range of protein and fatty degraders, will provide the major contribution of elements into the surrounding soil (Mn, K, C, O, H, N, S) (table 4.3).

Considering that 16 months have elapsed and the effects of decomposition are apparent, an important driving factor to the changes in the soil chemistry was internal soil temperature. In addition, air pockets in the soil were likely saturated with the toxic gases and compounds [26], which promoted less than optimal conditions for plant regrowth. Differences in soil chemistry and grass reflectance indicate that the compounds are near the surface, inhibiting microbe establishment and subsequent vegetation regrowth. The differences in soil chemistry were clearly expressed in the reflectance of the grass (fig. 4.6g). This inhibition of vegetation colonization may also indicate that decomposition was ongoing and has not yet reached the equivalent of the "remains" stage for surface decomposition, where plant growth is seen around the edges and in the interior of the decomposition island, corresponding to a second stage of slow cadaver mass loss [26].

The preceding discussion on the effects subsurface decomposition on the spectral response of grass is included in the discussion of this chapter not only because it presents a novel application of hyperspectral image analysis (i.e., location of clandestine graves) but also because cadaveric decomposition is an integral, albeit grossly overlooked, component of terrestrial ecology. For example, on Barro Colorado Island (BCI), Panama, Houston [47] estimates there are approximately 750 kg of mammalian cadavers per year, per square kilometer. In some ecosystems cadaveric decomposition may account for as much as 1% of the total organic matter input [26]. However, very little is understood regarding the influence of such decomposition on biodiversity and ecosystem functioning (both above and below ground) [26]. We have shown in the preceding section that changes to the vegetation following these processes are expressed in the spectral response and, in turn, should be considered as a local/regional factor that can affect vegetation reflectance within the same soil type and vegetation species.

4.5   CONCLUSIONS

From the nursery experiment we conclude that soil types with contrasting chemical and physical properties significantly affect plant growth. Even though the Inceptisol has the best chemical properties, it did not produce the best plant growth because of its low water-holding capacity. The Ultisol was systematically the worst substratum for plant growth due to its low fertility; however, once this soil is treated with NPK + $CaCO_3$, plant growth is one of the best among all treatments. The Andisol is a good substratum for plant growth and the addition of NKP improves its fertility level favoring plant vigor.

We also concluded that soil nutrient concentrations affect foliar nutrient concentrations and, for most part, the detected correlations are the result of expected logical relationships. Nevertheless, except for *T. grandis* seedlings, it was difficult to find a logical relationship for SPAD-relative chlorophyll values with either soil or foliar nutrient concentration. We concluded that this lack of correlation is due to the effect of plant water stress and leaf age. Consequently, SPAD-relative chlorophyll readings are not a useful approach to measure plant vigor in seedlings without prior calibration (i.e., species-specific chlorophyll calibration curves).

In this nursery experiment we compared *T. grandis* seedlings from two genetic sources: (a) genetically diverse uncontrolled seed sources and (b) less variable genetically superior clones. On average, plant growth in *T. grandis* seedlings was better than in *T. grandis* clones. However, growth differences could be associated with the fact that *T. grandis* seedlings are adapted better to the experimental conditions of this study because they came from the nursery with similar environmental conditions, while the clones came from the greenhouse. Hence, we expect these growth differences to disappear as time passes.

Because of the importance in nutrient cycling, the genetic source influence in foliar nutrient concentration deserves future research. In this study *T. grandis* clones apparently had lower foliar nutrient concentrations than *T. grandis* seedlings. It is particularly important to note that clones had notably less relative chlorophyll and foliar N, Ca, Fe, and Mn concentrations. This could imply that clones are more nutrient efficient (Ca and N) and less prone to absorbing toxic elements (Fe and Mn).

In general, differences in soil nutrient concentration and toxicity did affect the spectral response of plants. Specifically, in the cases where the photosynthetic and accessory leaf pigments were affected, the spectra of the leaves could be classified with good accuracy. Conversely, differences in soil nutrient concentration that either do not induce a difference in the photosynthetic or accessory pigments or potentially influence the vegetation in the same manner resulted in a classification problem that relied on the near-infrared wavelengths; consequently, the spectra were not readily separable. Importantly, the degree to which different soils affect leaf reflectance was found to be species specific.

Further research in the examination of the reflectance of soil types on leaf reflectance is needed across a wider range of species and under additional treatments. Castro-Esau et al. [37] show that overall climatic and edaphic conditions are extremely important in influencing the reflectance of mature trees; these effects greatly hinder regional classification efforts in the tropics. Currently, it is not possible to separate climatic and edaphic effects on mature trees growing in different conditions. With additional experimental consideration of climatic and edaphic conditions separately, perhaps it will be possible to expand classification efforts to regional scales.

ACKNOWLEDGMENTS

This work was supported by the National Sciences and Engineering Research Council of Canada (NSERC) Discovery Grant Program to Sanchez-Azofeifa and the Inter-American Institute for Global Change Research (IAI) under CRN-2, the NSERC postdoctoral fellowship program to M. Kalacska, the School of Criminology at Simon Fraser University, the University of Alberta, and the Instituto Tecnológico de Costa Rica. Logistical support was provided by the Interdisciplinary Research in the Mathematical and Computing Sciences Center (IRMACS) at Simon Fraser University. We thank Luis Coronado-Chacon, Enrique Salicetti, and Eva Snirer for their assistance in the data collection and organization as well D. Arias for participation in the experimental design.

REFERENCES

1. Carter, G.A. and Knapp, A.K., Leaf optical properties in higher plants: Linking spectral characteristics to stress and chlorophyll concentration, *American Journal of Botany*, 88, 677, 2001.
2. Clark, D.A., Roberts, D.A., and Clark, D.A., Hyperspectral discrimination of tropical rain forest tree species at leaf to crown scales, *Remote Sensing of Environment*, 96, 375, 2005.
3. Asner, G.P., Biophysical and biochemical sources of variability in canopy reflectance, *Remote Sensing of Environment*, 64, 234, 1998.
4. Castro-Esau, K.L., Sanchez-Azofeifa, G.A., and Caelli, T., Discrimination of lianas and trees with leaf-level hyperspectral data, *Remote Sensing of Environment*, 90, 353, 2004.
5. Jago, R.A., Cutler, M.E., and Curran, P.J., Estimating canopy chlorophyll concentration from field and airborne spectra, *Remote Sensing of Environment*, 68, 217, 1999.
6. Daughtry, C.S.T., Walthall, C.L., Kim, M.S., et al., Estimating corn leaf chlorophyll concentration from leaf and canopy reflectance, *Remote Sensing of Environment*, 74, 229, 2000.
7. Gitelson, A. and Merzlyak, M.N., Remote estimation of chlorophyll content in higher plant leaves, *International Journal of Remote Sensing*, 18, 2691, 1997.
8. Penuelas, J., Baret, F., and Filella, I., Semi-empirical indices to assess carotenoids/chlorophyll a ratio from leaf spectral reflectance, *Photosynthetica*, 31, 221, 1995.
9. Zarco-Tejada, P.J., Miller, J.R., Mohammed, G.H., et al., Vegetation stress detection through chlorophyll a+b estimation and fluorescence effects on hyperspectral imagery, *Journal of Environmental Quality*, 31, 1433, 2002.
10. Gamon, J.A., Field, C.B., Bilger, W., et al., Remote sensing of the xanthophyll cycle and chlorophyll fluorescence in sunflower leaves and canopies, *Oecologia*, 85, 1, 1990.
11. Carter, G.A., Cibula, W.G., and Miller, R.L., Narrowband reflectance imagery compared with thermal imagery for early of plant stress, *Journal of Plant Physiology*, 148, 515, 1996.
12. Carter, G.A., Ratios of leaf reflectance in narrow wavebands as indicators of plant stress, *International Journal of Remote Sensing*, 15, 697, 1994.
13. Carter, G. and Young, D.R., Responses of leaf spectral reflectance to plant stress, *American Journal of Botany*, 154, 239, 1993.

14. Jimenez, L.O. and Landgrebe, D., Supervised classification in high-dimensional space: Geometrical, statistical and asymptotic properties of multivariate data, *IEEE Transactions on Systems, Man and Cybernetics—Part C: Applications and Reviews*, 28, 39, 1998.

15. van der Heijden, F., Duin, R.P.W., de Ridder, D., et al., *Classification, parameter estimation and state estimation: An engineering approach using Matlab*. John Wiley & Sons, West Sussex, U.K., 2004.

16. Richards, A.J. and Jia, X., *Remote sensing digital image analysis*, 4th ed. Springer–Verlag, Berlin, 2005.

17. Hughes, G., On the mean accuracy of statistical pattern recognizers, *IEEE Transactions on Information Theory*, 14, 55, 1968.

18. Fielding, A.H., *Machine learning methods for ecological applications*. Kluwer Academic Press, Dordrecht, 1999.

19. Russell, S. and Norvig, P., *Artificial intelligence: A modern approach*, 2nd ed. Prentice Hall, Upper Saddle River, NJ, 2003.

20. Centro Cientifico Tropical (CCT), *Evaluacion Del Estado De Conservacion De La Caoba En Mesoamerica. Estudio Elaborado Para Comision Centroamericana Para El Ambiente Y El Desarrollo Y Proarca-Capas,* San Jose, Costa Rica, 2000.

21. Sanchez, P., *Properties and management of soils in the tropics*. John Wiley & Sons, New York, 1976.

22. Buol, S., Hole, F., and MacCracken, R., *Soil genesis and classification,* 2nd ed. Iowa State University Press, Ames, 1980.

23. Cochrane, T., Salinas, J., and Sanchez, P., An equation for liming acid mineral soils to compensate crops aluminum tolerance, *Tropical Agriculture*, 57, 18, 1980.

24. Kalacska, M., Bell, L.S., Sanchez-Azofeifa, G.A., et al., The application of remote sensing for detecting mass graves: An experimental animal case study from Costa Rica, *Journal of Forensic Sciences*, submitted.

25. Kalacska, M. and Bell, L.S., Remote sensing as a tool for the detection of clandestine mass graves, *Canadian Society of Forensic Science Journal*, 39, 1, 2006.

26. Carter, D.O., Yellowlees, D., and Tibbett, M., Cadaver decomposition in terrestrial ecosystems, *Naturwissenschaften*, 94, 12, 2007.

27. Emsley, J., *The elements*, 3rd ed. Clarendon Press, Oxford, 1998.

28. Kariya, K., Matsuzaki, A., and Machida, H., Distribution of chlorophyll content in leaf blade of rice plants, *Japanese Journal of Crop Science*, 51, 134, 1982.

29. Castro-Esau, K., Sanchez-Azofeifa, G.A., and Rivard, B., Comparison of spectral indices obtained using multiple spectroradiometers, *Remote Sensing of Environment*, 103, 276, 2006.

30. PP Systems [cited; available from: http://www.ppsystems.com].

31. Analytical Spectral Devices [cited; available from: http://www.asdi.com].

32. Statsoft Inc. *Statistica (Data Analysis Software System), Version 6.* 2001 [cited; available from: http://www.statsoft.com].

33. Karmoker, J., Clarkson, D., Saker, L., et al., Sulphate deprivation depresses the transport of nitrogen to the xylem and the hydraulic conductivity of barley (*Hordeum vulgare* L.) roots, *Plants*, 185, 269, 1991.

34. Hsiao, T. and Lauchli, A., Role of potassium in plant water relations, in *Advances in plant nutrition*, B. Tinker and A. Lauchli, eds. Praeger Scientific, New York, 1986, 281.

35. Gates, D.M., Keegan, H.J., Schleter, J.C., et al., Spectral properties of plants, *Applied Optics*, 4, 11, 1965.

36. Cho, M.A. and Skidmore, A., A new technique for extracting the red edge position from hyperspectral data: The linear extrapolation method, *Remote Sensing of Environment*, 101, 181, 2006.

37. Castro-Esau, K., Sanchez-Azofeifa, G.A., Rivard, B., et al., Variability in leaf optical properties of mesoamerican trees and the potential for species classification, *American Journal of Botany*, 93, 517, 2006.

38. Dannell, K., Berteaux, D., and Braathen, K.A., Effect of musk ox carcasses on nitrogen concentration in tundra vegetation, *Arctic*, 55, 389, 2000.

39. Vass, A., Bass, W.M., Wolt, J.D., et al., Time since death determinations of human cadavers using soil solution, *Journal of Forensic Sciences*, 37, 1236, 1992.

40. Rodriguez, W.C. and Bass, W.M., Decomposition of buried bodies and methods that may aid in their location, *Journal of Forensic Sciences*, 30, 836, 1985.

41. Hopkins, D.W., Wiltshire, P.E.J., and Turner, B.D., Microbial characteristics of soils from graves: An investigation at the interface of soil microbiology and forensic science, *Applied Soil Ecology*, 14, 283, 2000.

42. Towne, E.G., Prairie vegetation and soil nutrient responses to ungulate carcasses, *Oecologia*, 122, 232, 2000.

43. Vass, A., Beyond the grave—Understanding human decomposition, *Microbiology Today*, 28, 190, 2001.

44. Statheropoulos, M., Spiliopoulou, A., and Agapiou, A., A study of volatile organic compounds evolved from the decaying human body, *Forensic Science International*, 153, 147, 2005.

45. Moldrup, P., Olsen, T., Rolston, D.E., et al., Modeling diffusion and reaction in soils: VII. Predicting gas and ion diffusivity in undisturbed and sieved soils, *Soil Science*, 162, 632, 1997.

46. Fiedler, S. and Graw, M., Decomposition of buried corpses, with special reference to the formation of adipocere, *Naturwissenschaften*, 90, 291, 2003.

47. Houston, D.C., Evolutionary ecology of Afrotropical and neotropical vultures in forests, in *Neotropical Ornithology. American Ornithologists' Union Monograph No. 36*, M. Foster, ed. American Ornithologists Union, Washington, DC, 1985, 856.

5 Spectral Expression of Gender

A Pilot Study with Two Dioecious Neotropical Tree Species

J. Pablo Arroyo-Mora, Margaret Kalacska, Benjamin L. Caraballo, Jolene E. Trujillo, and Orlando Vargas

CONTENTS

5.1   INTRODUCTION

Few studies have considered effects of dioecy on plant ecophysiological functioning [1,2], even though a large proportion of species in the tropics are dioecious [1]. Since dioecious plants present separate male and female individuals, maintaining a proper ratio of gender ensures species survival, an important consideration in natural forest management, reforestation, and conservation projects. Nevertheless, it is a characteristic that has often been overlooked. Potentially, remote sensing may provide a set of tools to identify the gender of dioecious tree species, ensuring that male-to-female ratios are maintained when reforesting or during selective logging. Proper ratios must be maintained to ensure the reproduction and the population stability of species [3]. Our chapter aims to explore the spectral expression of dioecy of two neotropical species, *Hyeronima alchorneoides* and *Virola koschnyi*, through remote sensing techniques and leaf chemistry analysis.

H. alchorneoides (Euphorbiaceae) is a wet forest emergent tree with a range from Belize to the Amazon. It grows in humid and very humid forests where rainfall is between 3500 and 5000 mm, at elevations of 20–900 m, and in temperatures from 24–30°C and can grow to be 50 m in height with an extensive canopy. This species blooms twice a year with a primary flowering from May to July, peaking in June [4]. *V. koschnyi* (Myristicaceae) is found all over Central America [4]. It grows in moist to wet tropical forest with rainfall between 3500 and 5000 mm, between the elevations of 10 and 1200 m, and in temperatures from 24–35°C and can grow to be 30–45 m tall. *V. koschnyi*'s flowering period is from September to November, with fruiting in February and March and a smaller fruiting event in June [4].

H. alchorneoides and *V. koschnyi* are important for maintaining tropical biodiversity, but, as with the majority of dioecious species, gender is often difficult to ascertain. At the leaf level, spectroscopy is a remote sensing tool that potentially can be used to infer biophysical properties by analyzing characteristic light absorbance features. Reflectance, the ratio of incident to reflected light, has been widely used to infer leaf pigment contents and other biophysical parameters of interest [5]. Spectroscopy is an effective and noninvasive tool to identify certain vegetation characteristics and infer ecophysiological processes such as leaf pigmentation and drought stress and/or phenological differences between groups of species [2,5–7]. Spectral vegetation indices (SVIs) are functions calculated from reflectance spectra to infer vegetation characteristics. For example, Sims and Gamon [5] modified SVIs for pigment content so that the indices are species insensitive and can be applied to large-scale remote sensing studies. Others have focused on seasonal variations of plant pigments and applications of the photochemical reflectance index (PRI) to detect these differences [2,8,9], while Castro-Esau et al. [10] classified the spectral signatures of several species of dry forest tropical trees from leaf-level spectra. Similarly, the red edge parameter, the inflection point between high absorbance in the red region and high reflectance in the near infrared region of the spectra, has been used to infer leaf nitrogen content [11].

The main objectives of this chapter are to determine whether gender is expressed in the spectral signatures of *H. alchorneoides* and *V. koschnyi*, and to examine potential biophysical causes for any differences. Locating specific spectral features that distinguish gender at the leaf level is the first step in broader remote sensing applications such as scaling up to hyperspectral airborne imagery for gender identification of certain tree species.

5.2 METHODS

5.2.1 STUDY SITE

This study was conducted at the La Selva Biological Station (LS) in the Sarapiqui region of Costa Rica ($84°00'12.922''$ W, $10°25'52.61''$ N). Precipitation averages 4244 mm annually, with a relative dry season from January to April and a second, shorter dry season from August to October [12,13]. The old growth forest is classified as a tropical wet forest in the Holdridge Life Zone System and is characterized by a species-rich, multilayered community of trees, palms, lianas, and other terrestrial and epiphytic plants. There are at least 400 species of hardwood trees, of which over 10% are dioecious [12,14,15].

5.2.2 LEAF COLLECTION

Over the course of 1 week in July 2006, 20 leaves from nine *V. koschnyi* trees (five females and four males) were collected using a crossbow to avoid any major disturbance to the trees. All individuals were located in volcanic soils (Andisol) and all females sampled were producing fruits. A wide range of leaves was also collected comprising very young leaves and senescent leaves, specifically for the pigment-specific regression models discussed later. Leaves were placed in a plastic bag with water and brought back to the lab for the spectral analysis. The same sampling scheme was used for *H. alchorneoides* in July 2005. Leaf cores from the samples were frozen at $-80°C$ until the chemical extractions were performed (see later discussion). All leaves from the main sample were healthy, mature sun leaves and all individuals sampled were emergent canopy adult trees. Canopy position, leaf position in the canopy, and tree maturity were important considerations in sampling because it has been shown that canopy position and irradiance can affect leaf photosynthetic efficiency and morphology [16].

5.2.3 SPECTROSCOPY

Reflectance was measured between the midrib and leaf edge within 1 hour after collection using an ASD FieldSpec handheld spectrometer (Analytical Spectral Devices Inc., Boulder, Colorado). This spectrometer measures reflectance from 325 to 1075 nm. The resolution of full width at half maximum is 3.5 nm with a sampling interval of 1.5 nm. The spectrometer was fitted with a plant probe (with an internal light source) and leaf clip that holds the samples in place, excludes ambient light, and ensures a constant geometry for the light source and foreoptic. The integration time was automatically adjusted using a 99% reflective Spectralon™ white reference panel in the leaf clip. Subsequently, a dark current correction was performed to eliminate instrument noise from spectral measurements. White reference measurements with the Spectralon standard in the leaf clip were repeated at 5-minute intervals. Reflectance of the samples was computed as a ratio of each sample spectrum to the

---

TABLE 5.1

Formulas for the Spectral Vegetation Indices and Supporting References

Spectral vegetation index	Ref.
$SR_{680} = \dfrac{R_{800}}{R_{680}}$	5
$SR_{705} = \dfrac{R_{750}}{R_{705}}$	5
$ND_{680} = \dfrac{R_{800} - R_{680}}{R_{800} + R_{680}}$	5
$ND_{705} = \dfrac{R_{750} - R_{705}}{R_{750} + R_{705}}$	5
$mSR_{705} = \dfrac{R_{750} - R_{445}}{R_{705} - R_{445}}$	5
$mND_{705} = \dfrac{R_{750} - R_{705}}{R_{750} + R_{705} - 2(R_{445})}$	5
$SR_{GM} = \dfrac{R_{750}}{R_{700}}$	53
$SR_{V} = \dfrac{R_{740}}{R_{720}}$	54
$PRI = \dfrac{R_{531} - R_{570}}{R_{531} + R_{570}}$	8
D_{730} = first derivative at 730 nm	19

white reference spectrum. The areas of high instrument noise, below 400 nm and above 1000 nm, were removed from all the spectra.

5.2.4 SPECTRAL METRICS

Ten spectral vegetation indices (SVIs) were used to analyze the spectral data (table 5.1). These indices can be related to pigment content and leaf structure and used to explore the differences between species, vegetation types, life forms, etc. In general, wavebands in the 550- and 700-nm regions have been found to be optimal for vegetation indices because these regions require higher chlorophyll concentrations to saturate the absorbance features [5,6,17]. In contrast, only a small amount of chlorophyll is needed to obtain maximum absorption in the 660- to 680-nm region, which therefore does not provide an accurate measure of total chlorophyll [5]. At 445 nm, minimal reflectance of pigments occurs and therefore this correction offsets the effect of reflectance that may occur by the leaf tissue just below the epidermis [5,17].

5.2.5 CHLOROPHYLL AND CAROTENOID EXTRACTION

A subset of 48 leaves from each species was used in the dimethyl sulphoxide (DMSO) extraction developed by Hiscox and Israelstam [18] to remove chlorophyll and carotenoid pigments from leaf cores. With this method a leaf core is placed in a test tube with 10 mL of DMSO, which is then heated for 20 minutes, after which it is allowed to cool. Subsequently, the sample is vortexed and immediately placed in a 1-cm path length cuvette. The entire process takes place in subdued light to avoid pigment degradation.

This method has two advantages over other extractions, such as methanol, ethanol, or acetone. First, it is faster, largely because grinding and centrifuging are not required [19]. Second, the chlorophyll extracts are more stable in DMSO and do not break down as quickly as those in acetone. Richardson, Duigan, and Berlyn [19] reported that DMSO extracts are stable for up to 5 days, whereas with acetone extracts, for example, the measured levels of chlorophyll begin to drop off immediately.

Prior to any analysis, the spectrophotometer was calibrated using pure chlorophyll a and b and beta carotene standards to ensure that the optimal instrument-specific wavelength was used for each. From the extracts, the absorbance for chlorophyll a and b and carotenoids was measured with a Helios Beta Version 4.20 spectronic unicam spectrophotometer.

The chlorophyll and carotenoid concentrations of the extract were calculated using equations from Arnon [20] and Chappelle, Kim, and McMurtrey [21] and were then converted to leaf chlorophyll and carotenoid content expressed as milligrams per square centimeter (table 5.2). To infer chlorophyll and carotenoids for the leaves not used in the chemical extraction of pigments, regression models were derived from pigment content and spectral vegetation indices; the best-fit models are listed in table 5.3.

5.2.6 NITROGEN CONTENT

After the subset of 48 leaves from each of the species had been cored, all leaves, including those not used in the chemical extraction, were separated into groups of 5–20 leaves based on the similarity of their red-edge position regardless of gender, with a total of 16 groups for each of the species. Cutoff points for the red-edge value of the group memberships were determined by examining the quantiles of the distribution of the red-edge parameter for all leaves and also the logistic requirement of a minimum sample dry weight of 5 g. The leaves were placed in paper bags and dried at 70°C for 24 hours. The samples were then sent to the University of Costa Rica's Centro de Investigaciones Agronómicas to determine total nitrogen content.

5.2.7 ANALYSIS

5.2.7.1 Classification of Spectral Signatures

Reflectance spectra for both *H. alchorneoides* and *V. koschnyi* were analyzed using feature selection followed by pattern classifiers. Feature selection was used to (1) reduce the dimensions of the data, (2) determine the optimal wavelengths with

TABLE 5.2

Equations for Determining Chlorophyll and Carotenoid Content Following Pigment Extraction with DMSO

Equation	Ref.
$Chla = 0.0127A_{663} - 0.00269A_{645}$	20
$Chlb = 0.0229A_{645} - 0.00468A_{663}$	20
$Chltot = 0.0202A_{645} + 0.00802A_{663}$	20
$Carotenoids = \dfrac{1000A_{470} - 1.82chla - 85.02chlb}{198}$	21

Note: *Chla* = chlorophyll a content (g/L); *Chlb* = chlorophyll b content (g/L); *Chltot* = total chlorophyll content (g/L); A_{663}, A_{645}, and A_{470} are absorbance values at 663, 645, and 470 nm, respectively. The units (g/L) were converted to milligrams/area (mg/cm²), where the areas of the cores are 1.13 cm² for *H. alchorneoides* and 2.25 cm² for *V. koschnyi*. The units for the carotenoid formula are expressed as micrograms per milliliter. In the final calculations, carotenoids were converted to milligram per square centimeter. The absorbance wavelengths listed above are from the cited references. The actual absorbance wavelengths specific to the Helios Beta V. 4.20 Spectronic Unicam spectrophotometer were determined from pure standards of chlorophylls a and b and beta carotene.

the greatest difference between the genders, and (3) determine the optimal number of features (wavelengths) for classification with the lowest error. The maximum number of features that can be used without overfitting is $F = (n - g)/3$, where n is the number of spectra and g is the number of classes [22]. In order to determine the optimal number of features to use with each classifier, training and testing errors were plotted on a dual y-axis plot against number of features in increments of 10, from 10 to 70 features. The optimal number of features was where the two measurements of error were at global minima. The selected features were then used to classify the spectra of the male and female leaves for both species using parametric and nonparametric pattern classifiers (table 5.4) [23,24]. Analyses were performed using Matlab v6.5 (release 13).

5.2.7.2   Inter- and Intragender Spectral Metrics

Differences in spectral shape and amplitude were calculated between pairs of spectra between leaves at the tree level, between trees of the same gender, and between genders for both species. The D metric [25] describes differences in amplitude and the θ metric describes differences in the angle between spectra and can be thought of as a shape metric [25]. The spectral range used for these calculations was 400–1,000 nm (range with minimal instrument noise) with the spectra interpolated to 4 nm; S_1 and S_2 are pairs of spectra over a given wavelength region ($\lambda_a - \lambda_b$).

TABLE 5.3

Regression Models Used to Determine Chlorophyll a, Chlorophyll b, Total Chlorophyll, Carotenoids, and Nitrogen Content

Equations	SVI	R^2	Adj. R^2	RMSE	DF	SSE
Hyeronima alchorneoides						
Chlorophyll a = 0.0011785 (x − 3.75145)^2 + 0.007221 x + 0.001026	mSR	0.68	0.66	0.0036	39	0.0005
Chlorophyll b = 0.000921 (x − 3.77659)^2 + 0.0033432 x − 0.001335	mSR	0.64	0.62	0.0018	35	0.0003
Total chlorophyll = 0.0021113 (x − 3.77659)^2 + 0.0106496 x − 0.002516	mSR	0.67	0.65	0.0056	35	0.0010
Carotenoids = 10.872218 (x − 0.00721)^2 + 1.12710394 x − 0.001701	D_{730}	0.63	0.60	0.0008	29	0.0000
Nitrogen = 0.0131446 x − 9.12531	RE	0.74	0.72	0.0405	14	0.0213
Virola koschnyi						
Chlorophyll a − 0.4786093 (x − 0.04116)^2 + 0.1962289 x + 0.014507	PRI	0.65	0.63	0.0042	46	0.0008
Chlorophyll b = 1.0797677 x + 0.0021521	PRI	0.71	0.70	0.0014	35	0.0000
Total chlorophyll = 0.700922 (x − 0.03922)^2 + 0.2784824 x + 0.0210494	D_{730}	0.65	0.63	0.0062	41	0.0014
Carotenoids = −54.885376 (x − 0.00676)^2 + 0.3292282 x + 0.0044758	PRI	0.71	0.69	0.0006	35	0.0000
Nitrogen = 0.0103411 x − 7.058955	RE	0.70	0.68	0.0628	15	0.0552

Notes: SSE statistic is the least squares error of the fit. Adj. R^2 is the R-square statistic adjusted based on the residual degrees of freedom. RMSE is root mean square error. Units for chlorophyll and carotenoids are milligrams per square centimeter. Units for nitrogen are grams.

$$D = \left[\frac{1}{\lambda_b - \lambda_a} \int_{\lambda_a}^{\lambda_b} [S_1(\lambda) - S_2(\lambda)]^2 \partial\lambda \right]^{\frac{1}{2}} \tag{5.1}$$

$$\theta = \cos^{-1} \left[\frac{\int S_1(\lambda) S_2(\lambda) \partial\lambda}{\left[\int S_1(\lambda)^2 \partial\lambda \right]^{\frac{1}{2}} \left[\int S_2(\lambda)^2 \partial\lambda \right]^{\frac{1}{2}}} \right] \tag{5.2}$$

5.2.7.3 Hypothesis Tests

Leaves of each species were compared between genders for each of the following: chlorophyll a and b, total chlorophyll, carotenoids, red-edge, and all 10 SVIs. The

TABLE 5.4

Description of Pattern Classifiers

Classifier abbreviation	Description
logln	Log linear classifier uses a logistic function to separate classes
qdc	Quadratic classifier assumes normal densities
treec	Decision tree classifier
lmnc 2–5	Neural network classifier with Levenburg Marquardt optimization using two to five layers
kmnc 2–3	k nearest neighbor classifier with two and three neighbors

Sources: After [23, 24, 55].

samples were all tested for normalcy prior to using either a *t*-test (unequal variances) or a rank sum test. None of the populations met the assumptions of a normal distribution, except for SR_{680}; therefore, the nonparametric Wilcoxon rank sum test was used to test for differences. The *t*-test (unequal variances) was used for SR_{680}.

5.3   RESULTS

5.3.1   SPECTRAL METRICS AND INTER- AND INTRACLASS VARIABILITY

We found no significant difference ($P > .05$) between males and females for the ND_{680}, SR_{680}, and D_{730} for *H. alchorneoides* based on the *t*-test (SR_{680}) and Wilcoxon rank sums test for ND_{680} and D_{730}. The ND_{705}, SR_{705}, mSR, and mND, PRI, SR_V, and SR_{GM} were significantly different (Wilcoxon rank sum test, $P < .05$) between the males and females ($P < .05$).

For *V. koschnyi* the ND_{705}, SR_{705}, mSR, and SR_{GM} were not significantly different (Wilcoxon rank sum test, $P > .05$) between genders. Contrarily, the SR_{680}, ND_{680}, mND, SR_V, PRI, and D_{730} did show a significant difference based on the Wilcoxon rank sum and *t*-test (SR_{680}) for comparison ($P < .05$).

The red edge position was found to be at longer wavelengths for females of both species:

- For *H. alchorneoides*: males 709 nm (±4 nm); females 712 nm (±5 nm) (Wilcoxon rank sum test, $P = .0001$)
- For *V. koschnyi*: males 709 nm (±7 nm); females 712 nm (±5 nm) (Wilcoxon rank sum test, $P = .0073$)

The values for the shape and amplitude metrics are listed in table 5.5. No statistical difference (Wilcoxon rank sum test, $P > .05$) was found between any level of comparison (i.e., within tree, between trees of the same gender, between gender).

TABLE 5.5
Inter- and Intraspecies Variability in Spectral Shape and Amplitude

	D	θ	n
Virola koschnyi			
Within tree—male	0.0211 ± 0.0116	0.0349 ± 0.0245	760
Between tree—male	0.0237 ± 0.0131	0.0471 ± 0.0399	6
Within tree—female	0.0187 ± 0.0135	0.0261 ± 0.0131	928
Between tree—female	0.0234 ± 0.0123	0.0258 ± 0.0110	10
Between gender	0.0219 ± 0.0148	0.0332 ± 0.0294	20
Hyeronima alchorneoides			
Within tree—male	0.0160 ± 0.0104	0.0218 ± 0.0128	760
Between tree—male	0.0186 ± 0.0117	0.0213 ± 0.0096	6
Within tree—female	0.0161 ± 0.0096	0.0311 ± 0.0240	928
Between tree—female	0.0185 ± 0.0116	0.0132 ± 0.0062	10
Between gender	0.0244 ± 0.0140	0.0187 ± 0.0119	20

Notes: D = amplitude metric; θ = shape metric; n = number of total pairwise combinations.

5.3.2 Pattern Classifiers

We found specific wavelength regions with the greatest separability between spectra from leaves of male and female trees for both species (fig. 5.1). One of the regions with the greatest difference is 500–600 nm (green peak), where spectra from males exhibit a higher reflectance. The 800- to 1000-nm wavelength region (controlled by internal leaf structure), however, is not important in comparison for either species for separating spectra between the genders.

Specifically, for *H. alchorneoides* the Bhattacharyya test statistic [23,24,26] comparing the separability of each waveband found the green peak centered around 550 nm to be indicative of differences as well as the regions around 700 and 720 nm (fig. 5.1a). For *V. koschnyi*, the Bhattacharrya test statistic found the region between 500 and 650 nm and at 700 nm to be most indicative of differences (fig. 5.1b).

The pattern classifiers for *H. alchorneoides* found that 40–70 features (wavebands) were optimal for determining differences in the spectra; the exception was knnc2, with only 20 features. The best performing classifiers were lmnc2 (70 features), with an overall error of 7.1%; lmnc5 (60 features), with an overall error of 8.4%; and qdc (40 features), with an overall error of 8.4%. For *V. koschnyi* we found that 30–60 features were optimal for determining differences between the spectra; the exception was treec, with only 10 features. The best performing classifiers were lmnc2 (40 features), with an overall error of 5.6%; loglc (50 features), with an overall error of 5.6%; and lmnc4 (50 features), with an overall error of 5.6%.

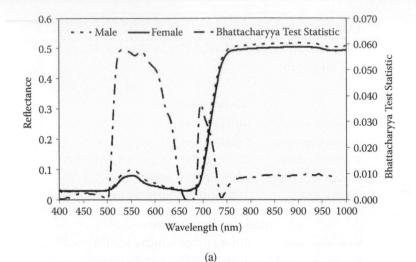

(a)

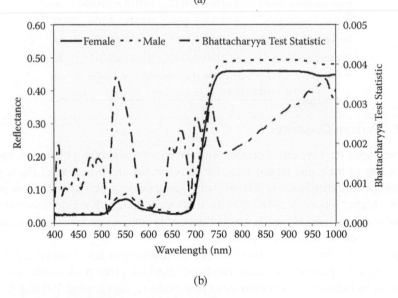

(b)

FIG. 5.1 Mean reflectance spectra for male and female trees superimposed with the Bhattacharyya test statistic for (a) *Virola koschnyi* and (b) *Hyeronima alchorneoides*. Peaks in the Bhattacharyya test statistic correspond to wavelength regions providing the greatest separability between genders.

5.3.3 LEAF CHEMISTRY

We found significant differences between chlorophyll a, chlorophyll b, and total chlorophyll concentrations and total leaf nitrogen from male and female trees of both species based on the Wilcoxon rank sum test ($P < .05$), with females having greater pigment and nitrogen concentrations. In addition, for *V. koschnyi*, a significant difference was also found for carotenoid concentration, also using the Wilcoxon rank sum test ($P < .05$), with females having more carotenoids (table 5.6).

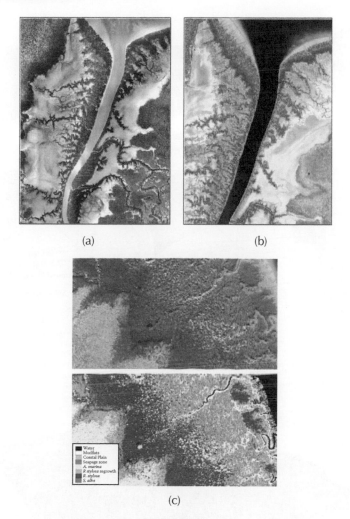

(a)   (b)

(c)

FIGURE 3.5

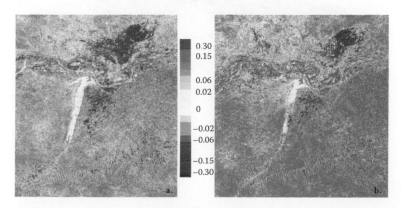

FIGURE 11.14

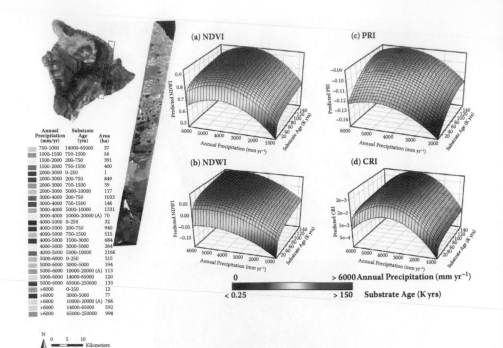

Annual Precipitation (mm/yr)	Substrate Age (yrs)	Area (ha)
750-1000	14000-65000	37
1000-1500	750-1500	56
1500-2000	200-750	391
1500-2000	750-1500	400
2000-3000	0-15	1
2000-3000	200-750	849
2000-3000	750-1500	39
2000-3000	5000-10000	117
3000-4000	200-750	1033
3000-4000	750-1500	146
3000-4000	5000-10000	1331
3000-4000	10000-20000 (A)	70
4000-5000	0-250	32
4000-5000	200-750	940
4000-5000	750-1500	515
4000-5000	1500-3000	684
4000-5000	3000-5000	264
4000-5000	5000-10000	2166
5000-6000	0-250	315
5000-6000	3000-5000	194
5000-6000	10000-20000 (A)	113
5000-6000	14000-65000	120
5000-6000	65000-250000	133
>6000	0-250	12
>6000	3000-5000	77
>6000	10000-20000 (A)	766
>6000	14000-65000	592
>6000	65000-250000	998

FIGURE 12.9

FIGURE 12.12

TABLE 5.6
Results of Hypothesis Tests with Chlorophyll a, Chlorophyll b, Total Chlorophyll, Carotenoids, and Nitrogen

	μ Male[a]	μ Female[a]	σ Male[b]	σ Female[b]	STE male[c]	STE female[c]
		Hyeronima alchorneoides				
Chlorophyll a[d]	0.0289	0.0308	0.0070	0.0058	0.0008	0.0006
Chlorophyll b[d]	0.0116	0.0132	0.0029	0.0031	0.0003	0.0003
Total chlorophyll[d]	0.0391	0.0427	0.0101	0.0085	0.0011	0.0009
Carotenoids	0.0078	0.0074	0.0019	0.0018	0.0002	0.0001
Nitrogen[d]	0.1941	0.2331	0.0521	0.0599	0.0067	0.0060
		Virola koschnyi				
Chlorophyll a[d]	0.0244	0.0257	0.0044	0.0043	0.0005	0.0004
Chlorophyll b[d]	0.0095	0.0109	0.0029	0.0018	0.0003	0.0002
Total chlorophyll[d]	0.0353	0.0370	0.0061	0.0062	0.0007	0.0006
Carotenoids[d]	0.0072	0.0070	0.0017	0.0006	0.0002	0.0001
Nitrogen[d]	0.2706	0.3013	0.0687	0.0551	0.0077	0.0055

Note: Units are milligrams per square centimeter for chlorophyll and carotenoids and grams for nitrogen.

[a] Mean.

[b] Standard deviation.

[c] Standard error of the mean.

[d] Significant difference between leaves from male and female trees with a Wilcoxon rank sum test $P < .05$.

5.4 DISCUSSION

A potential cause for the difference in spectral reflectance between genders is the stress incurred by female trees during reproduction. Reflectance spectra of leaves and fruits undergo numerous changes at various wavelengths [27,28] due to factors including deficiency of mineral nutrition [29], pollution [30], different stress conditions [31], and adaptation to solar irradiation [32–34] and in the course of leaf senescence [19,21,35–39].

The pattern classification has shown to be a useful analysis technique for locating the optimal wavebands in determining gender for each species and actually carrying out the classification with relatively low error. There were between 40 and 70 optimal wavebands for *H. alchorneoides* and between 30 and 60 wavebands for *V. koschnyi* (the exact number dependent on the classifier). The overall error was lower for *V. koschnyi* at 5.6% for three classifiers, compared to 7.1% with the best classifier for *H. alchorneoides,* which also required more (70 total) wavebands in the best classification in comparison to *V. koschnyi*, which optimally needed only 50. The lmnc 2-5 series of nonparametric classifiers and the qdc were the most effective in achieving the lowest overall error in both species. Potentially, they could be further applied to developing and refining applications of pattern recognition for the spectral separability of gender in dioecious species at larger scales.

Spectral vegetation indices also provide insight into additional vegetation characteristics such as photosynthetic efficiency. The PRI (table 5.1), which had the strongest relationship with chlorophyll a and chlorophyll b for *V. koschnyi*, has been found to have a positive correlation with photosystem II (PSII) photochemical efficiency, CO_2 uptake, and photosynthetic light use efficiency (LUE) and a negative correlation with nonphotochemical quenching (NPQ) [9]. Light use is a central component of photosynthesis that is, in turn, motivated by pigment concentrations. The PRI is calculated from wavelengths at 531 and 570 nm, which correspond to either side of the green peak, and has also been found to have a negative correlation with xanthophyll cycle activity [8,40]. Significantly higher PRI values in females of both species may be an indication of greater photosynthetic efficiency in females.

Reflectance in the visible range is determined by pigment concentrations in leaves [17,27,41], which are in turn related to photosynthesis [42]. The concentrations of the main photosynthetic pigments chlorophyll a, chlorophyll b, and carotenoids relate strongly to the photosynthetic potential of a plant and therefore give an indication of its overall physiological status [43,44]. It has been found in some dioecious species that females photosynthesize at a higher rate than males [45–48]. This is thought to occur because females can regulate photosynthesis to cover fruit production [46,49]. Our results show that there are significantly more chlorophyll a, chlorophyll b, and total chlorophyll in females from both species and this may be an indication of a possible trend also expressed in other dioecious species. The regulation of photosynthetic rate has been debated; Wheelwright and Logan [50] found that female *Ocotea tenera* trees, in fact, reduced their photosynthetic rate following the previous year's reproduction, a further indication of interspecies differences.

For *H. alchorneoides*, SVIs calculated using longer wavelengths (705–750 nm) were significantly different between males and females, corresponding to the peaks in the Bhattacharyya test statistic (fig. 5.1). In contrast the SVIs calculated from shorter wavelengths (680 nm) were not different, corresponding to the narrow valley shown by the Bhattacharyya test statistic centered at 680 nm (fig. 5.1). For *V. koschnyi* the indices with a significant difference are calculated from shorter wavelengths (i.e., 680- and 700-nm wavelengths), also corresponding to the peaks highlighted by the Bhattacharyya test statistic (fig. 5.1). These results reinforce that while gender may be expressed in the spectral reflectance and broad regions may be shared by multiple species, distinct wavelengths are most likely species specific.

Females of many dioecious plant species invest more nitrogen into reproduction than males [51]. Greater leaf nitrogen content has also been associated with increased chlorophyll content in female trees [50] and is also presumed to exert control over photosynthetic rate [52]. Nitrogen is an important element because it is a focal constituent of the porphyrin ring structure of chlorophyll and is an essential element in the enzyme rubisco. Increased nitrogen in females (table 5.6) may be the cause of the red edge being at statistically longer wavelengths for females of both species; the red edge is known to correlate to nitrogen concentrations within the leaves of plants [11].

Our results in this chapter show a definitive potential for the spectral differentiation of gender for dioecious species and possible biochemical explanations for these differences at the leaf level. To our knowledge, the spectral expression of dioecy has not been reported in the literature. Further exploration of this phenomenon in other dioecious species as well as the scaling up to canopy level needs to be investigated. In addition, an assessment of the differences in spectral reflectance during periods without fruits/flowers would be essential to understand whether the differences in the spectral properties of the leaves remain. An adequate ratio of male-to-female individuals is crucial to maintaining proper population densities and biodiversity. Therefore, spectroscopy as a means of rapid assessment of such ratios would be a useful tool to aid forest management practices that can facilitate an ecosystem conducive both to diversity and economic demands.

ACKNOWLEDGMENTS

We acknowledge the Organization for Tropical Studies Research Experience for Undergraduates (REU) funded by the National Science Foundation for funding throughout the two field seasons (2005 and 2006) and Titan Analysis Ltd. for the use of the field spectrometer. We thank Dr. Deedra McClearn and Leonel Campos for providing information on *Virola koschnyi* and Rigoberto Gonzalez for aiding in the collection of the leaves. We also thank Dr. Karen Castro for comments on an earlier version of the manuscript. We also extend a special recognition to the administrative and scientific staff at La Selva Biological Station in Costa Rica for their constant and excellent assistance for the duration of the project.

REFERENCES

1. Bawa, K.S. and Opler, P.A., Dioecism in tropical forest trees, *Evolution*, 29, 67, 1975.
2. Stylinski, C.D., Gamon, J.A., and Oechel, W.C., Seasonal patterns of reflectance indices, carotenoid pigments, and photosynthesis of evergreen chapparal species, *Oecologia*, 131, 366, 2002.
3. Endress, B.A., Gorchov, D.L., and Noble, R.B., Non-timber forest product extraction: Effects of harvest and browsing on an understory palm, *Ecological Applications*, 14, 1139, 2004.
4. Flores, E.M., *Tropical tree seed manual. Agricultural handbook 721.* Forest Service, U.S. Department of Agriculture, Washington, DC, 2002, 514.
5. Sims, D.A. and Gamon, J.A., Relationships between leaf pigment content and spectral reflectance across a wide range of species, leaf structures and developmental stages, *Remote Sensing of Environment*, 81, 337, 2002.
6. Gates, D.M., Keegan, H.J., Schleter, J.C., et al., Spectral properties of plants, *Applied Optics*, 4, 11, 1965.
7. Gamon, J.A., Kitajima, K., Mulkey, S.S., et al., Diverse optical and photosynthetic properties in a neotropical dry forest during the dry season: Implications for remote estimation of photosynthesis, *Biotropica*, 37, 547, 2005.
8. Gamon, J.A., Penuelas, J., and Field, C.B., A narrow-waveband spectral index that tracks diurnal changes in photosynthetic efficiency, *Remote Sensing of Environment*, 41, 35, 1992.

9. Guo, J.M. and Trotter, C.M., Estimating photosynthetic light-use efficiency using the photochemical reflectance index: Variations among species, *Functional Plant Biology*, 31, 255, 2004.

10. Castro-Esau, K.L., Sanchez-Azofeifa, G.A., Rivard, B., et al., Variability in leaf optical properties of mesoamerican trees and the potential for species classification, *American Journal of Botany*, 93, 517, 2006.

11. Cho, M.A. and Skidmore, A., A new technique for extracting the red edge position from hyperspectral data: The linear extrapolation method, *Remote Sensing of Environment*, 101, 181, 2006.

12. Frankie, G.W., Baker, H.G., and Opler, P.A., Comparative phenological studies of trees in tropical wet and dry forests in lowlands of Costa-Rica, *Journal of Ecology*, 62, 881, 1974.

13. OTS, *OTS Meteorological Data 1957–2003*. Organization for Tropical Studies, San Pedro, Costa Rica, 2003.

14. Hartshorn, G.S. and Hammel, B.E., Vegetation types and floristic patterns, in *La Selva: Ecology and natural history or a neotropical rain forest*, L.A. McDade, K.S. Bawa, H.A. Hepenheide, et al., eds. University of Chicago Press, Chicago, 1994, 73.

15. Vargas, O., *Sindromes De Dispersion Pollinizacion Y Sistemas Sexuales De Los Arboles Nativos De La Estacion Biologica La Selva Y Areas Circundantes*. La Flora Digital de La Selva, 2006.

16. Oberbauer, S.F. and Strain, B.C., Effects of canopy position and irradiance on the leaf physiology and morphology of *Pentaclethra macroloba* (Mimosaceae), *American Journal of Botany*, 73, 409, 1986.

17. Blackburn, G.A., Quantifying chlorophylls and carotenoids at leaf and canopy scales: An evaluation of some hyperspectral approaches, *Remote Sensing of Environment*, 66, 273, 1998.

18. Hiscox, J.D. and Israelstam, G.F., A method for the extraction of chlorophyll from leaf tissue without maceration, *Canadian Journal of Botany*, 57, 1332, 1979.

19. Richardson, A.D., Duigan, S.P., and Berlyn, G.P., An evaluation of noninvasive methods to estimate foliar chlorophyll content, *New Phytologist*, 153, 185, 2002.

20. Arnon, D.I., Copper enzymes in isolated chloroplasts. Polyphenoloxidase in *Beta vulgaris*, *Plant Physiology*, 24, 1, 1949.

21. Chappelle, E.W., Kim, M.S., and McMurtrey III, J.E., Ratio analysis of reflectance spectra (RARS): An algorithm for the remote estimation of the concentrations of chlorophyll a, chlorophyll b, and carotenoids in soybean leaves, *Remote Sensing of Environment*, 39, 239, 1992.

22. Defernez, M. and Kemsley, E.K., The use and misuse of chemometrics for treating classification problems, *Trac-Trends in Analytical Chemistry*, 16, 216, 1997.

23. Duin, R.P.W., *Prtools V.3.0: A Matlab toolbox for pattern recognition*. Pattern Recognition Group, Delft University, Delft, the Netherlands, 2000.

24. van der Heijden, F., Duin, R.P.W., de Ridder, D., et al., *Classification, parameter estimation and state estimation: An engineering approach using Matlab*. John Wiley & Sons, West Sussex, England, 2004.

25. Price, J.C., How unique are spectral signatures? *Remote Sensing of Environment*, 49, 181, 1994.

26. Chernoff, H., A measure of asymptotic efficiency for tests for a hypothesis based on the sum of observations, *Annals of Mathematical Science*, 23, 493, 1952.

27. Carter, G.A., Responses of leaf spectral reflectance to plant stress, *American Journal of Botany*, 80, 239, 1993.

28. Carter, G.A. and Young, D.R., Foliar spectral reflectance and plant stress on a barrier island, *International Journal of Plant Sciences*, 154, 298, 1993.

29. Kochubi, S.M., Koberts, N.I., and Shadchina, T.M., *Spektral'kye Svoistva Rastenii Kak Asnova Metodav Distantsionnoi Diagnostiki*. Naukova Durnka, Kiev, 1990.

30. Gitelson, A.A., Zur, Y., Chivkunova, O., et al., Assessing carotenoid content in plant leaves with reflectance spectroscopy, *Journal of Photochemistry and Photobiology*, 75, 272, 2002.
31. Merzlyak, M.N. and Solovchenko, A.E., Photostability of pigments in ripening apple fruit: Possible photoprotective role of carotenoids during plant senescence, *Plant Science*, 163, 881, 2002.
32. Penuelas, J. and Filella, I., Visible and near-infrared reflectance techniques for diagnosing plant physiological status, *Trends in Plant Science*, 3, 151, 1998.
33. Merzlyak, M.N., Gitelson, A.A., Chivkunova, O.B., et al., Non-destructive optical detection of pigment changes during leaf senescence and fruit ripening, *Physiologia Plantarum*, 106, 135, 1999.
34. Merzlyak, M.N. and Chivkunova, O., Light-stress induced pigment changes and evidence for anthocyanin photoprotection in apple fruit, *Journal of Photochemistry and Photobiology*, 55, 154, 2000.
35. Yoder, B.J. and Pettigrew-Crosby, R.E., Predicting nitrogen and chlorophyll concentrations from reflectance spectra (400–2500 nm) at leaf and canopy scales, *Remote Sensing of Environment*, 39, 199, 1995.
36. Lichtenthaler, H.K., Gitelson, A.A., and Lang, M., Non-destructive determination of chlorophyll content of leaves of a green and an aurea mutant of tobacco by reflectance measurements, *Journal of Plant Physiology*, 148, 483, 1996.
37. Merzlyak, M.N., Gitelson, A.A., Pogosyan, S.I., et al., Reflectance spectra of leaves and fruits during their development and senescence and under stress, *Russian Journal of Plant Physiology*, 44, 707, 1997.
38. Gitelson, A.A., Merzlyak, M.N., and Chivkunova, O., Optical properties and non-destructive estimation of anthocyanin content in leaves, *Journal of Photochemistry and Photobiology*, 74, 38, 2001.
39. Merzlyak, M.N., Solovchenko, A.E., and Gitelson, A.A., Reflectance spectral features and non-destructive estimation of chlorophyll, carotenoid and anthocyanin content in apple fruit, *Postharvest Biology and Technology*, 27, 197, 2003.
40. Bilger, W., Bjorkman, O., and Thayer, S.S., Light-induced spectral absorbance changes in relation to photosynthesis and the epoxidation state of xanthophyll cycle components in cotton leaves, *Plant Physiology*, 91, 542, 1989.
41. Turrell, F.M., Weber, J.R., and Austin, S.W., Chlorophyll content and reflection spectra of citrus leaves, *Botanical Gazette*, 123, 10, 1961.
42. Blackburn, G.A. and Steel, C.M., Towards the remote sensing of matorral vegetation physiology: Relationships between spectral reflectance, pigment and biophysical characteristics of semiarid bushland canopies, *Remote Sensing of Environment*, 70, 278, 1999.
43. Danks, S.M., Evans, E.H., and Whittaker, P.A., *Photosynthetic systems. Structure and assembly*. Wiley, New York, 1983.
44. Young, A. and Britton, G., Carotenoids and stress, in *Stress responses in plants: Adaptation and acclimation mechanisms*, R.G. Alscher and J.R. Cumming, eds. Wiley-Liss, New York, 1990, 87.
45. Dawson, T.E. and Bliss, L.C., Plants as mosaics, leaf-level variation in the physiology of the dwarf willow, *Salix artica, Functional Ecology*, 7, 293, 1993.
46. Dawson, T.E. and Ehleringer, J.R., Gender-specific physiology, carbon isotope discrimination, and habitat distribution in boxelder, *Acer Negundo, Ecology*, 74, 798, 1993.
47. Correia, O. and Barradas, M.C., Ecophysiological differences between male and female plants of *Pistancia lentiscus, Plant Ecology*, 149, 131, 2000.
48. Obeso, J.R., The costs of reproduction in plants, *New Phytologist*, 155, 321, 2002.
49. Obeso, J.R., Alvarez-Sanullano, M., and Retuerto, R., Sex ratios, size distributions, and sexual dimorphism in the dioecious tree *Ilex aquifolium* (Aquifoliaceae), *American Journal of Botany*, 85, 1602, 1998.

50. Wheelwright, N.T. and Logan, B.A., Previous-year reproduction reduces photosynthetic capacity and slows lifetime growth in females of a neotropical tree, *Proceedings of the National Academy of Sciences of the United States of America*, 101, 8051, 2004.
51. Gerber, M.A., Dawson, T.E., and Delph, L.F., *Gender and sexual dimorphism in flowering plants*. Springer, Berlin, 1999.
52. Field, C. and Mooney, H.A., The photosynthesis–nitrogen relationship in wild plants, in *On the economy of plant form and function*, T.J. Givnish, ed. Cambridge University Press, Cambridge, 1986, 23.
53. Gitelson, A.A. and Merzlyak, M.N., Remote estimation of chlorophyll content in higher plant leaves, *International Journal of Remote Sensing*, 18, 2691, 1997.
54. Vogalmann, J.E., Rock, B.N., and Moss, D.M., Red edge spectral measurements from sugar maple leaves, *International Journal of Remote Sensing*, 14, 1563, 1993.
55. Castro-Esau, K.L., Sánchez-Azofeifa, G.A., and Caelli, T., Discrimination of lianas and trees with leaf level hyperspectral data, *Remote Sensing of Environment*, 90, 353, 2004.

6 Species Classification of Tropical Tree Leaf Reflectance and Dependence on Selection of Spectral Bands

*Benoit Rivard, G. Arturo Sanchez-Azofeifa,
Sheri Foley, and Julio C. Calvo-Alvarado*

CONTENTS

6.1 INTRODUCTION

Few studies have focused on classification of tree species from spectral data, despite the many potential applications in conservation biology and forest conservation and management [1,2]. Fewer studies have examined classification of tropical tree species due to their high biodiversity [3–5].

Accurate classifications of boreal forest species and forest cover types using remote imagery are encouraging for continued research into the discrimination of tree species. Martin et al. [6], for example, were able to achieve an accuracy of 75% when classifying 11 North American forest cover types using airborne visible/infrared imaging spectrometer (AVIRIS) imagery. Other studies looking at boreal forest species differentiation include van Aardt and Wynne [7], who studied three pine and three deciduous tree species; Gong, Ru, and Yu [8], who investigated six coniferous species; and Shen, Badhwar, and Carnes [9], who studied three coniferous and two deciduous species.

It is clear that vegetation have dominant spectral properties controlled by pigment content, water content, and leaf structure [10], but it is not clear for tropical species which characteristics differ enough to enable remote detection. While it is understood that environmental variables, such as photon flux, temperature, nutrient availability, humidity, and precipitation can result in increased spectral variability within species groups [11–13], it is not yet determined whether or not intraspecific variation surpasses interspecific variation. Zhang et al. [5], working with airborne spectral data, and Castro-Esau et al. [3], who examined leaf spectral data, recently examined intra- and interclass spectral variability of tropical tree species. Their study shows that some species are separable at least locally. Strategies that have been taken to manage the large number of spectral bands of field and airborne data sets include preanalysis processing such as conversion to principle components [14], derivatives [15,16], wavelets [5,16,17], and filtered data [7]. Bands can also be chosen using statistical methods such as one-way analysis of variance (ANOVA) [18], correlation analyses [19], and linear discriminant analysis [8,20].

Discriminant analysis is particularly suited for classification of species because it not only reduces the number of bands, but also selects bands that contain maximal differences between species groups and minimal differences within those groups. It is more useful than principal component analysis (PCA), for example, which chooses eigenvectors containing the maximum amount of variability, with the drawback that results may vary depending on nondiagnostic features that can actually deter classification results.

The main objective of this study was to investigate the potential use of leaf reflectance to discriminate among 20 tropical rainforest tree species at the leaf scale. We hypothesized that the selected 20 tree species (table 6.1) have different leaf spectral properties and consequently can be differentiated using reflectance data. The relevance of these results for the interpretation of airborne imagery and the investigation of canopy-scale species differentiation represents the next step of this research [5].

In order to select and limit the number of bands used for classification of leaf spectral data collected from these 20 tropical species, we applied Wilk's stepwise discriminant analysis. An arbitrary number (four) of sample sets, each comprising a random selection of half of the original data set, were analyzed to test the robustness of this technique, as the number of bands/indices and species used far exceeded the assumptions of this analysis type. Five different band sets (band ranges and indices) were analyzed to determine how sensor limitations—caused by atmospheric interference, for example—affect band selection. The chosen bands were then compared against each other and evaluated based upon classification accuracy of a test data set.

TABLE 6.1
Tree Species Used in This Study and Number of Leaves Sampled

Name [ref.]	Genus	Species	Common name	No. of leaves	Site
Malpighiaceae	*Byrsonima*	*crispa* A. Juss	Nance	15	La Selva
Meliaceae	*Cedrela*	*odorata* L.	Cedro amargo	15	INBio
Euphorbiaceae	*Conceveiba*	*pleistemona* Donn. Sm	Algodón	15	La Selva
Fabaceae/Pap.	*Dipteryx*	*panamensis* (Pitter)	Almendro	13	INBio
Fabaceae/Pap.	*Dussia*	*macroprophyllata* (Donn. Sm) Harms	Sangrillo	17	La Selva
Meliaceae	*Guarea*	*guidonia* (L.) Sleumer	Cocora	15	La Selva
Fabaceae/Mim.	*Inga*	*pezizifera* Benth.	Guavilla	15	La Selva
Arecaceae	*Iriartea*	*deltoidea* Ruiz & Pav.	Palmito	17	La Selva
Flacourtiaceae	*Laetia*	*procera* (Poepp.) Eichler	Manga larga	15	La Selva
Lecythidaceae	*Lecythis*	*ampla* Miers	Jícarro	18	La Selva
Chrysobalanaceae	*Maranthes*	*panamensis* (Standl) Prance & F. Whi	Pejiballe	19	La Selva
Fabaceae/Mim.	*Pentaclethra*	*macroloba* (Willd.) Kuntze	Gavilán	15	INBio
Humiriaceae	*Sacoglottis*	*trichogyna* Cuartrec	Titor	15	La Selva
Elaeocarpaceae	*Sloanea*	*geniculata* Damon A. Sm.	Paleta	12	La Selva
Sterculiaceae	*Sterculia*	*apetala* (Jacq.) H. Karst.	Panama	15	INBio
Bignoniaceae	*Tabebuia*	*guayacan* (Seem.) Hemls.	Corteza	15	INBio
Meliaceae	*Trichilia*	*havanensis* Jacq.	Uruca	15	INBio
Myristicaceae	*Virola*	*koschnyi* Warb.	Fruta dorada	15	La Selva
Vocysiaceae	*Vochysia*	*ferruginea* Mart.	Botarrama	20	INBio
Arecaceae	*Welfia*	*regia* Mast.	Corozo	15	LaSelva

Notes: Only one tree per species was sampled. Species are sorted alphabetically by genus. Species descriptions can be found in references 40 and 41.

The tree species tested in this study all belong to habitat of the great green macaw (*Ara ambigua*) and are of importance to the survival of this bird species in the Caribbean lowlands of Costa Rica. The great green macaw is an umbrella species, meaning that it shares habitat with many other animals and its welfare is an indicator of the welfare of other coexisting species. Currently, this species is listed as vulnerable by the International Union for Conservation of Nature and Natural Resources [21] and is under severe pressure due to the extent of deforestation and forest fragmentation of its habitat. Current efforts to conserve this emblematic species include the establishment of a new national park, Maquenque, and the consolidation of the proposed San Juan-Braulio Carrillo biological corridor.

6.2 STUDY AREA

This study was conducted at the Costa Rican National Biodiversity (INBio) facilities in San Jose and at the La Selva biological station, Costa Rica. Both sites were chosen to acquire leaf samples due to their diverse vegetation and readily available facilities.

INBio has a botanical garden that contains trees representative of several ecosystems native to Costa Rica, such as the tropical dry forest, central valley forest, humid forest, tropical rain forest, and wetlands. La Selva is a biological station bordering the foothills of the central volcanic chain and the Caribbean coastal plain and is within the San Juan-Braulio Carrillo biological corridor that protects the habitat of the great green macaw.

6.3 METHODS

6.3.1 Sample Acquisition

The tree species (table 6.1) chosen for sample collection were selected as the most important species for the survival of the great green macaw (*A. ambigua*) [22]. Of these species, *Sacoglottis trichogyna* and *Dipteryx panamensis* trees are the most important food sources during the reproductive stage of this bird, particularly the *D. panamensis* tree, which also serves as the only tree species for nesting. A total of 20 trees, 1 per species, were selected within the botanical garden at INBio and at La Selva in May 2002.

Leaves were sampled from the well-lit upper portion of tree crowns to avoid sample variability introduced by light environment. Sun leaves are more relevant for the remote sensing context of this study because at the top of a tree canopy they strongly influence remote spectral measurements. Leaves at INBio were measured immediately in the laboratory while leaves collected from La Selva were put in bags and brought for measurement in the laboratory within 2 hours. Two hours was an acceptable time frame to preserve leaf freshness [23].

Based on expectation tests presented in the results section, a minimum of 15 leaves was collected from each tree at both sites. Exceptions include the *D. panamensis* tree (13 leaves), due to the small size of the tree and few leaves available, and *Sloanea geniculata,* due to holes and scars in the leaves, which were avoided during measurements but reduced the sample size to 12. As well, three of the sampled *D. panamensis* leaves appeared to be less than fully mature; this was considered in analysis steps. *Conceveiba pleiostemona* and *Inga pezizifera* leaves were damp from rain and the leaf surfaces were gently dried with gauze.

6.3.2 Spectral Measurements

Leaf sample reflectance was measured using a Fieldspec FR spectrometer (Analytical Spectral Devices, Boulder, Colorado), which is composed of three internal sensors ranging from 350 to 975 nm, 976 to 1770 nm, and 1771 to 2500 nm, with spectral resolutions of 3 nm at 700 nm and 10 nm at 1400 and 2100 nm. The sensor tip was positioned at nadir above each leaf, resulting in a 3.55-cm^2 field of view (FOV). The FOV for the *Lecythis ampla* leaves was smaller, 1.4 cm^2, in order to fit within the perimeter of the small leaves. A 50-W quartz halogen lamp was oriented at 45° from the normal to the surface and aimed directly at the FOV of the sensors. A black wooden panel measuring less than 5% reflectance was placed below each leaf to absorb light transmitted through the leaves. A 99% reflectance panel (Spectralon,

Labsphere, North Sutton) was used to calculate leaf reflectance by dividing leaf radiance by that of the panel.

Each leaf was measured in three to seven (three for most species) spots along an axis perpendicular to the main leaf vein, except for palm leaves that were measured on a central axis parallel with the length of the leaves. For small leaves the cumulative field of view from the three measurements covered most of the leaf surface area. For larger leaves the minimum of three spots per leaf was determined using expectation tests presented in the results section. At each spot 10 measurements were averaged and the average saved. For each leaf this process lasted a few seconds, precluding heating of the leaf surface.

6.3.3 LINEAR DISCRIMINANT ANALYSIS

The spectral measurements data were formatted for Wilk's linear stepwise discriminant analysis. This type of discriminant analysis involves first selecting the band with the lowest Wilk's lambda value, which indicates group separability; then, at each subsequent step the band that best increases species group separation is added to the model [24]. Four random samples containing 50% of leaf spot reflectance measurements were chosen in order to test the consistency among the four resultant models; in each case the remaining 50% was used to evaluate the classification accuracy of the models. SAS (statistical analysis system) and SPSS (statistical package for the social sciences) were used to perform linear discriminant analyses upon each 50% sample labeled A, B, C, and D. SAS was chosen because of its ability to run stepwise discriminant analysis with all 2150 bands as inputs. While SAS effectively ranked the input bands based upon Wilk's lambda, it did not select a subset of bands to be used eventually for classification. SPSS was used in a secondary step because of its ability to reduce the number of variables selected based upon significance and ability to classify test samples. The top 30 bands selected for each sample by SAS were grouped together, resulting in a maximum of 120 bands, and then entered into SPSS for a rerun of Wilk's stepwise discriminant analysis. Thirty of the top ranking bands from SAS were chosen because of the very small changes in Wilk's lambda for the subsequent bands selected and because the amount of 120 bands was very near the limit of input variables that SPSS would allow.

The preceding methodology was applied to vegetation and water spectral indices (table 6.2) and band sets (table 6.3) in order to capture the spectral range of a variety of sensors and to determine wavelength regions of importance for separating species. Band set 1 represents all available bands from 350 to 2500 nm; band set 2 is band set 1 minus bands that would be affected by atmospheric absorption in an airborne or satellite remote sensing situation; band set 3 includes bands from 1076 to 2470 nm without atmospheric bands; band set 4 includes bands that are encompassed by the Analytical Spectral Devices HH spectrometer (350–1075 nm); and, lastly, band set 5 includes both indices listed in table 6.2 and bands in band set 2.

Chlorophyll indices were selected from a recent comparative study by Le Maire, Fancois, and Dufrene [25], while water indices, the hyperspectral normalized difference vegetation index (NDVI), and hyperspectral simple ratio (SR) were selected from a comparative study by Sims and Gamon [26]. These indices were complemented by

TABLE 6.2

Indices Tested with Discriminant Analysis

Name	Index	Name [ref.]	Index
Chlorophyll index (DattA) [33]	$\dfrac{R_{780} - R_{710}}{R_{780} - R_{680}}$	PRI [28]	$\dfrac{R_{531} - R_{570}}{R_{531} + R_{570}}$
Chlorophyll index (DattB) [33]	$\dfrac{R_{850} - R_{710}}{R_{850} - R_{680}}$	Water index (WI_{1180}) [26]	$\dfrac{R_{900}}{R_{1180}}$
Chlorophyll index (SGA) [34]	$\dfrac{R_{750} - R_{705}}{(R_{750} + R_{705} - 2R_{445})}$	Water index (WI_{970}) [35]	$\dfrac{R_{900}}{R_{970}}$
Chlorophyll index (SGB) [34]	$\dfrac{R_{750} - R_{445}}{R_{705} - R_{445}}$	NDWI [29]	$\dfrac{R_{860} - R_{1240}}{R_{860} + R_{1240}}$
NDVI	$\dfrac{R_{800} - R_{680}}{R_{800} + R_{680}}$	MSI [30]	$\dfrac{R_{1600}}{R_{820}}$
SR	$\dfrac{R_{800}}{R_{680}}$	CAI [31]	$0.5(R_{2000} + R_{2200}) - R_{2100}$
Carotenoid chlorophyll a ratio (Car/Chl) [28]	$4.44 - 6.77\exp[-0.48\dfrac{R_{800} - R_{445}}{R_{800} - R_{680}}]$		

the photochemical reflectance index of Penuelas, Filella, and Gamon [27]; carotenoid–chlorophyll-a ratio of Penuelas, Baret, and Filella [28]; the normalized difference water index (NDWI) of Gao [29]; the moisture stress index (MSI) of Rock et al. [30]; and, finally, the cellulose absorption index (CAI) of Nagler et al. [31].

6.4 RESULTS

6.4.1 EXPECTATION TESTS

The minimum number of spots per leaf and leaf sample size per species was determined at INBioparque using expectation tests [32] for near-infrared (NIR) wavelengths (fig. 6.1). In this test, an adequate sample size is determined when the average reflectance approaches a constant value as more measurements are included. Figure 6.1 displays the average reflectance with increasing number of spots measured on one leaf for each of the six species. For small leaves (*D. panamensis*, *Cedrela odorata*, and *Tabebuia guayacan*), no more then three spots could be measured. With the exception of *Pentaclethra macroloba*, the average reflectance approached a constant value with the measurement of three spots; thus, this number was used as a minimum number of measurements per leaf. *P. macroloba* is a compound leaf presenting gaps in leaf surface area; therefore, finding locations where the leaf material filled the FOV was difficult and leaf material may have overlapped. In figure 6.2, expectation test results for leaf sample number are displayed for the same species.

TABLE 6.3

Spectral Bands Selected by Wilk's Stepwise Discriminant Analysis and Categorized by Input Band Sets and Input Sample Letter

Band set (nm)	Model name	Classification accuracy of test data	Bands selected (in order of selection)
1	1A	96.7	1442, 1316, 1662, 1729, 723, 707, 739, 694, 1397, 396, 1117, 526
350–2500	1B	91.2	1441, 1303, 1663, 1503, 1402, 723, 711, 739, 633, 390, 798, 497, 559, 606, 2185
	1C	90.9	1450, 1316, 1662, 1719, 735, 449, 390
	1D	81.1	1440, 1303, 1661, 1715, 744
2	2A	97.2	2326, 1089, 730, 707, 739, 691, 1500, 1661, 1723, 636, 557
350–1341,	2B	90.7	2313, 1090, 732, 707, 739, 694, 1503, 1256, 390, 1963, 2326
1500–1789,	2C	96.3	2326, 1090, 731, 708, 737, 1503, 1661, 1770, 445, 396, 1235
and 1954–2470	2D	96.0	2326, 1089, 734, 707, 747, 620, 1500, 1661, 1770, 698, 1235
3	3A	85.9	2326, 1089, 1959, 1500, 1661, 1729, 1226, 2022, 2259, 2053
1076–1341,	3B	81.9	2313, 1090, 1963, 1503, 1661, 1763, 1163
1500–1798,	3C	86.3	2326, 1090, 1503, 1661, 1729, 1959, 2259, 2030
and 1954–2470	3D	81.8	2326, 1089, 1963, 1500, 1661, 1753, 2022
4	4A	84.8	912, 737, 711, 747, 682, 666
350–1075	4B	87.7	911, 737, 711, 722, 693, 558, 571, 526
	4C	79.0	912, 736, 713, 747, 386, 429, 385, 377
	4D	82.5	912, 740, 711, 732, 678, 666
5	5A	93.1	MSI, DattB, SGA, WI970, DattA, 1787, 724, 750, 1500, 392, 483, 525, 696
Indices and	5B	95.7	MSI, DattB, SGA, Car/Chl, SGB, WI970, 1789, 637, 1500, 380, 664, 2156, 1723, 1341, 753, 733, 712, 702, 1961, 2296, SR680, 393
350–1341,			
1500–1789,	5C	93.6	MSI, DattB, SGA, Car/Chl, 1787, 722, 743, 1667, 1723, 712, 616, SR680, NDVI, 559, CAI
and 1954–2470	5D	94.7	MSI, DattB, SGA, WI970 Car/Chl, SGB, 1778, 1500, 2156, 2296, 1961, 559, 1723, 1628, SR680, 525, 380, 2141

Note: Bands and indices are listed in order of selection. Complete index names and abbreviations are shown in table 6.2.

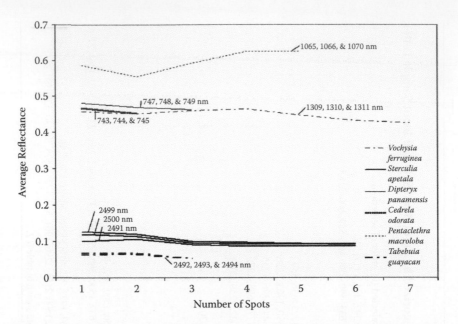

FIG. 6.1 Expectation test results used to determine the minimum number of measurements per leaf. Plot represents changing average reflectance values as sample spot number increases. For each species, results are shown for three wavelengths of highest standard deviation. Leaves of some species were too small for more than three measurements.

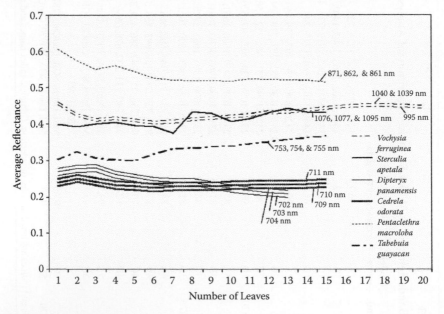

FIG. 6.2 Expectation test results used to determine the minimum leaf sample number for each species. For each species, results are shown for three wavelengths of highest standard deviation. Fifteen leaf measurements were chosen as the minimum number of measurements based upon slope values.

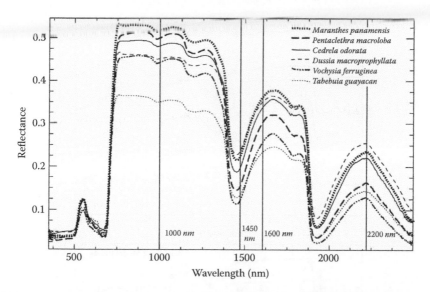

FIG. 6.3   Mean spectral reflectance for selected species. Note the fluorescence peak near 738 nm (red edge and near-infrared plateau convergence) on the *Maranthes panamensis* spectral line and the change in relative position of species reflectance between 1000, 1450, 1600, and 2200 nm. In order to preserve clarity, not all species are shown.

For all species, the change in the average reflectance is smaller than 1% following seven leaf measurements and less than 0.5% for five of the six species following 14 measurements. While seven leaf measurements may have been sufficient, the minimum number of leaf samples per tree was set to 15 to ensure a representative sample size. Expectation tests were performed for each species during sample measurement and collection as a precaution.

6.4.2 Species Mean Reflectance

Figure 6.3 shows similar mean reflectance for six species in the visible band (VIS) region and increased separation at longer wavelengths. Interspecies reflectance relationships change across the spectrum; for example, *Vochysia ferruginea* and *Dussia macroprophyllata* have very close mean reflectance at 900 nm but differ at 1660 nm. These results suggest graphically that NIR and shortwave infrared (SWIR) wavelengths may in some instances provide enhanced discriminating power.

6.4.3 Discriminant Analysis Models per Band Set

The bands selected by discriminant analysis for each sample and band set are summarized in table 6.3 and figure 6.4. In general, the bands selected early for each model are the most consistent among the four sample groups. Bands selected for all models were considered more robust to small changes between samples than bands selected less frequently. Band differences between band set models resulted in classification accuracy differences of up to 15%. The following paragraphs elaborate on the models obtained from each band set.

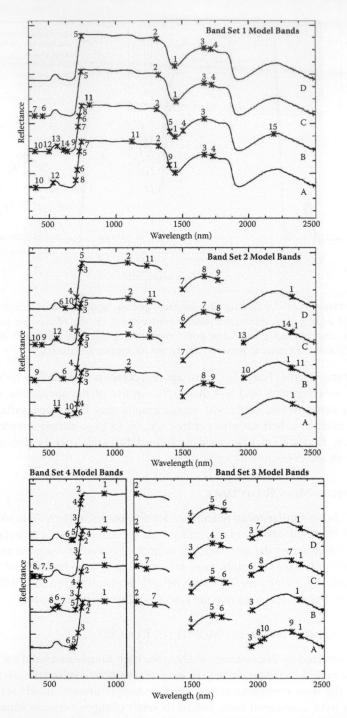

FIG. 6.4 Model bands selected from band sets 1 to 4. Numbers indicate order of band selection. Letters on the right side of each plot indicate the input sample for each model. The spectrum plotted in each case was chosen arbitrarily for illustrative purposes.

The models derived from wavelengths 350 to 2500 nm (band set 1) and for sample sets A, B, C, and D are composed of 5 to 15 bands. SPSS controlled the number of bands added via critical significance values that were the same for all samples. Models with the least number of bands (1C and 1D) also had the lowest classification accuracies. The first three bands added to each model were approximately 1440, 1316, and 1662 nm. Common bands that were not necessarily selected in the same order are 1720, 735, and 390 nm. The average classification accuracy of models 1A to 1D was 90%, the third highest value when compared to the averages for the other band sets.

For band set 2, bands were removed in order to approximate remote hyperspectral imagery affected by atmospheric absorption. The four resulting models had 11 to 14 bands; the first five bands added were approximately 2326, 1090, 730, 707, and 739 nm. Of the remaining selected bands 37% were VIS bands. Reoccurring bands or narrow wavelength regions (<20 nm) that were selected at different times for each model were approximately 1500 nm (100%), 1661 nm (75%), 1770 nm (50%), 1240 nm (75%), and 600 nm (100%). These models have a 95% classification accuracy average—the highest when compared to the other band set model averages.

For band set 3 atmospheric bands were removed again to represent realistic atmospheric effects on remote imagery and VIS bands were removed to observe if classification results were different without the VIS bands. The number of bands in each model ranged from 7 to 10. The first two bands selected for these models (2326 and 1090 nm) are the same as the bands selected for models from band set 2. The order of the bands selected is slightly variable but the next four bands included are approximately 1963, 1500, 1661, and 1729 nm. Average model classification accuracy was 84%, one of the two lowest averages when compared to the other band set model averages.

Band set 4 was intended to model data from sensors such as the Analytical Spectral Devices HH that only measure light from 350 to 1075 nm. Six to eight bands were chosen for each model. The first four bands selected in each model were approximately 911, 737, 711, and 722–747 nm. Out of the remaining 12 bands, 5 were between 650 and 700 nm, 2 were between 550 and 600 nm, 1 was 526 nm, and 4 were less than 450 nm. These models had a classification average of 84%, sharing the lowest average value with that of band set 3.

Several indices (table 6.2) were added to band set 2, creating band set 5, in order to determine if indices linked to biochemistry are important for differentiating between these species. The number of bands/indices selected (13 to 22) increased compared to the other model types and several different indices were selected before any individual bands. The MSI [30], chlorophyll index (DattB) from Datt [33], and chlorophyll index (SGA) from Sims and Gamon [34] were selected first for all four models. The fourth band selected for each model was either the water index (WI) (900/970 nm) [35] or Car/ChlA index [28]. Following the indices, the first band selected was approximately 1785 nm for all four models; then, the consistency between the models ceased and different bands were included in each model. The differences between each of the models have little impact on the overall classification results, which range between 93.1 and 95.7%. The average classification value is 94%, a slightly lower value than the average accuracy of band set 2.

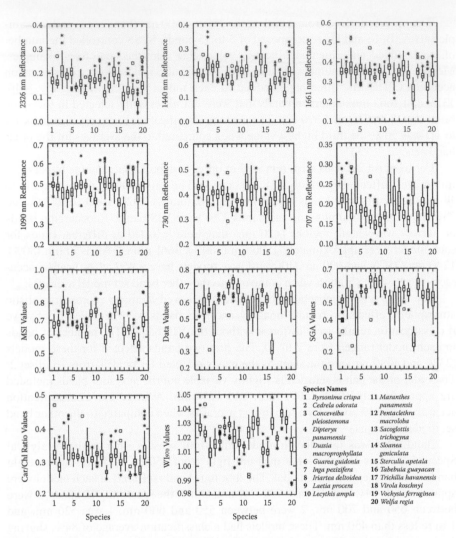

FIG. 6.5 Box-plots of selected variables showing relative distributions of species data. Species numbers correspond to x-axis numbers. Asterisks and squares represent outlying data points. Note similar positioning of species data for 2326 and 1440 nm and also note different positioning of species data for the other bands and indices shown here.

6.4.4 Selected Band Variance

The distribution of species data was compared to that of other species and described as interspecies relationships. Figure 6.5 depicts distributions of species data for several bands and indices belonging to the models (table 6.3). The first bands selected for band set 1, 2, and 3 models were 2326 or 1440 nm (fig. 6.5). These bands in addition to selected bands 1963, 1930, and 1500 nm show similar interspecies relationships; for example, in all of these bands the *Conceveiba pleiostemona* mean is slightly higher than the *D. panamensis* mean, while the *V. ferruginea* mean is the lowest of all 20 species. The second bands selected for band set 1, 2, and 3 models were NIR bands 1316 and 1090 nm, which, including 912 nm, are the second group sharing interspecies commonalities (1090 nm is plotted in fig. 6.5 as an example). Band 1661 nm was selected for many of the models and demonstrated an interspecies pattern similar to 1715 and 1787 nm, but different from previous bands mentioned (fig. 6.5). *V. ferruginea* and *T. guayacan* have distinct values for band 1661 nm compared to the rest of the species. Band 730 nm (fig. 6.5) contains the fourth interspecies pattern also demonstrated by band 747 and 737 nm. The 730-nm reflectance values appear (in fig. 6.5) to provide better species separability for more species subsets than 1661 nm (e.g., *Byrsonima crispa* vs. *Sterculia apetala*).

The models derived from indices and individual bands (band set 5) show the importance of specific spectral features for differentiating species. The first variable selected from band set 5 was MSI; thus, it has greater separation of species compared to the other input bands such as 2326 nm (fig. 6.5). The second variable selected was the chlorophyll index (DattB) [33], which clearly shows separation of *T. guayacan* from the other species (fig. 6.5). The chlorophyll index SGA [34] was selected third and when compared with DattB [33] shows a slight shift for *W. regia* and *I. pezizifera* species (fig. 6.5). These two indices are composed of different bands (table 6.2) but are highly correlated with one another ($r = 0.97$). The next variable chosen was either the carotenoid–chlorophyll ratio from Penuelas et al. [28] or WI_{970} [26]. WI_{970} shows greater group separation than the Car/Chl ratio, as shown in figure 6.5, and was found to have an inverse correlation coefficient of $r = 0.85$ with MSI. The subsequent variables were not as consistent among the four models (5A to 5D) but resulted in a maximum 2.6% difference in classification accuracy.

6.4.5   SPECIES-LEVEL CLASSIFICATION RESULTS

In general, all of the species, except for *D. panamensis*, were identified correctly most of the time (fig. 6.6). *D. panamensis* was confused mostly with *Laetia procera* by all of the models. As mentioned in section 6.4.2 and shown in table 6.2, models for band sets 2 and 5 have the highest overall accuracies. Species results show that they also have the highest accuracies for *D. panamensis*, especially for band set 5 models.

The classification of some species was not affected by the restriction to SWIR (band set 3) or VIS (band set 4) bands. The SWIR models (3A to 3D) maintained high accuracies for six species, while five different species had high accuracies when classified using VIS models (4A to 4D; bold on fig. 6.6). The high classification accuracy for these species indicates that they do not require more bands for differentiation. *S. trichogyna* was the most noticeably affected by the loss of SWIR bands, as its values dropped from 100% (model 2A) to 55% (model 4A; fig. 6.6). Other species classification accuracy values dropped for both SWIR (3A to 3D) and VIS (4A to 4D) band models (*L. procera* and *B. crispa*), suggesting at least some benefit from using bands from both regions.

The models derived from indices and bands classified the species very well (fig. 6.6e). Each of the species had classification accuracies greater than 94% for at least one sample—even *D. panamensis*, which had values ranging from 52 to 100%. The remaining species had values greater than 76%, with the sum of their four classification accuracies being greater than 345%.

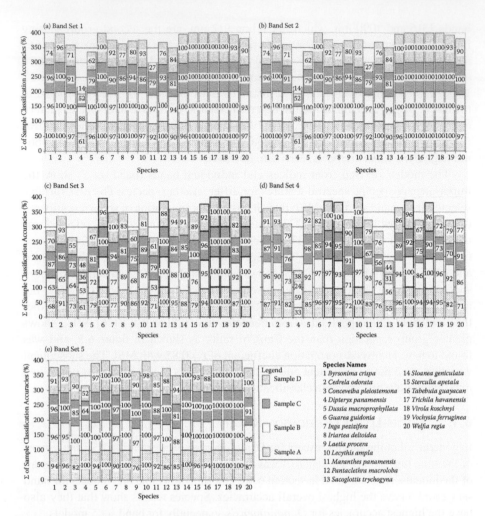

FIG. 6.6 Species-level differentiation results for band sets 1 to 4 and models A to D. Numbers within each bar represent the identification accuracy for that specified species and sample. Bars in bold highlight species that were classified accurately despite the limited range of bands belonging to those identification models.

6.5　DISCUSSION

The main objective of this study was to determine if 20 tropical tree species, important to the habitat of the great green macaw, could be differentiated using leaf spectral properties. All of the species were classified with greater than 90% accuracy by at least one model. One of the species, *D. panamensis*, was the most difficult to classify, but this was likely due to the possible immaturity (light coloring and structural flexibility) of 3 of the 13 sample leaves. Despite this difficulty the *D. panamensis* leaves were classified with 100% accuracy when indices were used as variables within the classification model. Other species such as *W. regia* and *T. guayacan* also showed VIS pigment variations but were classified accurately.

Four samples drawn from the same data pool were used to evaluate the robustness of Wilk's stepwise discriminant analysis for selecting variables from the wavelength region 350 to 2500 nm, a relatively large data set with 2150 input bands. Results indicate that the first bands selected were the most consistent between samples and overall classification results varied by a maximum of 15% between samples (table 6.3). Thus, stepwise discriminant analysis is robust to small amounts of variability for the first bands selected but remaining band combinations will likely be affected and can also affect classification results.

A shortcoming of this discriminant analysis is that, when applied to reflectance data, the spectral shape, which can include diagnostic features [36], is not exploited to its full extent. This is evident by the subtle increase in classification accuracy when indices, which exploit these diagnostic features, were added to the classification models. Also, indices were selected before individual bands, meaning that they had better discriminatory ability. Another shortcoming to this analysis was indicated by the selection of several correlated chlorophyll indices within single models. This problem can be managed by altering the tolerance criteria within SPSS; for this study the band selection constraints were kept constant for all of the models.

The spectral range of input bands was manipulated in order to create classification models more appropriate for data affected by atmospheric absorption and sensor wavelength limitations. The resulting models were used to interpret capabilities of indices and VIS and SWIR wavelengths separately or in combination for classification. When SWIR bands were included, resultant models first included bands associated with water absorption. In particular, bands were repeatedly selected near or at the 1430-nm water absorption feature. MSI, which was correlated with canopy moisture stress by Rock et al. [30], was selected first from all of the indices and bands listed for band set 5 (table 6.3). Thus, it seems that for the species of this study, leaf water content is the best individual variable for classification.

Similarly, Roche, Diaz-Burlinson, and Gachet [37] found that leaf water content (percent of total leaf weight or its inverse leaf dry matter content) proved to be the most reliable variable for differentiating 10 plant species at eight sites in southern Mediterranean France when compared to eight other leaf traits (area, fresh weight, dry weight, volume, density, thickness, specific leaf area, leaf nitrogen content). Leaf water content also had a significant correlation with average minimum temperature for each site. Leaf water content and climate variable relations may be different for tropical regions, although Geeske, Aplet, and Vitousek [38] showed that elevation affects leaf size and leaf mass per area (LMA) for trees in Hawaii.

Models with only VIS and NIR bands up to 1075 nm were insufficient to separate all 20 species studied, but despite the lack of strong water bands, some species were separated accurately (fig. 6.6). The best model derived from this region contained a band in the NIR plateau, several bands along the red edge, and some bands near the green peak. NIR and red edge bands were the most consistently selected and classifications without them had some of the lowest accuracies (models 1C, 1D, 3A, 3B, 3C, and 3D). NIR plateau bands, which were selected first from this band set and second from many other band sets, have been associated with scattering of light from within the leaf [39]; thus, species with leaf internal structure differences may be the cause of these band selections. It has been shown by Datt [33] that red edge bands are strongly associated with chlorophyll concentration; therefore, variation in chlorophyll concentrations caused by speciation may be the cause for selection of these bands. The inclusion of chlorophyll indices in other models reaffirms this point.

Models with only SWIR and NIR bands, like models with only VIS and NIR bands, also had lower classification accuracies, but again not for all species. In fact, the list of species with low classification accuracies was different for each model type; thus, when VIS, NIR, and SWIR bands were combined, models had higher overall classification accuracies (models 2A to 2D). Models derived from all bands and indices linked to biochemical features, including water content, and pigment content had the second highest classification accuracies and the highest accuracies for the otherwise poorest classifying species *D. panamensis*. It appears for this group of species that several bands or indices representing water absorption, pigment absorptions, and NIR scattering are necessary to classify all of the species in this study accurately, and that these representations can be accomplished with several combinations of bands and/or indices.

For other groups of species, studies have found different ranges of bands to be useful for classification. Fung and Siu [20] tested wavelengths of 400–900 nm for classifying tree leaves from 12 subtropical (Hong Kong) tree species and found that 7 out of 12 were classified with accuracies greater than 90%. Gong et al. [8] also tested VIS and NIR bands for classification, but of tree canopies for six conifer species, and found that VIS bands had higher discriminatory power than NIR bands. van Aardt and Wynne [7], who tested bands of 350–2500 nm, show different band sets useful for separating canopies of three pine tree species from each other and three hardwood tree species from each other, and for separating these pines from these hardwood trees. The hardwood trees, which are most similar to the 20 species used for this study compared to coniferous trees, were separated with bands largely in the VIS and NIR with few bands in the SWIR regions; notably, there was not a preference for water bands. These classification examples, in addition to the results from this study, demonstrate several wavelength combinations that are possibly only relevant for the species under investigation. Some of the wavelength selection differences could also result from differences between leaf and canopy measurements.

6.6   CONCLUSIONS

The results from this leaf scale investigation indicate that Wilk's linear discriminant analysis is an effective technique for selecting reliable bands/indices for classification of leaf data. Classification of 20 tropical tree species using these selected bands/indices resulted in high classification accuracies ranging from 79 to 97%. It was found that models with the highest accuracies included bands or indices correlated with water content (especially 1450-nm band and MSI), followed by pigmentation properties (red edge bands, green peak, and chlorophyll indices were selected). While classification accuracies were high when the 350- to 2500-nm spectrum was exploited, some species were classified accurately (up to 100%) using only NIR and SWIR bands (1076–2500 nm; without bands affected by atmospheric absorption) or using only VIS and NIR bands (350–1075 nm). Furthermore, published investigations of other species groups show different band/indices significance patterns for classification, suggesting that spectral differences can differ between species groups. Future studies investigating species differentiation should analyze bands/indices within the entire 350- to 2500-nm region, as this study suggests that exclusion of any of these wavelengths may reduce classification accuracy.

It was found that indices linked to leaf water content and chlorophyll concentration had better discriminating potential than individual bands between 350 and 2500 nm. Index-based models resulted in improved classification for some species and high overall classification accuracies. Consequently, continued research focusing on leaf/canopy biochemical and biophysical indices/variables should also increase species detection capabilities. The ability to detect species would have significant ramifications for the fields of ecology, conservation biology, and forest conservation and management.

REFERENCES

1. Penuelas, J. and Filella, I., Visible and near-infrared reflectance techniques for diagnosing plant physiological status, *Trends in Plant Science*, 3, 151, 1998.
2. Turner, W., Spector, S., Gardiner, N., et al., Remote sensing for biodiversity science and conservation, *Trends in Ecology and Evolution*, 18, 306, 2003.
3. Castro-Esau, K., Sanchez-Azofeifa, G.A., Rivard, B., et al., Variability in leaf optical properties of mesoamerican trees and the potential for species classification, *American Journal of Botany*, 93, 517, 2006.
4. Clark, D.A., Roberts, D.A., and Clark, D.A., Hyperspectral discrimination of tropical rain forest tree species at leaf to crown scales, *Remote Sensing of Environment*, 96, 375, 2005.
5. Zhang, J., Rivard, B., Sanchez-Azofeifa, G.A., et al., Intra- and interclass spectral variability of tropical tree species at La Selva, Costa Rica: Implications for species identification using HYDICE imagery, *Remote Sensing of Environment*, 105, 129, 2006.
6. Martin, M.E., Newman, S.D., Aber, J.D., et al., Determining forest species composition using high spectral resolution remote sensing data, *Remote Sensing of Environment*, 65, 249, 1998.
7. van Aardt, J.A. and Wynne, R.H., Spectral separability among six southern tree species, *Photogrammetric Engineering and Remote Sensing*, 67, 1367, 2001.
8. Gong, P., Ru, R., and Yu, B., Conifer species recognition: An exploratory analysis of in situ hyperspectral data, *Remote Sensing of Environment*, 62, 189, 1997.
9. Shen, S.S., Badhwar, G.D., and Carnes, J.G., Separability of boreal forest species in the Lake Jennette area, Minnesota, *Photogrammetric Engineering and Remote Sensing*, 51, 1775, 1985.
10. Gates, D.M., Keegan, H.J., Schleter, J.C., et al., Spectral properties of plants, *Applied Optics*, 4, 11, 1965.
11. Curran, P.J., Dungan, J.L., Macler, B.A., et al., Reflectance spectroscopy of fresh whole leaves for the estimation of chemical concentration, *Remote Sensing of Environment*, 39, 153, 1992.
12. Carter, G. and Young, D.R., Responses of leaf spectral reflectance to plant stress, *American Journal of Botany*, 154, 239, 1993.
13. Cao, K.F., Leaf anatomy and chlorophyll content of 12 woody species in contrasting light conditions in a Bornean heath forest, *Canadian Journal of Botany*, 78, 1245, 2000.
14. Bell, I.E. and Baronoski, G.V.G., Reducing the dimensionality of plant spectral databases, *IEEE Transactions on Geoscience and Remote Sensing*, 42, 570, 2004.
15. Tsai, F. and Philpot, W., Derivative analysis of hyperspectral data, *Remote Sensing of Environment*, 66, 41, 1998.

16. Kaewpijit, S., Le Moigne, J., and El-Ghazawi, T., Automatic reduction of hyperspectral imagery using wavelet analysis, *IEEE Transactions on Geoscience and Remote Sensing*, 41, 863, 2003.

17. Kalacska, M., Sanchez-Azofeifa, G.A., Rivard, B., et al., Ecological fingerprinting of ecosystem succession: Estimating secondary tropical dry forest structure and diversity using imaging spectroscopy, *Remote Sensing of Environment*, 108, 82, 2007.

18. Luther, J.E. and Carrol, A.L., Development of an index of balsam fir vigor by foliar spectral reflectance, *Remote Sensing of Environment*, 69, 241, 1999.

19. Shaw, D.T., Malthus, T.J., and Kupiec, J.A., High-spectral resolution data for monitoring Scots pine (*Pinus Sylvestris* L) regeneration, *International Journal of Remote Sensing*, 19, 2601, 1998.

20. Fung, T. and Siu, W.L., Hyperspectral data analysis for subtropical tree species recognition, in *IGARSS '98 Sensing and Managing the Environment, IEEE International Geoscience and Remote Sensing*, Seattle, WA, 1998.

21. BirdLife International, *Ara ambigua*, in *IUCN red list of threatened species*, IUCN, Cambridge, 2004.

22. Powell, G., Wright, P., Guindon, C., et al., *Research findings and conservation recommendations for the great green macaw (Ara ambigua) in Costa Rica*, unpublished report, 1999.

23. Foley, S., Rivard, B., Sanchez-Azofeifa, G.A., et al., Foliar spectral properties following leaf clipping and implications for handling techniques, *Remote Sensing of Environment*, 103, 265, 2006.

24. Huberty, C.J., *Applied discriminant analysis: Wiley series in probability and mathematical statistics*. John Wiley & Sons, New York, 1994.

25. Le Maire, G., Fancois, C., and Dufrene, E., Towards universal broad leaf chlorophyll indices using prospect simulated database and hyperspectral reflectance measurements, *Remote Sensing of Environment*, 89, 1, 2004.

26. Sims, D.A. and Gamon, J.A., Estimation of vegetation water content and photosynthetic tissue area from spectral reflectance: A comparison of indices based on liquid water and chlorophyll absorption features, *Remote Sensing of Environment*, 84, 526, 2003.

27. Penuelas, J., Filella, I., and Gamon, J., Assessment of photosynthetic radiation use efficiency with spectral reflectance, *New Phytologist*, 131, 291, 1995.

28. Penuelas, J., Baret, F., and Filella, I., Semi-empirical indices to assess carotenoids/chlorophyll a ratio from leaf spectral reflectance, *Photosynthetica*, 31, 221, 1995.

29. Gao, B.C., NDWI: A normalized difference water index for remote sensing of vegetation liquid water from space, *Remote Sensing of Environment*, 58, 257, 1996.

30. Rock, B.N., Vogelmann, J.E., Williams, D.L., et al., Remote detection of forest damage, *BioScience*, 36, 439, 1986.

31. Nagler, P.L., Inoue, Y., Glenn, E.P., et al., Cellulose absorption index (CAI) to quantify mixed soil–plant litter scenes, *Remote Sensing of Environment*, 87, 310, 2003.

32. Richard, A.G., *An introduction to probability and statistics using Basic*. Marcel Dekker, New York, 1979.

33. Datt, B., Visible/near infrared reflectance and chlorophyll content in *Eucalyptus* leaves, *International Journal of Remote Sensing*, 20, 2741, 1999.

34. Sims, D.A. and Gamon, J.A., Relationships between leaf pigment content and spectral reflectance across a wide range of species, leaf structures and developmental stages, *Remote Sensing of Environment*, 81, 337, 2002.

35. Penuelas, J., Filella, I., Biel, C., et al., The reflectance at the 950–970 nm region as an indicator of plant water status, *International Journal of Remote Sensing*, 14, 1887, 1993.

36. Curran, P.J., Dungan, J.L., and Peterson, D.L., Estimating the foliar biochemical concentration of leaves with reflectance spectrometry testing the Kokaly and Clark methodologies, *Remote Sensing of Environment*, 76, 349, 2001.

37. Roche, P., Diaz-Burlinson, N., and Gachet, S., Congruency analysis of species ranking based on leaf traits: Which traits are the more reliable? *Plant Ecology*, 1174, 37, 2004.
38. Geeske, J., Aplet, G., and Vitousek, P., Leaf morphology along environmental gradients in Hawaiian *Metrosideros polymorpha, Biotropica*, 26, 17, 1994.
39. Slaton, M.R., Hunt, E.R., Jr., and Smith, W.K., Estimating near-infrared leaf reflectance from leaf structural characteristics, *American Journal of Botany*, 88, 278, 2001.
40. Gentry, H.A., *A field guide to the families and genera of woody plants of Northwest South America, Columbia, Ecuador, Peru.* Conservation International, Arlington, VA, 1993.
41. Mabberly, D.J., *The plant book.* Cambridge University Press, Cambridge, 1990.

7 Discriminating *Sirex noctilio* Attack in Pine Forest Plantations in South Africa Using High Spectral Resolution Data

Riyad Ismail, Onisimo Mutanga, and Fethi Ahmed

CONTENTS

7.1 INTRODUCTION

The wood-boring pest *Sirex noctilio* Fabricius (Hymenoptera: Siricidae) [1] is causing mortality along the heavily afforested eastern regions of South Africa, with recent reports indicating that mortality might be as high as 30% in some forestry compartments [2]. *S. noctilio* affects all commercial pine species in South Africa with none of the species showing a high resistance to attack [3]. Based on recent bioclimatic studies [4], the wasp is likely to spread farther north, where the majority of South Africa's commercial pine forests are located (fig. 7.1).

Management strategies by South African forest companies now focus on the combined use of remote sensing, silvicultural treatments, and biological control to reduce *S. noctilio* population numbers and to minimize the potential economic threat to the industry [5]. Remote sensing is a key component of the integrated management

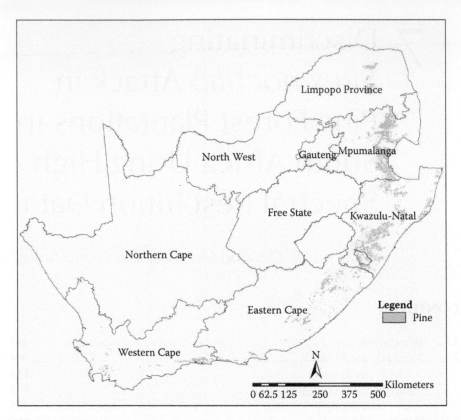

FIG. 7.1 Map showing the distribution of pine (all commercial species) in South Africa. There are approximately 721,000 ha of commercial pine plantations, with plantations concentrated in Limpopo Province (3.9%), Western Cape (12.1%), Eastern Cape (14.6%), Kwazulu-Natal (25.4%), and Mpumalanga (44%) [38]. See CD for color image.

strategy and remains crucial for the detection and monitoring of the wasp and for the effective deployment of appropriate suppression activities [6].

However, there are operational limitations that restrict the successful implementation of remote sensing by South African forestry companies. These limitations are primarily due to classification errors that arise due to the inability of broadband multispectral sensors to discriminate between the different damage classes associated with *S. noctilio* attack. *S. noctilio* symptoms can be represented on a damage scale as the green, red, and gray stages of attack (fig. 7.2).

Initial evidence of attack, or the green stage of attack, includes the appearance of resin droplets and the presence of ovipositors on the bark with a dark fungal stain appearing along the cambium. There is minimal needle loss and the canopy appears green and healthy. The red stage of attack occurs approximately 3 months later when the canopy of the attacked tree wilts and changes color from green to yellow to reddish brown [7]. Ultimately, during the gray stage of attack, the canopy defoliates and round exit holes appear on the bark. A new generation of adult wasp emerges, resulting in a compartment of scattered pattern of dead or dying trees [7–9]. During the gray stage of attack the wood is totally desiccated and the timber is nonutilizable

Stages of Attack		Symptoms
Green	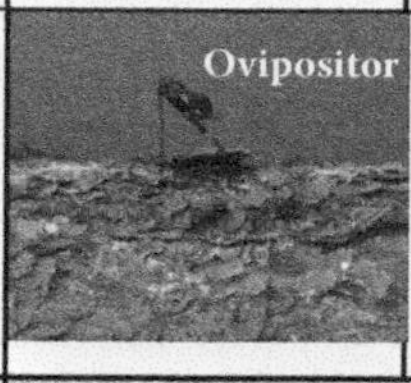 Ovipositor	Green, healthy crown, presence of resin droplets, cambium stain, ovipositors found on the trunk, and no substantial needle loss
Red		Severe chlorosis, reddish-brown canopy, and high needle loss
Gray		Emergence holes, no canopy, and 100% needle loss

FIG. 7.2 Description of the damage symptoms due to *S. noctilio* attack. See CD for color image.

and economic losses are incurred. It has been shown that the spectral separability between these different classes of damage remains elusive when high spatial resolution broadband multispectral sensors are used [6].

The limitations associated with the use of broadband multispectral channels are primarily due to classification errors that arise due to the inability of multispectral sensors to accurately discriminate between healthy, green-, and red-stage trees. For example, it has been shown that the spectral separability between healthy-green and green-red trees remains problematic when using high spatial resolution (0.5 m) multispectral sensors [6]. Given these limitations, there is strong optimism that with the new generation of hyperspectral sensors, significantly higher quality data will be available [10], thus allowing for the detection of detailed features using many narrow bands that would have been otherwise masked by broadband sensors [11].

Internationally, the number of airborne and space-borne hyperspectral sensors has increased [12,13]. However, due to cost, availability, and accessibility of hyperspectral imagery in South Africa, only a few studies have investigated the potential of using high spectral resolution data. For example, Mutanga [14] and Mutanga and Skidmore [15] successfully used HYMAP imagery to map grass quantity and quality in savanna ecosystems. Using the same imagery, Ferwerda [16] mapped the total content of polyphenols and condensed tannins in *Colophospermum mopane* trees in an effort to understand herbivore distributions in the Kruger National Park. Studies focusing on the use of hyperspectral imagery for forestry purposes are limited with the exception of Ahmed and Mthembu [17], who calculated the leaf area index (LAI) of *Eucalyptus grandis* trees using space-borne hyperspectral imagery (HYPERION) and narrowband vegetation indices.

However, there should be an increased interest in using high spectral resolution data for a wide variety of environmental applications due to the future availability and accessibility of hyperspectral sensors in the southern African region. It is envisaged that the South African satellite, Sumbandila (ZASat-002), is due for launch in 2007 from a Russian submarine, followed by ZASat-003 to be launched in 2008 [18]. ZASat-003 will carry a full multisensor microsatellite imager (MSMI) instrument as well as a hyperspectral sensor with a 14.9-km swath and 14.5-m ground sampling distance. This hyperspectral sensor will slice the spectrum between 400 and 2350 nm into 200 bands, each 10 nm wide [18,19]. The question that then arises is whether, with the future availability and accessibility of high spectral resolution data in South Africa, is there potential to successfully discriminate between healthy and *S. noctilio* attacked pine trees?

Therefore, the preliminary aim of this study was to use high spectral resolution data to identify diagnostic spectral features of *Pinus patula* needles showing varying degrees of *S. noctilio* attack. The results obtained from this study could thus form the basis of future algorithms or spectral indices capable of discriminating *S. noctilio* attacked trees from healthy trees at either airborne or space-borne platforms. More specifically, the objectives of this chapter are (1) to determine whether there is a significant difference between the mean reflectance (percent) at each measured band (from 400 to 1300 nm) for the three stages of *S. noctilio* attack (green, red, and gray), and (2) for the wavelengths that are spectrally different in this region, to test whether some bands have more discriminating power than others in the detection of *S. noctilio* induced stress. Reflectance measurements were taken from three groups of *Pinus patula* trees, classified according to the severity of attack by *Sirex noctilio* in the forestry compartments of Kwazulu-Natal Province, South Africa, using a field spectrometer.

7.2 MATERIALS AND METHODS

7.2.1 FOLIAR SAMPLES

During April 2006, needle samples from healthy, green-, and red-stage *Pinus patula* trees were collected from a known *S. noctilio* attacked compartment at the Pinewoods Plantation (centroid 30°4'13.83" E and 29°38'36.06" S) Kwazulu-Natal, South Africa. All pine species are susceptible to attack [3]; however, only *P. patula* trees have been attacked in Kwazulu-Natal. Before any spectral measurements were taken, the trees were carefully examined with the assistance of experienced foresters and classified into mutually exclusive classes (i.e., healthy, green, or red). Green-stage trees were checked for the presence of ovipositors, resin droplets, and the cambium staining. Additionally, trees that were classified as red stage were destructively sampled to evaluate the presence or absence of *S. noctilio* larvae. The gray stage of attack was excluded from this study since gray-stage trees are completely defoliated and therefore no needle samples can be collected.

Five trees from each stage, including the healthy trees, were then sampled. Pine needles were collected from three branches (upper, middle, and lower crowns) with two needle samples from the same branch. Sampled needles used for spectral measurements were from the same age trees and no other damaging agents were observed on the pine needles. The samples ($n = 90$) were immediately sealed in resealable plastic bags and placed in coolers on ice and taken to the laboratory at the University of Kwazulu-Natal for spectral measurements within 4 hours of collection.

7.2.2 Spectral Data Acquisition

Spectral measurements of pine needles were acquired using Analytical Spectral Devices' (ASD) FieldSpec Pro FR spectroradiometer, which senses in the spectral range 350–2500 nm at a spectral bandwidth of 1.4–2.0 nm and a spectral resolution of 3–10 nm [20]. This spectral range incorporates the visible (400–700 nm), near infrared (NIR) (700–1200 nm), and the shortwave infrared (SWIR) (1200–2500 nm). The ASD spectroradiometer, equipped with a field of view of 25°, was mounted on a tripod and positioned 0.5 m above the needle sample at the nadir position. A 150-W halogen bulb was used as the light source to illuminate the pine needle leaves. Reflectance spectra were obtained by calibrating the radiance of the target pine needles with the radiance of a standard (white reference panel, Spectralon™) of known spectral characteristics. Needle samples for the different classes (healthy, green, and red) were randomly stacked on a target platform. Ten reflectance measurements were taken while the needles were shuffled and this process was repeated for each sample ($n = 90$). The entire experiment was conducted under controlled laboratory conditions (i.e., dark room, 25°C) in order to avoid ambient light sources unrelated to the true spectral signal of the needles [21]. In total, 300 reflectance values were acquired for each class but these were later averaged in order to reduce within-class variability.

7.2.3 Data Analysis

The hypothesis that the mean reflectances between healthy, green, and red stages were significantly different at each measured wavelength in a 350- to 1300-nm region was tested using one-way analysis of variance (ANOVA) with a Tukey's HSD post hoc test. The ANOVA with the post hoc test was calculated at each measured wavelength for the respective class pair (i.e., healthy-green, green-red, and healthy-red, hereafter referred to as H-G, G-R, and H-R) and then summarized using a histogram. The histogram was calculated by counting the number of significant bands at each wavelength for all class pairs. The histogram then indicates the frequency of significant wavelengths and which wavelengths are relatively more important for discriminating all classes (i.e., healthy, green, and red, hereafter referred to as H-G-R). Additionally, to identify wavelengths most responsive to *S. noctilio* attack, sensitivity analyses were undertaken following the procedure described by Carter [22] and Cibula and Carter [23]

$$R_\lambda = (R_{\lambda i} - R_{\overline{\lambda h}}) / R_{\overline{\lambda h}} \tag{7.1}$$

where $(R_{\lambda i} - R_{\overline{\lambda h}})$ is the spectral reflectance difference curves calculated by subtracting the mean reflectance of the healthy pine needles at each wavelength from the mean reflectance of the green- or red-stage needles.

When the reflectance of healthy needles is subtracted from the reflectance of the green- or red-stage needles, the resulting difference curves indicate the wavelengths in which reflectance changed greatly with stress [24]. When this difference curve is divided by the mean reflectance of the healthy needles, the results yield the relative change in reflectance or reflectance sensitivity [25]. Sensitivity analysis therefore shows the wavelengths at which reflectance was strongly affected by *S. noctilio* attack. We also calculated the coefficient of variation (CoV) at each wavelength for reflectance values across all three classes. The CoV is the ratio of the standard deviation to the mean reflectance [25]. Higher CoV values result in greater variability and hence discriminatory power of the respective wavelength.

Finally, we tested the hypothesis that some bands have more discriminatory power than others by calculating the Jeffries–Matusita (JM) distance [26]. The JM distance calculation delivers a value between 0 and $\sqrt{2}$ ($\approx$1.414), with higher values representing better separability of class pairs [26]. Therefore, the band or band combinations producing the highest JM distance averaged over each class pair can be considered the best for discriminating *S. noctilio* attack. Although previous studies used JM distance thresholds of ≥95% [21] to indicate separability, in this study we opted to use higher separability values (≥99%) largely due to the potential of upscaling from leaf to tree canopy and the variability associated with canopy reflectance when compared to leaf samples [27].

The Jeffries–Matusita distance is formally stated as

$$\alpha = \frac{1}{8}\left(\mu_i - \mu_j\right)^T \left(\frac{C_i + C_j}{2}\right)^{-1} \left(\mu_i - \mu_j\right) + \frac{1}{2}\ln\left(\frac{\left(\left|C_i + C_j/2\right|\right)}{\sqrt{\left|C_i\right| * \left|C_j\right|}}\right) \quad (7.2)$$

For normally distributed classes, this distance becomes the Bhattacharyya (BH) distance [26] and is stated as

$$JM_{ij} = \sqrt{2\left(1 - e^{\alpha}\right)} \quad (7.3)$$

where

  i and j = the two classes being compared
  C_i = the covariance matrix of signature i
  μ_i = the mean vector of signature i
  ln = the natural logarithm function
  $\left|C_i\right|$ = the determinant of C_i (matrix algebra)

7.3   RESULTS

The results of the ANOVA for individual class combinations (H-G, G-R, and H-R) are shown in figure 7.3 (a, b, and c). The shaded areas indicate the reflectance wavelengths where the class pairs show a significant statistical difference in reflectance ($p < .001$).

Table 7.1 shows the frequency of statistically significant wavelengths for spectral regions as defined by Gong, Pu, and Heald [28]. These spectral regions were

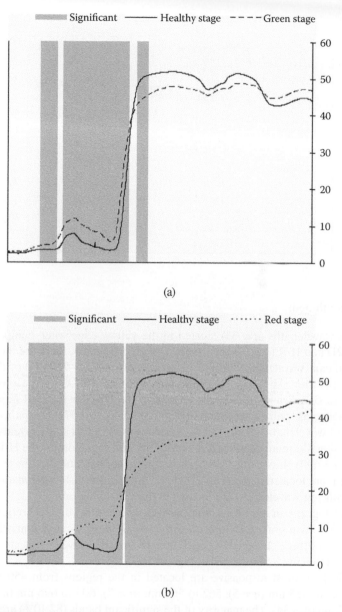

FIG. 7.3 ANOVA results for individual class pairs: (a) healthy green; (b) healthy red; and (c) green-red. The gray shades indicate regions of electromagnetic spectrum where there were significant differences.

selected to simplify the spectral interpretation between individual class pairs. There are no significant wavelengths located in the blue range; however, the blue edge is very responsive to the G-R and H-R class pairs. Additionally, the H-R combination has more significant wavelengths ($n = 60$) located in the green range than any other class pairs (H-G, $n = 40$, and G-R, $n = 25$), whereas the H-G combination has more

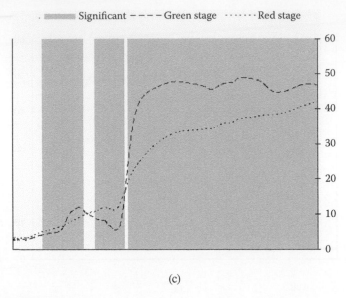

(c)

FIG. 7.3 (continued)

significant wavelengths ($n = 33$) located in the yellow edge when compared to the G-R ($n = 21$) and H-R ($n = 19$) class pairs. For all class pairs, there are more statistically significant wavelengths located in the red region (600–700 nm) of the electromagnetic spectrum (H-G, $n = 100$; H-R, $n = 100$; and G-R, $n = 93$). However, there are far more significant wavelengths located in the NIR for the G-R ($n = 594$) and H-R ($n = 450$) class pairs when compared with the H-G ($n = 57$) class pair. An examination of all wavelengths (from 350 to 1300 nm) used in this study reveals that there are proportionally more significant wavelengths in this region for the H-R ($n = 814$) and G-R ($n = 699$) class pairs. For the H-R combination, 62.68% of these significant wavelengths are located in the visible region, while for the G-R combination, 73.50% of the significant wavelengths are located in the visible region.

The histogram in figure 7.4 summarizes the results of the frequency table (table 7.1). The histogram then indicates which wavelengths can potentially discriminate among all three classes (H-G-R). These responsive wavelengths are defined by regions with the maximum gray shading shown in figure 7.4.

Wavelengths most responsive are located in the regions from 450 to 500 nm ($n = 51$), 521 to 525 nm ($n = 5$), 562 to 568 nm ($n = 7$), 604 to 696 nm ($n = 93$), and 750 to 783 nm ($n = 34$). The majority of the significant bands (82.10%) are located in the visible part (300–700 nm) of the electromagnetic spectrum, with the remainder of the significant wavelengths located in the shoulder region of the NIR.

7.3.1 SENSITIVITY ANALYSIS

Sensitivity analysis was used to reduce the number of significant bands prior to the JM calculation. Therefore, sensitivity analysis was restricted to regions that were identified as significant for all three classes. These significant areas are shaded in gray in figure 7.5. It was not possible to calculate the JM distance by using all significant

TABLE 7.1

Frequency Table of Statistically Significant Bands for the Spectral Regions[a]

Wavelength region (nm)	Description	No. of bands	H–G	%	G–R	%	H–R	%
350–400	Blue range	51	0	0.00	0	0.00	0	0.00
490–530	Blue edge	41	21	51.22	38	92.68	41	100.00
501–560	Green range	60	40	66.67	25	41.67	60	100.00
550–582	Yellow edge	33	33	100.00	21	63.64	19	57.58
640–680	Red well	41	41	100.00	41	100.00	41	100.00
670–737	Red edge	68	54	79.41	58	85.29	58	85.29
700–900	NIR	201	58	28.86	191	95.02	194	96.52
350–700	Visible	351	231	65.81	248	70.66	220	62.68
350–1300	All wavelengths	951	288	30.28	699	73.50	814	85.59

Note: The table shows the results for individual class pairs.

[a] Defined by Gong, P. et al., *International Journal of Remote Sensing*, 23, 1827, 2002.

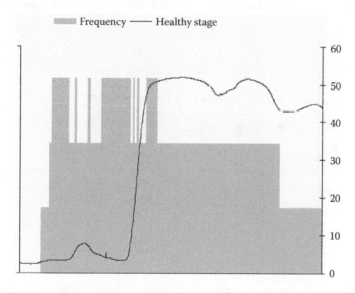

FIG. 7.4 Frequency of statistically significant differences for all classes (healthy, green, and red). The maximum gray shadings indicate the wavebands that could discriminate between all damage class combinations. The spectral signature for the healthy needle is included for comparative purposes.

bands ($n = 190$) because of the singularity problem of matrix inversion [21]. Additionally, spectral information needed to discriminate among healthy, green, and red stages could be extracted from a few optimal bands since many bands are highly correlated and often redundant [13].

Figure 7.5 shows the results of the sensitivity analysis. The wavelengths most responsive to *S. noctilio* attack are shown using black arrows, indicating wavelengths

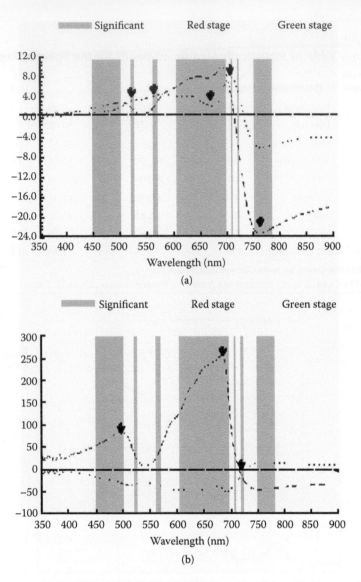

FIG. 7.5 Sensitivity analysis. (a) Reflectance difference curves for the red and green stages of attack. The curves were calculated by subtracting the mean reflectance of the red- as well as the green-stage needles from the reflectance of healthy needles. (b) The result of the sensitivity analysis. The curves were calculated by dividing the reflectance difference curves of the red- and green-stage needles from the reflectance of healthy needles.

that were selected for further analysis. Sensitivity maxima located at 500 and 695 nm followed by a wavelength located at 720 nm were identified from the sensitivity curves (fig. 7.5b), while wavelengths located at 521, 565, and 707 nm were identified from the spectral difference curves (fig. 7.5a). Minima located at 685 and 760 nm were also identified from these difference curves. Table 7.2 shows the results of the CoV calculated for the wavelengths. Noticeable are the high CoV values for the 500-,

TABLE 7.2
Coefficient of Variation (CoV) Results for the Nine Selected Bands

Band	nm	Description	CoV
1	500	Blue edge	30.10
2	521	Blue edge	24.21
3	565	Yellow edge	25.12
4	685	Red edge	56.26
5	690	Red edge	51.79
6	695	Red edge	42.86
7	707	Red edge	24.59
8	720	Red edge	19.38
9	760	NIR	25.01

TABLE 7.3
Results of the Average Jefferies–Matusita Distance Analysis for the Three Classes (Healthy, Green, and Red)

Best combination	500	521	565	685	690	695	707	720	760	J–M value	%
Single band				×						1.186	83.88
Two band							×	×		1.354	95.76
Three band	×	×	×							1.392	98.44
Four band	×	×	×						×	1.404	99.29
Five band	×	×		×	×				×	1.410	99.72
Six band	×	×	×	×	×				×	1.412	99.86
Seven band	×	×	×	×	×	×			×	1.413	99.93
Eight band	×	×	×	×	×		×	×	×	1.413	99.93
Nine band	×	×	×	×	×	×	×	×	×	1.413	99.93

Note: The symbol "×" indicates the optimal bands that were selected in each band combination.

685-, 690-, and 695-nm wavelengths located in the red edge region of the electromagnetic spectrum.

7.3.2 DISTANCE ANALYSIS

Based on the results from the sensitivity and CoV analysis, nine spectral bands (table 7.3) were selected for further analysis. Jefferies–Matusita distances [26] were then calculated to determine the best combinations of bands for separating the classes (i.e., healthy, green, and red) from each other. Jefferies–Matusita distances range from 0 to 1.414, with higher values representing better separability of all class pairs (H-G, G-R, and H-R). Therefore, the bands or band combinations producing the highest JM distance averaged over all class pairs can be considered optimal for discrimination.

TABLE 7.4

JM Values for Individual Class Pairs Using All Possible Band Combinations

Best combination	H-G	%	H-R	%	G-R	%
Single band	1.240	87.69	1.404	99.29	1.224	86.56
Two band	1.339	94.70	1.414	100.00	1.368	96.75
Three band	1.370	96.89	1.414	100.00	1.395	98.66
Four band	1.400	99.01	1.414	100.00	1.403	99.22
Five band	1.408	99.58	1.414	100.00	1.408	99.58
Six band	1.410	99.72	1.414	100.00	1.411	99.79
Seven band	1.411	99.79	1.414	100.00	1.413	99.93
Eight band	1.412	99.86	1.414	100.00	1.413	99.93
Nine band	1.413	99.93	1.414	100.00	1.414	100.00

Table 7.4 shows the resulting JM distance averaged for all combined class pairs. The best separability obtained when using a single band located at 685 nm produces an unacceptable JM value of 1.186. However, using more bands considerably improves the separability of the classes. The best average JM values (1.413) are reached when using a seven-band combination or greater, with individual bands located at 500, 521, 565, 685, 690, 695, and 760 nm. However, using a four-band combination comprising bands located at 500, 521, 685, and 760 nm produces an acceptable JM value of 1.401 or 99.29% separability. Additionally, the frequencies of these bands, as shown in table 7.4, are relatively high when compared to other band combinations.

7.4 DISCUSSION

Numerous studies have examined the ability of hyperspectral data to discriminate the damage caused by forest pests and pathogens [25,27,29–34]. However, this study was the first to report on *S. noctilio* attacking *P. patula* foliage, with results indicating that hyperspectral reflectance data measured in a laboratory environment can successfully discriminate varying needle-level damage. Results show that there is a significant difference between the mean reflectance for all three classes (H-G-R), with a large number of significant wavelengths located in the visible region of the electromagnetic spectrum. More specifically, results show that at least 77% of significant wavelengths are located from 450 to 500 nm ($n = 51$) and from 604 to 696 nm ($n = 93$). The results obtained in this study corroborate previous leaf-level studies that have shown that reflectance measured at visible wavelengths (400–700 nm) is generally the most consistent response to stress [22,24,32].

Although chlorophyll content was not directly measured in this study, it is plausible that the prevalence of significant wavelengths in the visible region, especially in the red region, is due to the effects of chlorophyll. Wavelengths within the 690- to 700-nm regions are particularly sensitive to decreases in leaf chlorophyll content and represent the blue shift of the red edge that frequently accompanies stress [22,23]. There is evidence that the combined efforts of a phytotoxic mucus and the wood-decaying fungus *Amylostereum areolatum* result in the breakdown of chlorophyll during initial *S. noctilio* attack [1,35]. It is therefore assumed that the chlorophyll concentrations will vary with the different stages of *S. noctilio* attack. This hypothesis is still speculative and would have to be further tested using foliar biochemistry analysis.

While pigments may control the spectral responses of leaves in the visible wavelengths, it is the cellular structure and water content of leaves that are the main determinants in the near- and mid-infrared regions of the electromagnetic spectrum [30]. Similar NIR spectral responses are prevalent during the later stages of attack (i.e., red stage). The red stage is characterized by the collapse of vascular tissue, which ultimately leads to water relations breaking down completely and irreversibly due to the growth of *A. areolatum* [35,36]. More specifically, results indicate that there are more significant bands located in the NIR region for the G-R ($n = 594$) and H-R ($n = 450$) class pair when compared with the H-G ($n = 57$) combination.

Sensitivity analysis was used to reduce the number of significant bands prior to the JM calculation and nine spectral bands were selected for further analysis. Two bands from the blue edge, one band from the yellow edge, five bands from the red edge, and one band from the NIR were selected. JM distances indicate only a few optimal bands are needed to produce acceptable separability results. Although no single band is capable of total separability, spectral separability between all the classes (healthy, green, and red) is possible when using a four-band combination. Bands located at 500, 521, 685, and 760 nm provided the best average separability (99.01%) for all stages of *S. noctilio* attack. Similar band combinations (500, 521, 685, and 760 nm) also produce the best separability results for individual class pairs. The results are consistent with previous studies that state that only three or four well-placed bands were needed for a good classification [37]. The results therefore encourage further investigation into the capability of using airborne and satellite hyperspectral sensors for mapping *S. noctilio* attack.

7.5 SUMMARY

To summarize, a better understanding has been gained about specific regions of the electromagnetic spectrum that offer the maximum information content for discriminating *S. noctilio* attack. It has been shown that there is a significant difference between the mean reflectance for all three classes with a large number of significant wavelengths located in the visible region of the electromagnetic spectrum. We therefore established an important prerequisite (i.e., band selection) for the potential upscaling of results to either an airborne or a space-borne platform. Although no single band can discriminate between all the stages of *S. noctilio* attack, bands located at 500, 521, 685, and 760 nm show the greatest potential. Overall, the results provide optimistic evidence that encourages canopy-scale investigation into the capability of high spectral resolution data for discriminating *S. noctilio* attack in pine forest plantations in South Africa.

REFERENCES

1. Tribe, G.D. and Cillie, J.J., The spread of *Sirex noctilio* Fabricius (Hymenoptera: Siricidae) in South African pine plantations and the introduction and establishment of its biological control agents, *African Entomology*, 12, 9, 2004.

2. Slippers, B., The *Sirex* epidemic in Kwazulu-Natal and its control, *Wood and Timber Times Southern Africa*, 31, 24, 2006.

3. Anon., *Sirex* woodwasp. FABI: Tree Protection Cooperative Program: Pretoria, 2, 2004.

4. Carnegie, A., Matsuki, M., Haugen, D.A., et al., Predicting the distribution of *Sirex noctilio* Fabricius (Hymenoptera: Siricidae), a significant exotic pest of *Pinus* plantations, *Annals of Forest Science*, 63, 119, 2006.

5. Ismail, R., Brink, A., Verleur, M., et al., An integrated approach to managing *Sirex noctilio* infestations, *The Leaflet* (Sappi Forest quarterly newsletter), 41, 7, 2005.

6. Ismail, R., Mutanga, O., and Bob, U., The use of high resolution airborne imagery for the detection of forest canopy damage by *Sirex noctilio*, in *Precision Forestry in Plantations, Seminatural Areas and Natural Forest: Proceedings of the International Precision Forestry Symposium*, Stellenbosch University, Stellenbosch, South Africa, 2006.

7. Ciesla, W.M., European woodwasp: A potential threat to North America's conifer forests, *Journal of Forestry*, 101(22), 18, 2003.

8. Haugen, D.A. and Underdown, M.G., *Sirex noctilio* control program in response to the 1987 green triangle outbreak, *Australian Forestry*, 53, 33, 1990.

9. Haugen, D.A., Bedding, R.A., Underdown, M.G., et al., National strategy for control of *Sirex noctilio* in Australia, *Australian Forest Grower*, 13, 8, 1990.

10. Kumar, L., Schmidt, K.S., Dury, S., et al., Review of hyperspectral remote sensing and vegetation science, in *Imaging spectrometry: Basic principles and prospective applications*, F. van der Meer and S.M. de Jong, eds. Kluwer Academic Press, Dordrecht, 2001.

11. Schmidt, K.S. and Skidmore, A.K., Exploring spectral discrimination of grass species in African rangelands, *International Journal of Remote Sensing*, 22, 3421 2001.

12. Wulder, M.A., Optical remote sensing techniques for the assessment of forest inventory and biophysical parameters, *Progress in Physical Geography*, 22, 449, 1998.

13. Lucas, R., Rowlands, A., Niemann, O., et al., Hyperspectral sensors and applications, in *Advanced image processing techniques for remotely sensed hyperspectral data*, P.K. Varshney and M.J. Arora, eds. Springer–Verlag, New York, 2004, 11.

14. Mutanga, O., Discriminating tropical grass canopies grown under different nitrogen treatments using spectra resampled to HyMap, *International Journal of Geoinformatics*, 1, 21, 2005.

15. Mutanga, O. and Skidmore, A., Integrating imaging spectrometry and neural networks to map grass quality in the Kruger National Park, South Africa, *Remote Sensing of Environment*, 90, 104, 2004.

16. Ferwerda, J.G., *Charting to quality of forage: Mapping and measuring the chemical composition of foliage using hyperspectral remote sensing*, ITC, Wageningen, Enschede, 2005, 183.

17. Ahmed, F. and Mthembu, I., Assessing the utility of Hyperion in extracting vegetation indices and leaf area index, in *AARSE 2006: Proceedings of the 6th AARSE International Conference on Earth Observation and Geoinformation Sciences in Support of Africa's Development*, Cairo, Egypt, 2006.

18. Scholes, B. and Annamalai, L., CSIR imaging expertise propels SA to a science high, in *Aerospace Science Scope*, 2006.

19. van Aardt, J. and Coppin, P., Current state and potential of the Is-Hs project: Integration of in situ data and hyperspectral remote sensing for plant production modeling, in *Precision Forestry in Plantations, Seminatural and Natural Forests. Proceedings of the International Precision Forestry Symposium*, Stellenbosch University, Stellenbosch, South Africa, 2006.
20. Analytical Spectral Devices. *Fieldspec Pro Users Guide.* 2002 [cited; available from: http://www.asdi.com/products-FS3.asp].
21. Vaiphasa, C., Ongsomwang, S., Vaiphasa, T., et al., Tropical mangrove species discrimination using hyperspectral data: A laboratory study, *Estuarine, Coastal, and Shelf Science*, 65, 371, 2005.
22. Carter, G.A., Ratios of leaf reflectance in narrow wavebands as indicators of plant stress, *International Journal of Remote Sensing*, 15, 697, 1994.
23. Cibula, W.G. and Carter, G.A., Identification of a far-red reflectance response to Ectomycorrhizae in slash pine, *International Journal of Remote Sensing*, 13, 925, 1992.
24. Carter, G.A. and Knapp, A.K., Leaf optical properties in higher plants: Linking spectral characteristics to stress and chlorophyll concentration, *American Journal of Botany*, 88, 677, 2001.
25. Stone, C., Chisholm, L., and McDonald, S., Spectral reflectance characteristics of *Pinus radiata* needles affected by *Dothistroma* needle blight, *Canadian Journal of Botany*, 81, 560, 2003.
26. Richards, J.A. and Jia, X., *Remote sensing digital image analysis*, 3rd ed. Springer, Berlin, 1999.
27. Stone, C., Chisholm, L., and Coops, N., Spectral reflectance characteristics of Eucalypt foliage damaged by insects, *Australian Journal of Botany*, 49, 687, 2001.
28. Gong, P., Pu, R., and Heald, R., Analysis of in situ hyperspectral data for nutrient estimation of giant sequoia, *International Journal of Remote Sensing*, 23, 1827, 2002.
29. Pontius, J., Hallet, R., and Martin, M., Assessing hemlock decline using visible and near-infrared spectroscopy: Indices comparison and algorithm development, *Applied Spectroscopy*, 59, 836, 2005.
30. Coops, N., Dury, S., Smith, M.L., et al., Comparison of green leaf Eucalypt spectra using spectral decomposition, *Australian Journal of Botany*, 50, 567, 2002.
31. Coops, N.C., Stanford, M., Old, K., et al., Assessment of *Dothistroma* needle blight of *Pinus Radiata* using airborne hyperspectral imagery, *Phytopathology*, 33, 1524, 2003.
32. Carter, G.A., Cibula, W.G., and Miller, R.L., Narrowband reflectance imagery compared with thermal imagery for early detection of plant stress, *Journal of Plant Physiology*, 148, 515, 1996.
33. Ahern, F., The effects of bark beetle stress on the foliar spectral reflectance of lodgepole pine, *International Journal of Remote Sensing*, 63, 61, 1988.
34. Ruth, B., Hoque, E., Weisel, B., et al., Reflectance and fluorescence parameters of needles of Norway spruce affected by forest decline, *Remote Sensing of Environment*, 38, 35, 1991.
35. Neumann, F.G. and Minko, G., The *Sirex* woodwasp in Australian *Radiata* pine plantations, *Australian Forestry*, 44, 46, 1981.
36. Slippers, B., Coutinho, T.A., Wingfield, B.D., et al., A review of the genus *Amylostereum* and its association with woodwasps, *South African Journal of Science*, 99, 70, 2003.
37. Leckie, D.G., Cloney, E., and Joyce, S., Automated detection and mapping of crown discoloration caused by jack pine budworm with 2.5-m resolution multispectral imagery, *International Journal of Earth Observation and Geoinformation*, 7, 61, 2005.
38. DWAF, *Commercial timber resources and primary roundwood processing in South Africa 2003/2004.* Department of Water Affairs and Forestry, South Africa, 2005.

8 Hyperspectral Remote Sensing of Exposed Wood and Deciduous Trees in Seasonal Tropical Forests

Stephanie Bohlman

CONTENTS

8.1 INTRODUCTION

Most methods for analyzing remote sensing to gain ecological insight about forested landscapes were developed in temperate, not tropical, forests. Until recently the image processing methods used to quantify ecologically meaningful, spatial, and temporal spectral variation across forested landscapes were not appropriate for mature, undisturbed tropical forests. For example, geometric-optical models that quantify variation in shadows and sunlit vegetation are ideally suited for conifer forests [1–3]. In general, tropical forests have much flatter crowns than conifers, reducing the dramatic difference in sunlit and shaded crowns that is apparent in conifer forests. Spectral mixture analysis and vegetation indices are used to detect differences in soil and green vegetation found in open forests, woodlands, and savanna [4–6]. Most humid and wet mature tropical forests have continuous vegetation coverage such that the soil is not exposed except for a brief time after gap-forming events. Leaf area indices (LAIs) in mature tropical forests are generally above the threshold where vegetation indices, such as the normalized difference vegetation index (NDVI), are responsive to leaf density changes [7]. The aforementioned techniques have been used extensively to classify the early regeneration stages of tropical forests, but, infrequently to analyze mature tropical forests.

Improvements in sensor technology and analytical methods in the past 10 years, in combination with an increased realization that tropical forests are not spectrally invariant, have provided important advances in quantifying biophysical information for mature tropical forests. Using broadband multispectral data, quite large variation in LAI has been inferred in the Amazon using MODIS data [8,9] and subtle spatial variation in reflectance from Landsat images has been correlated with differences in species diversity between sites [10]. Hyperspectral remote sensing, because of its greater spectral range and detail, provides an even better opportunity to detect subtle spectral differences relevant to species composition, biomass, or canopy functioning in mature tropical forests. On an individual crown level, high resolution hyperspectral data have shown promise in distinguishing individual species [11,12] and quantifying high and low liana coverage [13]. There is also evidence that the narrowband features of hyperspectral data can be used to detect drought stress and changes in leaf nitrogen and water content in mature tropical forests [14,15].

Spectral mixture analysis (SMA) is one method used to gain ecological information from remotely sensed images that has been little used for mature tropical forests. SMA quantifies the proportion of different material present in a landscape (called endmembers) rather than using an index or correlative approach. Because of its strong physical basis, it can be more easily compared directly against field data and extrapolated to areas outside the study site [16]. Mixture model analysis has been used to successfully classify different regeneration categories of tropical forests [17–20], but there has been limited use of SMA in mature, undisturbed tropical forest [21].

Human-altered landscapes that include pasture and young second-growth forest typically have a wide array of endmembers. Logging or slash-and-burn agriculture removes the high leaf density canopy of mature forest, revealing bare soil and wood. Pasture or crops also contain exposed soil and, especially in periods of low rainfall, senescent vegetation. Typical endmember spectra used in tropical land use studies are green vegetation, soil, and shade/shadow [18,19] and, additionally in some cases, nonphotosynthetic vegetation (NPV) [17,20], which refers to senescent vegetation and woody material.

On the other hand, mature evergreen tropical forest has much more limited range of materials visible from above—mostly green vegetation and shade/shadow [21]. Soil or NPV is only potentially visible in canopy gaps [21]. However, seasonal, semi-deciduous tropical forests contain partially or fully deciduous trees in the dry season and thus have a large amount of wood exposed at the top of the canopy. But it is uncertain if there is enough exposed wood to be detected in a high leaf density system like a tropical forest. Models suggest the contribution of wood to high LAI ecosystems is negligible [22,23]. This study addresses the question of whether NPV in the form of exposed canopy branches is an important component of seasonal tropical forests, and if it can be quantified with spectral mixture analysis.

In this study, spectral mixture analysis of hyperspectral, hyperspatial images were used to test if exposed wood and the percentage of the forest that is deciduous can be accurately quantified in a seasonal tropical forest.

8.2 METHODS

8.2.1 STUDY SITE

Parque Natural Metropolitano (8°56' N, 79°33' W) is located within the city limits of Panama City on the Pacific side of the Isthmus of Panama. It has a strong dry season from January to April. The average annual rainfall is 1850 mm with 90% on average falling in the wet season from June to November. During the dry season, 24% of canopy trees were deciduous in a nearby forest at Cocoli [26]. The dominant species under Parque Metropolitano crane are *Anacardium exelsum* and *Luehea seemanii*. There are many species that lose their leaves for the entire dry season, including *Bursera simarouba*, *Cavanillesia platanifolia*, *Calycophyllum candidissimum*, *Guazuma ulmifolia*, *Pseudobombax septenatum*, and *Spondias mombin*. In 1991, the Smithsonian Tropical Research Institute erected a 50-m-tall construction crane in the park for accessing and studying the forest canopy. Numerous studies of canopy physiology and ecology have been conducted on the tree crowns at the crane site (e.g., Meinzer et al. [27] and Mulkey, Kitajima, and Wright [28]).

8.2.2 IMAGE DATA COLLECTION

All spectral analyses were performed on images of Parque Metropolitano collected by the airborne Hyperspectral Digital Imagery Collection Experiment (HYDICE) sensor. These images, taken in the March 1998 dry season, have 1-m spatial resolution and contain 210 1.0-µm-wide wavelength bands in the range of 0.4–2.5 µm (fig. 8.1). During the flight, four wooden (painted) and four canvas calibration panels were placed in a clearing within 100 m of the crane to calibrate the images from radiance to apparent spectral reflectance using the empirical line method [29,30]. Average radiance values of each panel were extracted from the image and used to develop regression equations between image radiance and reflectance for each panel measured in the field during the HYDICE mission using a portable ASD FieldSpec spectrometer (Analytical Spectral Devices, Boulder, Colorado). The slope and intercept (gain and offset) of the regression equations were applied to each pixel in the image.

Interpretation of vegetation cover types of parts of the study area was aided by color aerial photographs taken on five dates, four in the dry season and one in the

(a)　　　　　　　　　　　　　　　　(b)

FIG. 8.1 HYDICE images of a 5-ha section of Parque Natural Metropolitano, Panama, taken in the dry season of 1998. Green polygons are nondeciduous vegetation, red polygons are deciduous crowns, and blue polygons are intercrown shade/shadows. (a) Visible color image displayed as red: 684 nm, green: 571 nm, blue: 484 nm. (b) Spectral mixture analysis results with green displayed as green vegetation, red as NPV, and blue as shade/shadows. See CD for color image.

wet season, in 2001 and 2002. These photos were taken from a helicopter flown at approximately 1000 feet and a spatial resolution of approximately 0.2 m per pixel.

8.2.3 GROUND DATA COLLECTION

For a subset of crowns visible in the HYDICE images, the species identity, the state of deciduousness, and/or percent leaf cover was determined in the field. Individual crowns visible in the images were located on the ground by taking a three-band color composite of the images in the field. The many man-made features, such as roads and buildings, visible in the images allowed navigation to individual crowns. Once a crown was co-located in the image and on the ground, it was identified to species and recorded as being deciduous or nondeciduous. Overall, 368 crowns from 29 species were identified. For 70 of these crowns that were under and near the canopy crane, the deciduousness of each crown as measured by percent leaf cover was also estimated following Condit et al. [26]. Over the whole area the locations of large patches that were where individual tree crowns were not identified to species, the locations of large patches that were predominately composed of deciduous trees, grass, *A. excelsum* trees, or *L. seemanii* trees were marked on the images.

8.2.4 DATA ANALYSIS

Reflectance data for the crowns identified in the field were extracted from the images by drawing polygons of 10–100 pixels over individual crowns. *A. excelsum,* *L. seemanii,* and *Ficus* sp. crowns were considered separately, while the crowns of other fully-leaved species were lumped together and are referred to as "other non-deciduous trees." Of the fully-leaved species, *A. excelsum, L. seemanii,* and *Ficus* sp. crowns cover the most canopy area in the study area, comprising approximately 20, 10, and 1% of the study area, respectively. All deciduous crowns together were lumped together, as previous experience has shown that individual deciduous species are not separable in the images. The average spectra for *A. excelsum, L. seemanii, Ficus* sp., deciduous, and nondeciduous trees were calculated from 115, 80, 25, 52, and 38 crowns, respectively.

For the spectral mixture analysis, three endmembers—green vegetation, non-photosynthetic vegetation (NPV), and shade—were chosen to represent the materials visible at the top of the canopy. Soil was not included because the canopy at "Parque Metropolitano is closed and even where there are gaps from tree falls, vegetation grows to completely cover the gap within a year" [31]. For the green vegetation and NPV endmembers, tree crown spectra were used rather than spectra of plant components or partial canopies taken with a ground-based spectrometer. Using ground-based spectra of plant components as endmembers introduces major errors in spectral mixture analysis [21,23] and it is difficult to get spectra of whole tree crowns from above using a handheld spectrometer, even from the canopy crane. Even from the canopy crane, only spectra of sections of most crowns, not the whole crown, can be obtained with a handheld spectrometer.

The green vegetation endmember was taken from a fully-leaved *A. excelsum* tree, which is the species with the highest leaf density in the study area [32]. The NPV endmember was taken from an area that was a mixture of *B. simarouba* and *S. mombin* crowns with almost no leaf density in either the overstory or understory. The shade endmember was taken from a black test panel that was included in the HYDICE image for calibration. All endmember spectra were taken from crowns outside the 5-ha area used to test the mixture model.

The linear mixture model using these endmembers was constrained to a fractional sum of one for each pixel. The mixture model was applied to the entire HYDICE image at 1-m resolution. Endmember composition for individual crowns was then extracted from the endmember fraction images derived from the full resolution images. For the 5-ha test area described later, SMA was also applied to the HYDICE image degraded to a pixel resolution of 10 and 30 m. All image processing was done with the Environment for Visualization of Images (ENVI) software package (ITT Visual Information Solutions, Boulder, Colorado).

A 5-ha test area that contained a mixture of deciduous and nondeciduous trees was chosen to test the accuracy of the mixture model classification compared to a classification based on field observations. For the field observations, the entire area was designated into three land-surface materials—fully-leaved tree crowns, deciduous tree crowns, and intercrown shadows. Species identifications of individual tree crowns were used to classify 30% of the 5-ha area. The rest of the area, which lacked individual species identifications, was classified based on field observations of the locations of large deciduous patches and patches of grass, and the color aerial photographs. Intercrown shadows were quantified as the area between crowns remaining after individual crowns were delineated (fig. 8.1).

8.3 RESULTS

The three endmembers used in this study had clearly contrasting spectra in the visible and infrared bands (fig. 8.2). The NPV endmember lacked the chlorophyll absorption features in the blue and red bands found in the green vegetation endmember. The strong contrast between red and the near-infrared (NIR) plateaus was much less for NPV than for green vegetation. Reflectance in the SWIR bands (SWIR1 = 1.55–1.69 µm and SWIR2 = 2.04–2.23 µm) was much greater for the NPV endmember than for the green vegetation endmember. The shade endmember was flat spectra across all bands. The slight features and some negative values for the shade endmember were a result of imperfect calibration of the image.

The average spectra for the different species with fully-leaved crowns (*A. excelsum, L. seemanii,* and *Ficus* sp.) during the dry season showed strong similarity to the green vegetation endmember (fig. 8.2). All three species had strong chlorophyll absorption features and high NIR reflectance. In the SWIR, *A. excelsum* and *Ficus* sp. strongly resembled the green vegetation endmember, but the *L. seemanii* spectrum strongly resembled the NPV endmember.

The average spectra for deciduous tree crowns showed both similarities to and differences from the NPV spectra. In the NIR wavelengths (0.90–1.39 µm), the average deciduous crown and the NPV endmember were very similar. However, deciduous tree crowns showed chlorophyll absorption features like the green vegetation endmember, although the absorption troughs were not as deep as in the spectra for the fully-leaved species. In contrast, the NPV endmember had no absorption troughs in the blue and red bands. In the SWIR1 (1.55–1.69 µm), the average deciduous

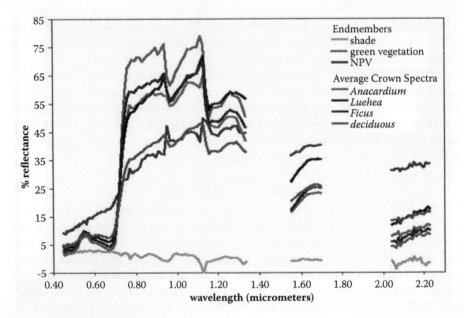

FIG. 8.2 Endmember spectra used for linear mixture model of HYDICE images of Parque Natural Metropolitano, Panama. Also, average spectra for three nondeciduous tree species, and all deciduous species combined.

crown spectra were similar to the green vegetation endmember. In this part of the spectrum, *L. seemanii* was actually more similar to the NPV endmember than the average fully-leaved crown spectra. In the SWIR2 (2.04–2.23 µm), both the average deciduous and *L. seemanii* spectra were between the green vegetation and NPV endmember spectra, and distinct from the spectra of *A. excelsum* and *Ficus* sp. for statistical comparison of different species and deciduous crowns, see Bohlman and Lashlee [33].

Overall, the average spectrum of deciduous crowns indicates these crowns were composed of a mixture of NPV and green vegetation. Images of a deciduous tree crown taken from a multispectral camera indicated that the deciduous tree crown spectra were composed of a wood signal originating from the upper tree crowns, combined with a green vegetation tree crown spectra originating from the understory tree crowns (fig. 8.3). In the red band (0.680 µm), the exposed wood in the overstory tree crown caused some reflectance back to the sensor, whereas the green understory crown strongly absorbed in this wavelength band. In the NIR band (0.85 µm), the understory tree crown was strongly reflecting light back to the sensor, dominating the signal.

8.3.1 ENDMEMBER MIXTURES

Overall, the mixture model approach clearly separated deciduous from nondeciduous tree crowns, and even separated different nondeciduous species. For individual tree crowns, the percentage of the green vegetation endmember increased and the

(a)   (b)

FIG. 8.3   Red and near-infrared images of the same branch of a deciduous tree crown at Parque Natural Metropolitan, Panama. The images were taken looking straight down on the canopy from a gondola suspended from a construction crane. The camera consisted of a charge coupled device (CCD) video camera (Marshall Electronics Model 1050A) fitted with 10-nm filters (Oriel optics) centered around 0.68 µm (red) and 0.85 µm (near infrared). The crown is a 25-m-tall *P. septenatum* (deciduous) overtopping a 15-m-tall *A. exelsum* (high leaf density).

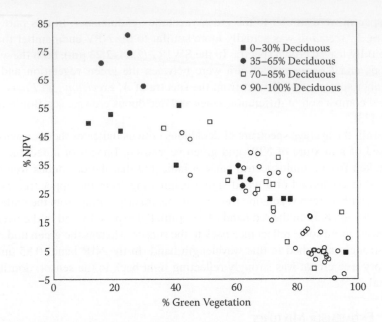

FIG. 8.4 Variation in endmember composition for tree crowns with different levels of deciduousness as measured in the field for Parque Natural Metropolitano, Panama.

percentage of NPV decreased with the level of deciduousness measured in the field (fig. 8.4). The amount of shade did not change with different levels of deciduousness per crown. Overall, the mixture model approach with three endmembers produced good fits for all of the tree species, except *L. seemanii*, whose crowns had negative fraction of the shade/shadow component (fig. 8.5).

Crowns of deciduous species had a different endmember mixture compared to fully-leaved crowns during the dry season (fig. 8.5). Deciduous crowns always had a lower fraction of the green vegetation endmember than fully-leaved trees. With the exception of *L. seemanii* crowns, the deciduous species' crowns had a lower fraction of the NPV endmember than crowns of fully-leaved species. In the range where *L. seemanii* and deciduous species' crowns had the same NPV (40–50%), *L. seemanii* crowns always had a higher green vegetation component and generally a lower shade/shadow component. Crowns of fully-leaved species showed a gradient of green vegetation/NPV ratios (fig. 8.6). *A. excelsum* and *Ficus* sp. had the greatest amounts of green vegetation and the lowest amounts of NPV. *Ficus* sp. had lower shade/shadow than *A. excelsum* (fig. 8.5). *L. seemanii* was intermediate between the other fully-leaved species and the deciduous species in terms of green vegetation/NPV ratio (fig. 8.6), but had lower intercrown shade and shadowing than any crown type (fig. 8.5).

8.3.2 LANDSCAPE CLASSIFICATION

The endmember mixture of a green vegetation, NPV, and shade/shadow was successful in determining the percentage of crown types in the 5-ha test area of Parque

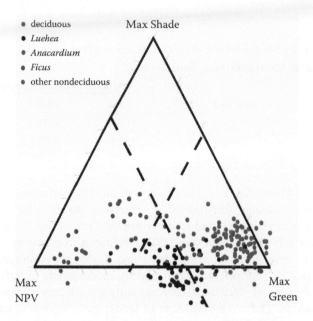

FIG. 8.5 Ternary diagram of the endmember mixtures of tree crowns in Parque Natural Metropoliano, Panama. Each symbol is the mean of all pixels covering one crown. The two dashed lines show the upper bound of green vegetation and the lower bound of NPV for deciduous crowns.

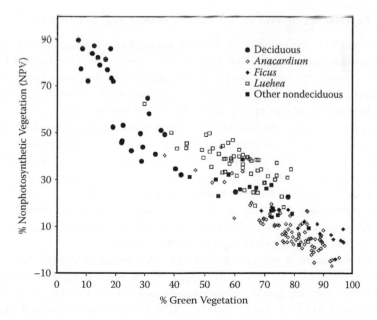

FIG. 8.6 The percentage of green vegetation and NPV endmembers for different tree crowns at Parque Natural Metropolitano, Panama. Each symbol represents the mean of all pixels for one tree crown.

TABLE 8.1
Percent Coverage of Different Material in a 5-ha Test Area of Parque Natural Metropolitano, Panama

		Mixture model image resolution		
	Measured	1 m	10 m	30 m
Deciduous	31.5%	31.9%	33.1%	31.9%
Nondeciduous	52.6%	49.3%	48.3%	54.4%
Shadow/shade	14.5%	17.8%	16.9%	12.2%

Natural Metropolitano (fig. 8.1). This area contained a mixture of different non-deciduous trees, including fully-leaved *A. excelsum* crowns with dense crowns, senescing *L. seemanii* crowns, and sparsely leaved *Cecropia* sp. crowns that comprised just over half the test area (table 8.1). Deciduous tree crowns covered about one-third of the area, and intercrown shade and shadowing coverage, about 15%. At full resolution of 1 m² per pixel, the mixture model nearly exactly determined the correct coverage of deciduous crowns, fully-leaved crowns, and intercrown shade and shadows. The correct coverage of landscape components was also accurately determined when the images were degraded to 10- and 30-m pixel resolution (table 8.1).

8.4 DISCUSSION

8.4.1 Importance of Wood in Tropical Forest Spectral Signal

Nonphotosynthetic vegetation in the form of exposed wood at the top of the canopy is an important component of the land surface spectra in this seasonal tropical forest. Deciduous tree crown spectra were a mixture of the green vegetation and NPV spectra, indicating that NPV is affecting the spectra for these tree crowns. The contrast between deciduous crowns and fully-leaved crowns such as *A. excelsum* is not merely the result of lower leaf density in the deciduous trees. Since NPV and soil spectra are similar, one could argue that the deciduous tree crown spectra are a result of a mixture of soil and green vegetation, not NPV and green vegetation. However, the understory in the Panama Canal area where Parque Metropolitano is located remains green throughout the dry season, even though the canopy is semideciduous [26], which precludes soil being visible from above in all but the most deciduous forests. Also, the three-endmember mixture model very accurately determined the percent cover of deciduous and fully-leaved tree crowns.

In this landscape, a mixture model that does not include NPV may lead to incorrect interpretation of land cover types. Many spectral mixture analyses in tropical forests use only green vegetation, soil, and shade endmembers [18,19]. If a soil endmember had been used instead of NPV, the deciduous crowns would have erroneously contained a high fraction of the exposed soil. A high fraction of soil might be incorrectly interpreted as either a canopy gap or as pasture or another secondary land use, if a human-altered landscape were being considered. In a landscape that includes both deciduous forest and pasture, it is important then to include both NPV and soil as endmembers for spectral mixture analysis, in order to separate high-NPV deciduous trees from high-NPV pastures, which also include exposed soil. Also, differences in texture may be useful in distinguishing between the different high-NPV land cover types [4,34].

8.4.2 ACCURACY OF MIXTURE MODEL TO PREDICT THE PERCENTAGE OF DECIDUOUS TREES

The high accuracy of the mixture model in predicting the percentage of deciduous trees was surprising for several reasons. First, the complex vertical structure of this forest, with varying amounts of exposed wood at the top of the canopy overtopping a green mid-story and understory, was well predicted by a linear SMA approach that assumes essentially a two-dimensional surface of green vegetation, NPV, and shade/shadows. The linear mixture model may work well because endmembers were derived from whole crowns, rather than from spectra of tree components such as wood and leaves. The green vegetation and NPV endmembers derived from the images already account for intracrown shadowing and the three-dimensional interaction of green leaves and wood, thus reducing the complexity of this forested landscape.

The linear mixture model also worked well despite the high spectral heterogeneity within the green vegetation and NPV endmembers. Sources of spectral variation among crowns may be generated by a number of factors, including variation in leaf spectra, which is well quantified at this site [7,39,40], and variation in wood reflectance among deciduous species, which has not been measured. Additional sources of spectral variation in tropical forest include variation in canopy structure and leaf phenology among species, as well as presence of fruits, flowers, and epiphylls [41,42]. Indeed, the mixture model used here provided a poor fit for one of the species, *L. seemanii*, and there was large variation in the endmember mixtures within the fully-leaved and deciduous categories. Despite the variation, there was a distinct threshold in endmember mixtures between fully-leaved and deciduous crowns and the SMA very accurately quantified the relative cover of the deciduous and nondeciduous forest.

The accuracy of the mixture model in predicting percentage of deciduous crowns may result because the spectral contrast between NPV and green vegetation is much greater than the subtle variation caused by the factors mentioned earlier. Also, species diversity in the upper canopy at Parque Metropolitano is quite low in comparison to other tropical forests. Under the canopy crane, *L. seemanii* and *A. excelsum* compose about 50% of the upper canopy. In a more diverse forest, the spectral variation among species might reduce the ability of a single spectrum to represent each endmember and necessitate an approach that accounts for spectral variability within each endmember [43–45].

8.4.3 SPATIAL SCALE

The three endmember mixture model works well at different spatial scales, including the 1-m full resolution of the HYDICE images, which is actually a finer resolution than the objects being resolved—namely, whole tree crowns. At the coarser spatial resolutions of the degraded images (10 and 30 m), single pixels in general contain at least one whole tree crown. The fact that this simple approach worked at all three spatial scales is probably due to the strong spectral contrast between NPV and green vegetation at different spatial scales. At the 1-m subcrown spatial resolution, the pixels contain a mixture of leaves, wood, and intracrown shadows. At the 30-m supracrown resolution, pixels are composed of a mixture of deciduous crowns, nondeciduous crowns, and intercrown shadows. The success of the SMA with the 30-m resolution images indicates that it could be used widely with image data such as ASTER or Landsat, although these sensors contain less spectral resolution and breadth than the HYDICE images.

8.4.4 INDIVIDUAL SPECIES SPECTRA AND ENDMEMBER MIXTURES

The endmember mixtures of the different species matched field observations of the species' crown structure and phenology and gave information about why different species may be separable spectrally in this tropical forest [33]. Deciduous trees had low, but still substantial, vegetation fractions (less than 40%), which is consistent with three-dimensional composition of semideciduous forests. In the dry season, deciduous overstory crowns have a fully green mid-canopy and understory trees. Also, liana growth on deciduous crowns can increase the green vegetation signal, as lianas tend to maintain a higher leaf density in the dry season than host trees [35–37].

In general, deciduous crowns showed a higher fraction of intracrown shading and shadows than fully-leaved crowns. In a deciduous tree crown, the space that had been filled with leaves will be empty, allowing greater light penetration to lower levels of the forest, and thus creating more intracrown shadows when viewed from above.

A. excelsum and *Ficus* sp. both had high leaf area during the dry season, but have different branch and canopy structures [32]. *A. excelsum* has more vertically oriented leaves than *Ficus* sp., which creates greater shadowing at the branch scale than *Ficus* sp. Also, *Ficus* sp. branches are oriented to create a crown that consists of three to four discreet, relatively flat layers of leaves. *A. excelsum* branches are oriented more vertically, so the surface of *A. excelsum* has more microtopographic variation and thus greater shade and shadowing.

There are several possible explanations for the high NPV and low green vegetation fractions for *L. seemanii*. *L. seemanii* has lower leaf density in the dry season than in the wet season, which may lead to more exposed branches. *L. seemanii* also has a high load of woody seeds born directly above the canopy in March, which may obscure the green vegetation signal from the green leaves below. Also, the woody seeds should have similar spectra to wood, which would increase the NPV fraction. Also, *L. seemanii* leaves are senescing at the end of the dry season [7,38], which may cause the spectra to look more like wood than green leaves. The leaves and branches of *L. seemanii* are more horizontal than the leaves and branches of *A. excelsum* [32], which is consistent with the higher intracrown shadow fraction for *A. excelsum* vs. *L. seemanii*.

Given the importance of species diversity in tropical ecosystems and the emerging evidence that at least some species can be distinguished spectrally in tropical forests using hyperspectral data [11,12], the spectral mapping of individual species will probably be the focus of more attention in the future. As in this study, the mixture model approach may give additional insight into compositional features of different species, which allow some species to be distinguished while others cannot be separated.

8.4.5 IMPLICATIONS

Increasing evidence shows that tropical forest canopies should not just be viewed as invariant high leaf density systems. The phenology of seasonal ecosystems leads to important variation in leaf quantity, leaf age, and materials that are exposed at the top of the canopy. This study has shown that exposed wood at the top of the canopy significantly affects the spectra of the canopy and can be quantified accurately using a mixture model approach. Determining the amount of deciduous vs. nondeciduous trees across a landscape can potentially be important for studies of landscape patterns of carbon uptake, hydrology, and species diversity.

Increasing evidence shows that tropical forest canopies should not be viewed just as invariant high leaf density systems. The phenology of seasonal ecosystems leads to important variation in leaf quantity, leaf age, and materials that are exposed at the top of the canopy. This study has shown that exposed wood at the top of the canopy significantly affects the spectra of the canopy, and can be quantified accurately using a spectral mixture model approach. The SMA approach can be used to explore ecology and biogeography of seasonal tropical forests. At a local scale, SMA can provide important insights into temporal patterns of the landscape, showing the degree to which the species in the forest are synchronous in the transition between deciduousness in the dry season and full leaf content in the wet season. The amount of leaf density throughout the year has direct effects on carbon uptake, rainfall interception, and evapotranspiration. At a regional scale, SMA can be used to map semi- and fully-deciduous tropical forests and their changes through time. Forty percent of the global distribution of lowland tropical forest is semi-evergreen, semi-deciduous, or deciduous [46]. In areas with increasing frequency of drought, seasonal forests may become more widespread. SMA provides a tool to accurately quantify the distribution of this type of forest spatially and temporally from a wide range of remotely sensed images with different spatial resolutions. Finally, the SMA approach may be useful in any high leaf density forests where forest damage from wind, crown fires, or pathogen outbreaks has caused wood to be exposed at the top of the canopy.

REFERENCES

1. Hall, F.G., Knapp, D.E., and Huemmrich, K.F., Physically based classification and satellite mapping of biophysical characteristics in the southern boreal forest, *Journal of Geophysical Research*, 102, 29567, 1997.
2. Kobayashi, H., Suzuki, R., and Kobayashi, S., Reflectance seasonality and its relation to the canopy leaf area index in an eastern Siberian larch forest: Multisatellite data and radiative transfer analyses, *Remote Sensing of Environment*, 106, 238, 2007.
3. Peddle, D.R., Hall, F.G., and LeDrew, E.F., Spectral mixture analysis and geometric-optical reflectance modeling of Boreal forest biophysical structure, *Remote Sensing of Environment*, 67, 288, 1999.

4. Asner, G.P., Archer, S., Hughes, F.R., et al., Net changes in regional woody vegetation cover and carbon storage in Texas drylands, 1937–1999, *Global Change Biology*, 9, 316, 2003.
5. Asner, G.P. and Heidebrecht, K.B., Imaging spectroscopy for desertification studies: Comparing AVIRIS and EO-1 Hyperion in Argentina drylands, *IEEE Transactions on Geoscience and Remote Sensing*, 41, 1283, 2003.
6. Harris, A.T., Asner, G.P., and Miller, D.M., Changes in vegetation structure after long-term grazing in pinyon-juniper exosystems: Integrating imaging spectroscopy and field studies, *Ecosystems*, 6, 368, 2003.
7. Gamon, J.A., Kitajima, K., Mulkey, S.S., et al., Diverse optical and photosynthetic properties in a neotropical dry forest during the dry season: Implications for remote estimation of photosynthesis, *Biotropica*, 37, 547, 2005.
8. Myneni, R.B., Yang, W., Nemani, R.R., et al., Large seasonal swings in leaf area of Amazon rainforests, *Proceedings of the National Academy of Sciences*, 104, 4820, 2007.
9. Xiao, X., Hagen, S., Zhang, Q., et al., Detecting leaf phenology of seasonally moist tropical forests in South America with multi-temporal MODIS images, *Remote Sensing of Environment*, 103, 465, 2006.
10. Tuomisto, H., Ruokalainen, K., Aguilar, M., et al., Floristic patterns along a 43-km long transect in an Amazonian rain forest, *Journal of Ecology*, 91, 743, 2003.
11. Clark, D.A., Roberts, D.A., and Clark, D.A., Hyperspectral discrimination of tropical rain forest tree species at leaf to crown scales, *Remote Sensing of Environment*, 96, 375, 2005.
12. Zhang, J., Rivard, B., Sanchez-Azofeifa, G.A., et al., Intra- and interclass spectral variability of tropical tree species at La Selva, Costa Rica: Implications for species identification using HYDICE imagery, *Remote Sensing of Environment*, 105, 129, 2006.
13. Kalacska, M., Bohlman, S.A., Sanchez-Azofeifa, G.A., et al., Hyperspectral discrimination of tropical dry forest lianas and trees: Comparative data reduction approaches at the leaf and canopy levels, *Remote Sensing of Environment*, 109, 406, 2007.
14. Asner, G.P., Nepstad, D., Cardinot, G., et al., Drought stress and carbon uptake in an Amazon forest measured with space-borne imaging spectroscopy, *Proceedings of the National Academy of Sciences*, 101, 6039, 2004.
15. Asner, G.P. and Vitousek, P.M., Remote analysis of biological invasion and biogeochemical change, *Proceedings of the National Academy of Sciences*, 102, 4383, 2005.
16. Adams, J.B. and Gillespie, A.R., *Remote sensing of landscapes with spectral images: A physical modeling approach*. Cambridge University Press, Cambridge, 2006.
17. Adams, J.B., Sabol, D., Kapos, V., et al., Classification of multispectral images based on fractions of endmembers; application to land-cover change in the Brazilian Amazon, *Remote Sensing of Environment*, 52, 137, 1995.
18. Lu, D., Moran, E., and Batistella, M., Linear mixture model applied to Amazonian vegetation classification, *Remote Sensing of Environment*, 87, 456, 2003.
19. Shimabukuro, Y.E., Batista, G.T., Mello, E.M.K., et al., Using shade fraction image segmentation to evaluate deforestation in Landsat thematic mapper images of the Amazon Region, *International Journal of Remote Sensing*, 19, 535, 1998.
20. Souza, C.M., Carlos, Firestone, L., et al., Mapping forest degradation in the Eastern Amazon from spot 4 through spectral mixture models, *Remote Sensing of Environment*, 87, 494, 2003.
21. Asner, G.P., Knapp, D.E., Cooper, A.N., et al., Ecosystem structure throughout the Brazilian Amazon from Landsat observations and automated spectral unmixing, *Earth Interactions*, 9, 1, 2005.
22. Huemmrich, K.F. and Goward, S.N., Vegetation canopy par absorptance and NDVI: An assessment of 10 tree species with the sail model, *Remote Sensing of Environment*, 61, 254, 1997.

23. Asner, G.P., Biophysical and biochemical sources of variability in canopy reflectance, *Remote Sensing of Environment*, 64, 234, 1998.

24. Asner, G.P., Wessman, C.A., Bateson, C.A., et al., Impact of tissue, canopy and land-scape factors on the hyperspectral reflectance variability of arid ecosystems, *Remote Sensing of Environment*, 74, 69, 2000.

25. Asner, G.P. and Heidebrecht, K.B., Spectral unmixing of vegetation, soil and dry carbon cover in arid regions; comparing multispectral and hyperspectral observations, *International Journal of Remote Sensing*, 23, 3939, 2002.

26. Condit, R.C., Watts, K., Bohlman, S.A., et al., Quantifying the deciduousness of tropical forest canopies under varying climates, *Journal of Vegetation Science*, 11, 649, 2000.

27. Meinzer, F.C., Andrade, J.L., Goldstein, G., et al., Control of transpiration from the upper canopy of a topical forest: The role of stomatal, boundary layer and hydraulic architecture components, *Plant Cell and Environment*, 20, 1242, 1997.

28. Mulkey, S.S., Kitajima, K., and Wright, S.J., Plant physiological ecology of tropical forest canopies, *Trends in Ecology and Evolution*, 11, 408, 1996.

29. Hall, F.G., Strebel, D.E., Nickeson, J.E., et al., Radiometric rectification: Toward a common radiometric response among multidate, multisensor images, *Remote Sensing of Environment*, 35, 11, 1991.

30. Perry, E.M., Warner, T., and Foote, P., Comparison of atmospheric modeling versus empirical line fitting for mosaicking HYDICE imagery, *International Journal of Remote Sensing*, 21, 799, 2000.

31. van der Meer, P.J. and Bongers, F., Formation and closure of canopy gaps in the rain forest at Nouragues, French Guiana, *Vegetation*, 126, 167, 1996.

32. Kitajima, K., Mulkey, S.S., and Wright, S.J., Variation in crown light utilization characteristics among tropical canopy trees, *Annals of Botany*, 95, 535, 2005.

33. Bohlman, S.A. and Lashlee, D., *High spatial and spectral resolution remote sensing of Panama Canal Zone forests: An applied example mapping tropical tree species.* Kluwer Press, Dordrecht, 2005.

34. Hudak, A.T. and Wessman, C.A., Textural analysis of high resolution imagery to quantify bush encroachment in Madikwe Game Reserve, South Africa, 1955–1996, *International Journal of Remote Sensing*, 22, 2731, 2001.

35. Kalacska, M., Calvo-Alvarado, J.C., and Sanchez-Azofeifa, G.A., Calibration and assessment of seasonal changes in leaf area index of a tropical dry forest in different stages of succession, *Tree Physiology*, 25, 733, 2005.

36. Opler, P.A., Baker, H.G., and Frankie, G.W., *Seasonality of climbers: A review and example from Costa Rican dry forest.* Cambridge University Press, Cambridge, 1991.

37. Schnitzer, S. and Bongers, F., The ecology of lianas and their role in forests, *Trends in Ecology and Evolution*, 17, 223, 2002.

38. Kitajima, K., Mulkey, S.S., and Wright, S.J., Seasonal leaf phenotypes in the canopy of a tropical dry forest: Photosynthetic characteristics and associated traits, *Oecologia*, 109, 490, 1997.

39. Castro-Esau, K.L., Sanchez-Azofeifa, G.A., and Caelli, T., Discrimination of lianas and trees with leaf-level hyperspectral data, *Remote Sensing of Environment*, 90, 353, 2004.

40. Castro-Esau, K., Sanchez-Azofeifa, G.A., Rivard, B., et al., Variability in leaf optical properties of mesoamerican trees and the potential for species classification, *American Journal of Botany*, 93, 517, 2006.

41. Andrew, M.E. and Ustin, S.L., Spectral and physiological uniqueness of perennial pepperweed (*Lepidium latifolium*), *Weed Science*, 54, 1051, 2006.

42. Roberts, D.A., Nelson, B.W., Adams, J.B., et al., Spectral changes with leaf aging in Amazon caatinga, *Trees—Structure and Function*, 12, 315, 1998.

43. Asner, G.P. and Lobell, D.B., A biogeophysical approach for automated SWIR unmixing of soils and vegetation, *Remote Sensing of Environment*, 74, 99, 2000.

44. Bateson, C.A., Asner, G.P., and Wessman, C.A., Endmember bundles: A new approach to incorporating endmember variability in spectral mixture analysis, *IEEE Transactions on Geoscience and Remote Sensing*, 38, 1083, 2000.
45. Smith, M.O., Adams, J.B., and Sabol, D.E., *Spectral mixture analysis—New strategies for the analysis of multispectral data*. ECSC, Brussels, 1994.
46. United Nations Environment Programme, World Conservation Monitoring Center, *Global Distribution of Forest Types*, 2000.

9 Assessing Recovery Following Selective Logging of Lowland Tropical Forests Based on Hyperspectral Imagery

J. Pablo Arroyo-Mora, Margaret Kalacska, Robin L. Chazdon, Daniel L. Civco, German Obando-Vargas, and Andrés A. Sanchún Hernández

CONTENTS

9.1 INTRODUCTION

Development of effective, long-term plans to conserve and manage biodiversity requires an understanding of the distribution, diversity, and abundance of species in both protected and unprotected areas. More than ever, due to extensive fragmentation and other anthropogenic effects, forest fragments in unprotected areas must serve a dual function: providing a sustainable source of timber and nontimber products and providing essential intact forest habitat for biodiversity conservation within the landscape matrix. According to the Food and Agricultural Organization (FAO) [1], sustainable forest management is a land use with similar goals to those that focus on biodiversity conservation principles. However, factors such as the initial structure of the forest and the scale and intensity of the logging operations [2], not to mention the short- and long-term effects of postharvesting [3], are important when forest management and biodiversity conservation goals converge. Therefore, monitoring of natural forest areas subject to forest management (selective logging) is necessary in order to assess forest recovery processes after disturbance.

From the ecological standpoint, forest recovery processes are highly variable depending upon factors such as disturbance gap size, seed dispersal limitations, disturbance intensity, light availability [4], and forest type, among others. Ecological studies of recovery following selective logging have generally focused on the long-term effects of recovery [5], regeneration in extraction roads [5,6], and species composition [7]. Nevertheless, an ongoing major limitation for assessing forest recovery after selective logging is the lack of information representing a chronological sequence, precluding solid forecasting of regeneration trajectories. A rapid, reliable method to detect, assess, and monitor areas over large expanses that had formerly been selectively logged is required. Remotely sensed data in the form of airborne or satellite imagery is a candidate technology to meet these needs provided that management units undergoing regeneration for various time periods are readily identifiable (regardless of unique microsite differences).

Remote sensing has been used for monitoring and mapping selective logging [8,9], forest recovery postlogging activities [10], and canopy spatial patterns [11] with some success using low spectral and spatial resolution imagery (i.e., Landsat and SPOT sensors). In addition, the potential of remote sensing at large spatial scales is only recently being realized to describe leaf biochemical characteristics of forest canopies [12] and canopy species detection [13]. Due to the potential variability in regeneration following selective logging, assessment and monitoring techniques must be a balance between generalization (i.e., able to capture the general regeneration trend) and specialization (i.e., able to detect subtle difference in overall canopy reflectance). The use of hyperspectral data with multitemporal management information is a novel application of such data, addressing the need for studies focusing on the assessment of forest recovery essential to monitoring efforts.

Few tropical regions have the economic resources to acquire multitemporal hyperspectral data; however, forest management data are available, even if they may not be methodologically organized. In addition, newer hyperspectral data analysis techniques are constantly being developed to exploit the wealth of spectral information within the data—beyond the commonly used indices (i.e., normalized difference vegetation index [NDVI]) and statistics (i.e., ordination methods). These new analyses include machine learning (ML) techniques that provide more flexible analytic power for data exploration and pattern recognition [14]. Many of these methods originate from the artificial intelligence, pattern recognition, and signal processing communities and are powerful tools that can be adapted to support the synergy of hyperspectral remotely sensed data with ecological applications [15].

Of the various types of machine learning techniques, we focus on "generalist" pattern classification with well-known *nonparametric learning* classifiers such as "decision tree," "neural network," and "k-nearest neighbor" classification and also explore the use of a soft (probabilistic) unsupervised clustering technique with the expectation maximization (EM) algorithm [16]. The nonparametric classifiers derive the hypotheses from the data rather than assuming a particular form for the class density function [17].

This chapter takes advantage of a multitemporal forestry geographical information system (FGIS) pre- and postlogging data set provided by the Foundation for the Central Volcanic Mountain Range Conservation Area (FUNDECOR), a hyperspectral airborne data set (HyMap II-HyVista, Castle Hill, New South Wales, Australia) from the year 2005 obtained through a collaborative imaging effort between the Costa Rican government and NASA, and a set of ML techniques for an integrative analysis of both data sets. Our objectives are:

- to identify the most significant wavelengths (features) for separating the spectra of recovery following selective logging of natural forest areas, thereby exploring the most likely forest characteristics (i.e., soil exposure, canopy structure, gap vegetation recovery) responsible for this separability
- to determine the accuracy with which the spectral signature of different selectively logged natural forest areas can be classified
- to determine how the spectral separability of the managed forest areas is affected by spatial resolution and in turn how this affects the accuracy of the different classifiers

This chapter aims to increase the knowledge base surrounding the application of hyperspectral data to assess tropical forest recovery after management (i.e., selective logging). Most of the previous studies applying remote sensing to forest management have been carried out in the Amazon basin (Brazil and Bolivia), while other tropical regions have been neglected. In addition, none of those studies utilized machine learning techniques. In addition, this study has the potential to motivate the development of a research agenda in forest management in Costa Rica specifically utilizing remotely sensed data. In the field of hyperspectral data analysis we are contributing to the knowledge base through a direct application of machine learning techniques to explain patterns of forest recovery after logging and its linkage with changes in forest biophysical properties through time. This multidisciplinary approach can strengthen both forestry and remote sensing applications aiming to improve long-term conservation goals.

9.2 METHODOLOGY

This study was carried out for tropical wet lowland forest areas located along the proposed San Juan–La Selva biological corridor, Costa Rica (fig. 9.1). This biological corridor is partially located within the Central Volcanic Mountain Range Conservation Area (ACCVC). The study area encompasses an agricultural matrix of farms with diverse agriculture, reforestation activities, cattle pastures, extensive fields of cash crops, and patches of remnant forests of variable size. A significant land use in the last decade has been forest management (selective logging) for which an FGIS is

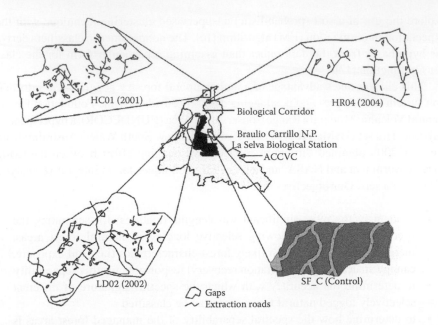

FIG. 9.1 Management units with postlogging gap density within the San Juan–La Selva biological corridor.

currently under development. The following sections briefly describe the FGIS and the airborne hyperspectral data set used in the study.

9.2.1 Natural Forest Management and Hyperspectral Data

This study considered four natural forest management units within the biological corridor where forest management was planned and supervised by FUNDECOR. Three out of the four management units were logged in different years (1–4 years prior to image acquisition), while the fourth area was not logged and is used as a control. The unlogged control site (referred to as SF_C) is representative of relatively undisturbed forest in the study area. A direct chronological trajectory between the management units is not assumed. All four management units are located within the tropical wet forest life zone [18]. The mean slope for all the units is less than 6%; therefore, we focused our analysis in topographically homogeneous areas. The mean elevation ranges from 117 m above sea level (m.a.s.l.) in SF_C, followed by LD02 (78 m.a.s.l.), HR04 (60 m.a.s.l.), and HC01 (50 m.a.s.l.). The total number of trees above 30 cm in diameter measured at breast height (DBH) per hectare ranges from 73 trees/ha in LD02 to 100 trees/ha in SF_C (92 and 75 trees/ha in HR04 and HC01, respectively). For the four units the species richness ranges from 47 species above 30 cm DBH in SF_C to 15 species in HR04 (42 and 35 species in LD02 and HCO1, respectively). Although the units have different species richness, a common characteristic among them is the dominance of *Pentaclethra macroloba*, which accounts for 38% of the total number of trees above 30 cm DBH in LD02, 39% in SF_C, and 49 and 57% in HC01 and HR04, respectively.

TABLE 9.1

Management Units, Management Period, and Total Gap Area Based on a Postlogging Field Survey

Management unit	Management period	Trees/ha logged	Unit area (ha)	Total gap area (%)	Road area (%)	Disturbance (%)
HR04	2004	5.5	10.9	1.17	3.62	4.79
LD02	April–Nov. 2001	6.4	36.5	3.32	2.53	5.95
HC01	Jan. 2001; Feb. 18, 2002	7.5	11.4	5.14	3.62	8.76
SF_C	Unmanaged forest	NA	30.5	NA	NA	NA

In each management unit we used a geographic information system (GIS) that integrated information about farm location, size, number, and location of remnant and harvested trees during the management period, as well as biophysical characteristics of the area (hydrology and topography). For our analysis we focused in areas within the management unit where postlogging, extraction roads, and gap location information were surveyed in the field after selective logging activities. Table 9.1 illustrates the logging period, the management unit area, gap area (percent), extraction road area (percent), and total disturbance area (percent).

The hyperspectral imagery was collected by the HyMap II airborne sensor in March and April 2005 as part of the CARTA II mission (see chapter 1) at 16-m resolution. The imagery consists of 125 bands spanning 458–2491 nm. Of the data products available from this mission (e.g., radiance, reflectance), we used the atmospherically corrected reflectance data, which had been processed in house with the ATREM algorithm followed by EFFORT polishing by HyVista prior to the release of the imagery. The four management areas examined were located on four flight lines (table 9.1).

9.2.2 Data Analysis

We followed three broad categories of analysis: data preparation, feature selection, and pattern classification; figure 9.2 illustrates the overall analysis approach.

9.2.2.1 Hyperspectral Data Preparation

Based on the management units chosen for the study we located the flight lines that completely contained the management units without cloud cover. Subsequently, the shape file of the farms was used to clip the HyMap II image for each management unit. Because the farm/management unit boundaries were clearly defined in the imagery, any minor misregistration between the image and the shape files was manually corrected. Once we had the clipped management units, the spectra from all pixels representing the four managed units were extracted at the original 16-m resolution in ENVI 4.1. The imagery was subsequently resampled to a 30-m spatial resolution and once again the pixels representing the managed units were extracted.

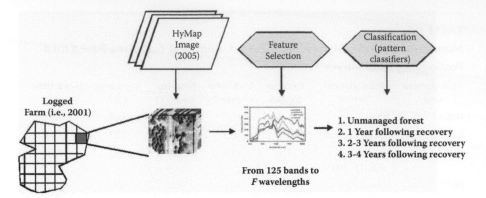

FIG. 9.2 Overall approach for analyzing natural forest recovery based on multitemporal logging data and hyperspectral airborne imagery. See CD for color image.

The resampling was done to examine the effect of spatial resolution on the separability of the spectra and also because the emergent tree crowns in this lowland forest are commonly greater than 15 m in diameter. The 30-m pixel size was also chosen to match that of the commercially available hyperspectral satellite Hyperion. These data were then assigned to training and testing data sets (i.e., 50% for training and 50% for testing).

9.2.2.2   Feature Selection and Classification

A forward feature selection algorithm was used to reduce the dimensions of the data, determine the optimal wavelengths with the greatest difference between classes (managed forest at time t_1 and t_n), and determine the optimal number of wavelengths (features) for classification with the lowest error. This algorithm initially chooses the best wavelength for separating the classes based on a criterion (nearest neighbor in this case) and then iteratively adds additional wavelengths in the order in which they most improve the separability. Once the optimal number of wavelengths has been reached, the remainder are ranked based on their ability to separate the classes and are added to the list of wavelengths. Generally, the maximum number of wavelengths that can be used without overfitting is $F = (n - g)/3$, where n is the number of spectra and g is the number of classes [19].

In order to determine the optimal number of wavelengths to use with the classifiers we selected the number of features with the best separability criterion [14]. These optimal wavelengths were subsequently used to train the classifiers to separate the spectra of the selectively logged from the unlogged old-growth forest. The classifiers we examined were nonparametric (k-nearest neighbor "knnc"), neural network (feed forward with a Levenberg–Marquardt optimization—"lmnc"), and decision tree "treec."

We also examined a probabilistic clustering with the EM algorithm [16]. With this technique, rather than learning/constructing decision boundaries to separate the classes from a previously labeled training data set, likelihood values are used to assign the data to each class (i.e., each pixel to each of the four management units). The final classification label is that of the class it most closely resembles (i.e., the most likely management unit) as opposed to the nonparametric classification, where data are assigned to only one class based on the derived decision boundaries. This approach is more flexible and guarantees to converge on an optimum result. In order to minimize the possibility of convergence at a local minimum, we ran 1000 iterations with various initialization points [14,16]. All ML analyses were run in Matlab 7.4 with PRTools v2004 [16]. To assess intrafarm variability vs. interfarm separability we examined the spectral angle for each pixel [20] in each management unit at both resolutions using ENVI 4.1.

9.3 RESULTS

9.3.1 OVERALL ANALYSIS AT 16- AND 30-M SPATIAL RESOLUTION

The forward feature selection process highlighted the bands that provided the best separability for the four farms at both resolutions. At 16-m spatial resolution, eight spectral bands were optimal (based on the nearest-neighbor criterion [nnc = 0.985]— b1–b5: 459.8–518.3 nm; b12: 621.9 nm; b15: 665.2 nm; and b122: 2443.6 nm). Similarly, for the imagery resampled to 30-m resolution, eight spectral bands showed the most separability among farms (b1–b3: 473.9–488.8 nm; b11: 607.5 nm; b70: 1503.6 nm; b96: 1989.1 nm; b98: 2027.3 nm; b113: 2295.8 nm); the criterion value was 0.980. Figure 9.3 illustrates the separability of the four farms based on the eight dimensions (bands) for both resolutions.

At 16-m resolution, the spectral angle indicates the greatest difference between HR04 and the control farm SF_C, and the greatest similarity between LD_02 and SF_C is the narrowest (fig. 9.3a). Nevertheless, all four farms tend to form well-defined clusters with minimal overlap. At 30-m resolution (fig. 9.3b) the same pattern can be seen in the eight-dimensional space, but the clusters are not as well defined and there is greater overlap, especially between LD_02 and SF_C (fig. 9.3b). Results from the classifiers at both resolutions using the optimal eight spectral bands indicate that the spectral response of the four farms can be reliably separated—that is, low training and validation errors (<2%) with all classifiers except lmnc2 and lmnc3 (66.58% training/testing errors) at 16-m resolution and treec (6.7% training error) at 30-m resolution (table 9.2).

9.3.2 INTERFARM PAIRWISE COMPARISON

We also examined interfarm separability through forward feature selection for a matrix of pairwise analyses at both spatial resolutions. With only two farms per comparison, the classification problem (definition of the decision boundaries) was simpler than the overall analysis of four management units. For each interfarm pairwise analysis, a number of bands representing the entire spectral range presented perfect separability (based on the feature selection process). The spectral regions highlighted by the feature selection are illustrated for each farm pair in figures 9.4 and 9.5. A subset of the optimal bands was chosen as input for the classifiers. At 16-m resolution, the visible region was among the wavelength regions chosen for each pairwise comparison except for the LD02–SF_C comparison (fig. 9.4e); these

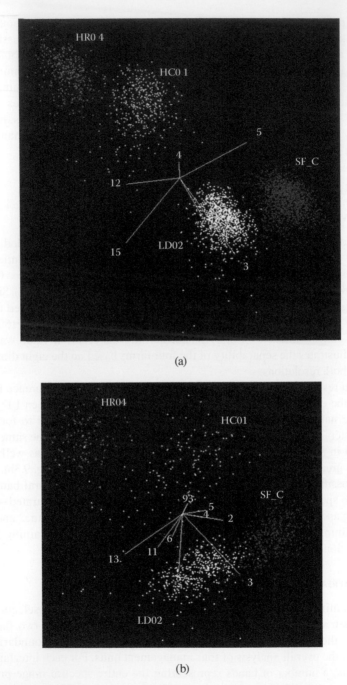

FIG. 9.3 *n*-Dimensional visualization of the four management units at (a) 16-m spatial resolution and (b) data resampled to 30-m resolution. See CD for color image.

TABLE 9.2

Training and Testing Errors at 16- and 30-m Resolutions
(All Classifiers) for Discriminating Spectra of All Four
Management Data Sets

	16-m Resolution		30-m Resolution	
	Training error (%)	Testing error (%)	Training error (%)	Testing error (%)
	Nonparametric classifiers			
treec	0	1.84	0	6.7
lmnc 2	66.58	66.58	0	0
lmnc 3	66.58	66.58	0	0.67
lmnc 4	0	0.1	0	1.67
lmnc 5	0.1	0.1	0	0.84
knnc 2	0.31	0.7	0	0.67
	EM clustering			
emclust-nmc	0.11	0.11	0.45	0.45

two farms are the closest in n-dimensional spectral space (fig. 9.4e, inset). Similarly, the near infrared was selected for each pairwise comparison except HR04 vs. HC01 (fig. 9.4b).

At 30-m resolution a similar trend is seen where the visible wavelengths are limited in the LD02–SF_C comparison (fig. 9.5e) and no near-infrared wavelengths were chosen for HR04–HC01 (fig. 9.5b). Results from the 16-m resolution were similar; LD02 and SF_C are the closest in n-dimensional spectral space (fig. 9.5e, inset). Table 9.3 illustrates the classification results (testing and training errors) for each comparison pair. It can be seen that all classifiers with the exception of lmnc3 for HC02 vs. LD_06 (16 m) and HR04 and SF_C (16 m) have less than 1% training and testing errors. At 30 m both training and testing errors are less than 5% for all pairs, except lmc3 for HR04 vs. SF_C (19.78%).

Figure 9.6 illustrates the four management units with the pixels color coded to represent the management units to which they are more similarly based on the spectral angle. At both resolutions pixels most similar to HC01 are found in the HR04 management unit, while pixels most similar to LD02 are found in the control management unit SF_C. At both resolutions LD02 is homogeneous with only a few pixels more similar to SF_C.

9.4 DISCUSSION

In this study we compared the spectral signatures of four management units that had been logged at different times (0.3, 2, and 3 years since management and one unmanaged unit) based on hyperspectral data using machine learning techniques. Our results indicate that it is possible to separate and accurately classify these management units with less than 5% error (table 9.2 and fig. 9.3). The management

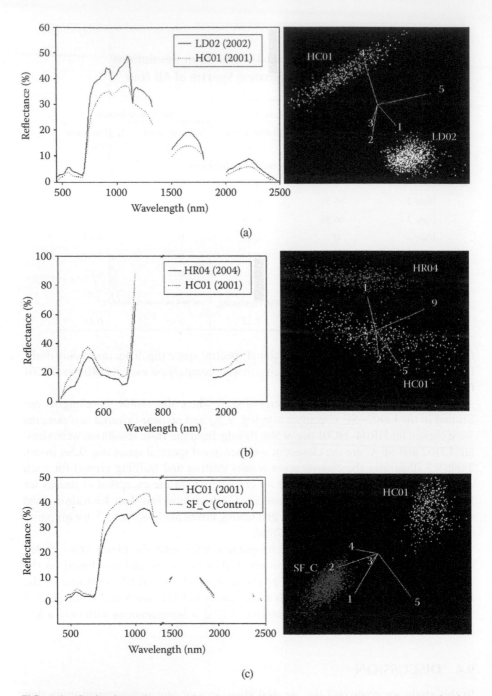

FIG. 9.4 Optimal wavelength regions for separability based on the forward feature selection for each interfarm comparison pair at 16-m spatial resolution. Insets represent *n*-dimensional visualization of each management unit pair (five bands shown). (a) HC01 vs. LD02; (b) HC01 vs. HR04; (c) HC01 vs. SF_C; (d) HR04 vs. LD02; (e) LD02 vs. SF_C; (f) HR04 vs. SF_C.

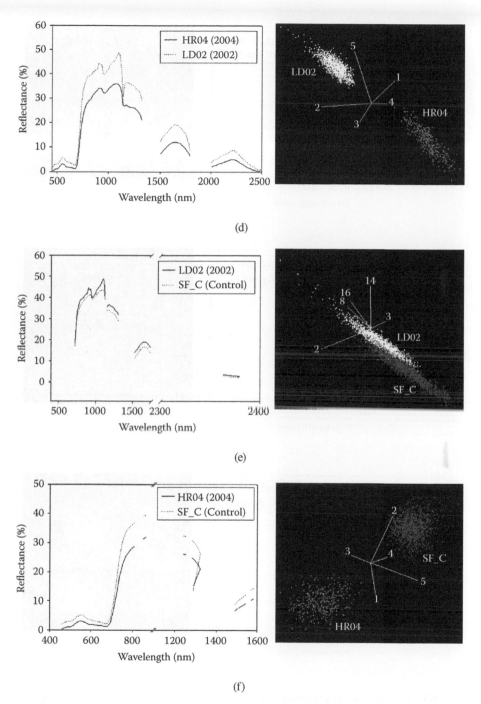

(d)

(e)

(f)

FIG. 9.4 *continued.*

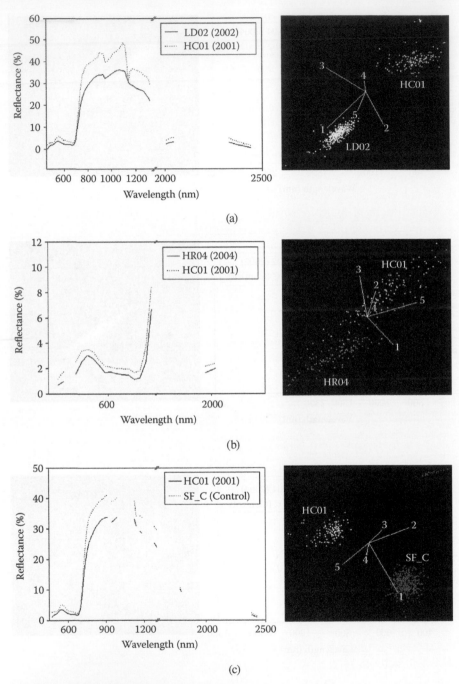

FIG. 9.5 Optimal wavelength regions for separability based on the forward feature selection for each interunit comparison pair at 30-m spatial resolution. Insets represent *n*-dimensional visualization of each management unit pair (five bands shown). (a) HC01 vs. LD02; (b) HC01 vs. HR04; (c) HC01 vs. SF_C; (d) HR04 vs. LD02; (e) LD02 vs. SF_C; (f) HR04 vs. SF_C.

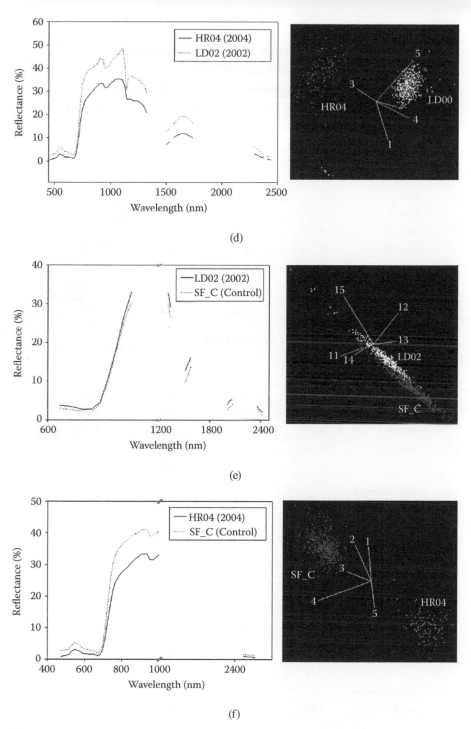

(d)

(e)

(f)

FIG. 9.5   *continued.*

TABLE 9.3

Training and Testing Errors at 16- and 30-m Spatial Resolution for Interunit Analysis

	16-m Resolution		30-m Resolution	
	Training error (%)	Testing error (%)	Training error (%)	Testing error (%)
HC01 vs. LD02				
Nonparametric classifiers				
treec	0	0	0	0.06
lmnc 2	0	0	0	0
lmnc 3	0	97.37	0	0
lmnc 4	0	0	0	0
lmnc 5	0	0	0	0.06
knnc 2	0	0	0	0
EM clustering				
emclust-nmc	<0.01	<0.01	0.06	0.06
HC01 vs. HR04				
Nonparametric classifiers				
treec	0	0.8	0	0
lmnc 2	0.22	0.22	0	2.56
lmnc 3	0	0.45	0	4.27
lmnc 4	0	0.22	0	1.71
lmnc 5	0	0.45	0	1.71
knnc 2	0	0	0.85	0.85
EM clustering				
emclust-nmc	0.09	0.08	0.88	0.88
HC01 vs. SF_C				
Nonparametric classifiers				
treec	0	0.35	0	1
lmnc 2	0	0.09	0	0
lmnc 3	0	0	0	0
lmnc 4	0	0	0	0
lmnc 5	0	0	0	0
knnc 2	0	0	0	0
EM clustering				
emclust-nmc	0.09	0.09	0.03	0.03

TABLE 9.3 (continued)
Training and Testing Errors at 16- and 30-m Spatial
Resolution for Interunit Analysis

	16-m Resolution		30-m Resolution	
	Training error (%)	Testing error (%)	Training error (%)	Testing error (%)
		LD02 vs. HR04		
Nonparametric classifiers				
treec	0	0	0	0
lmnc 2	0	0	0	0
lmnc 3	0	0	0	0
lmnc 4	0	0	0	0
lmnc 5	0	0	0	0
knnc 2	0	0	0	0
EM clustering				
emclust-nmc	0.11	0.11	0.01	0
		LD02 vs. SF_C		
Nonparametric classifiers				
treec	0	0.13	0	0.91
lmnc 2	0.2	0.13	0	0
lmnc 3	0	0	0	0
lmnc 4	0	0.13	0	0
lmnc 5	0	0.07	0	0
knnc 2	0	0.73	0	0
EM clustering				
emclust-nmc	0.07	0.07	0.31	0.31
		HR04 vs. SF_C		
Nonparametric classifiers				
treec	0	0	0	0
lmnc 2	0	0	0	0
lmnc 3	0	97.37	19.78	19.78
lmnc 4	0	0	0	0
lmnc 5	0	0	0	0
knnc 2	0	0	0	0
EM clustering				
emclust-nmc	0.19	0.19	0.04	0.04

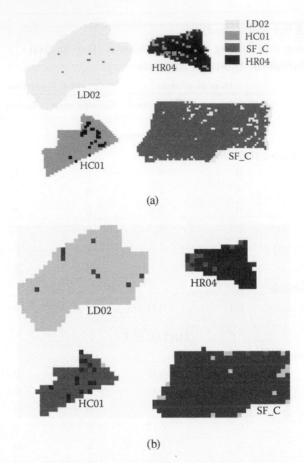

FIG. 9.6 Similarity of pixels based on the spectral angle using the optimal eight bands chosen with the forward feature selection. Red represents pixels most similar to HR04; green: HC01; yellow: LD01; and blue: SF_C at (a) 16-m spatial resolution and (b) 30-m spatial resolution. See CD for color image.

units considered in this study fall between reduced impact logging (RIL) based on disturbance (percent) compared to those in the Amazon forest (unit HR04 = 4.79%) and conventional logging (CL) (unit HC01 = 8.76%), with the third unit intermediate between these two management intensities (unit LD02 = 5.96%). Asner et al. [9], Matricardi et al. [21], and Pereira et al. [22] have shown at broad scales that changes in canopy texture (derived from imagery) due to logging disturbance are a driver for differences seen in the spectral reflectance of management units. The management units assessed in this study have undergone various intensities of logging resulting in a range of disturbances (table 9.1). The resulting differences in canopy degradation may be responsible for the clear differences in the spectral signatures of the units (<5% error). Recovery in both gaps and extraction roads [23] also influences the spectral signature of the management units. While a certain amount of intra-unit spectral variation is expected, figure 9.6 illustrates that intrafarm variation is minimal in comparison to differences between the units.

At 16 m the eight wavelengths (features) driving these differences are found primarily in the visible range (459.6–665.2 nm), with one band in the shortwave (2243.6 nm). Kalacska et al. [24] and Asner [25] showed that for high leaf area index (LAI) values reflectance can be influenced by pigment content (biochemical features are enhanced by the multiple scattering between the leaves). The one band in the shortwave is indicative of differences primarily in canopy architecture and water content [15,25,26]. Variations in canopy leaf angle may also be responsible for differences in the spectral response in both the visible and shortwave regions [26]. Similar results were also seen in the optimal bands selected at 30 m with the addition of two bands primarily influenced by water absorption (1200 and 1900 nm). However, at both scales, to discern the quantitative contribution of each canopy element to the separability of the four management units, an unmixing analysis of the spectra is required [27] (chapter 8).

For the between-management unit comparison our results indicate a lower separability for the management units that do not include wavelengths from the visible (insets in figs. 9.4e and 9.5e). A potential explanation for the aforementioned trend is that LD02 and SF_C (table 9.1) have a similar canopy structure following the recovery of LD02 after logging. LD02 underwent overall what is most similar to RIL (5.95% disturbance) compared to HC01 (8.76%). Although HR04 also underwent RIL (4.97%), this unit only had 0.3 years of recovery following logging, resulting in a canopy that was still disturbed at the time of image acquisition.

The unusually high error in table 9.3 for the lmnc3 classifier is an artifact of the neural network having been overfitted in the training process [14,28]—a common tendency of neural networks. The higher impact of the conventional logging on HC01 (i.e., leading to a large variety of canopy characteristics) is also expressed in the dispersion pattern of the spectra in multidimensional space. In figures 9.3–9.5 the class shape of HC01 (defined by the covariance matrix) is less defined, resulting in greater dispersion. Another trend seen in the results may also be explained by the disturbance and recovery processes: In figures 9.4(a), 9.4(c), 9.4(d), and 9.4(e), the units with the most mature and potentially advanced canopy recovery (SF_C and LD02) have considerably higher reflectance in the near infrared (750–1100 nm) and may be due to the effect of sunlit pixels at the subcrown scale (i.e., 16 m). In comparison, HR04 and HC01, which are presumed to have ongoing disturbance effects (i.e., several canopy gaps), would be subject to shading effect at the sub-canopy scale [13].

At 30-m resolution we expect the reflectance to be more indicative of canopy level effects. For example, in figures 9.5(a) and 9.5(d) the management unit LD02 has a lower reflectance in the near infrared (750–1100 nm). With the sunlit effect partially negated at this resolution, canopy water content as shown by Matricardi et al. [21] may account for this difference. The canopy in LD02 (in an advanced stage of recovery) is presumed to have a higher canopy water content than HR04 or HC01 because of its close resemblance to a closed canopy. HR04 and HC01 in comparison are recently and highly disturbed, respectively. The potential explanations of the trends seen in our results will require additional field data for confirmation, as was done by Asner et al. [9]. Nevertheless, we show the benefit of hyperspectral data is that it is a tool that can provide an integrated picture of different recovery patterns following selective logging.

Forest management units in the study are conducted on a smaller scale (10–100 ha) compared to the logging operations in other tropical areas (i.e., Brazil). Nonetheless, these management units have both temporal (i.e., time since logging) and disturbance (low-impact vs. conventional logging) variability that can be used to study the general patterns for assessing forest recovery after logging using hyperspectral data and machine learning techniques. Both the data and the analytical techniques have shown to be general (i.e., examine overall recovery trends), yet specific enough to be able to detect differences between the management units. This study is a positive first step toward examining the utility of such data for monitoring forest management of smaller units at the landscape level. A follow-up of this study will examine in greater detail the contribution of additional logging variables (e.g., extraction roads, specific tree canopy effects, species richness, and topography, among others) at the management level through endmember extraction and pixel unmixing. Understanding of logging impacts at the landscape-scale level is key to enhancing conservation practices outside protected areas. Our study is the first attempt to implement these techniques in a biological corridor in Costa Rica.

ACKNOWLEDGMENTS

We would like to acknowledge the Center for Environmental Sciences and Engineering at the University of Connecticut, the Bamford Research Award (Ecology and Evolutionary Biology Department) at the University of Connecticut, and the National Science Foundation. Logistical support for data analysis was provided by the Interdisciplinary Research Center in the Mathematical and Computing Sciences (IRMACS) at Simon Fraser University. FUNDECOR kindly provided the forest management data and the Costa Rican National Center for High Technology provided the hyperspectral data. Also, we would like to express our gratitude to the Organization for Tropical Studies for the field logistic and informatics support while in Costa Rica.

REFERENCES

1. Food and Agricultural Organization (FAO), Global forest resources assessment 2005: Progress towards sustainable forest management, FAO Forestry paper 147, 2005, 348.
2. Bawa, K.S. and Seidler, R., Natural forest management and conservation biodiversity in tropical forests, *Conservation Biology*, 12, 46, 1998.
3. Asner, G.P., Broadbent, E.N., Oliveira, P.J.C., et al., Condition and fate of logged forests in the Brazilian Amazon, *Proceedings of the National Academy of Sciences*, 103, 12947, 2006.
4. Nicotra, A.B., Chazdon, R.L., and Iriarte, S.V.B., Spatial heterogeneity of light and woody seedling regeneration in tropical wet forests, *Ecology*, 80, 1908, 1999.
5. Guariguata, M.R. and Dupuy, J., Forest regeneration in abandoned logging roads in lowland Costa Rica, *Biotropica*, 29, 15, 1997.
6. Nabe-Nielsen, J., Severiche, W., Fredericksen, T., et al., Timber tree regeneration along abandoned logging roads in a tropical Bolivian forest, *New Forests*, 34, 31, 2007.

7. Hall, J.S., Medjibe, V., Berlyn, G.P., et al., Seedling growth of three co-occurring *Entandrophragma* species (Meliaceae) under simulated light environments: Implications for forest management in Central Africa, *Forest Ecology and Management*, 179, 135, 2003.

8. Souza, C., Firestone, L., Moreira Silva, L., et al., Mapping forest degradation in the Eastern Amazon from spot 4 through spectral mixture models, *Remote Sensing of Environment*, 87, 494, 2003.

9. Asner, G.P., Keller, M., Pereira, R., et al., Remote sensing of selective logging and textural analysis, *Remote Sensing of Environment*, 80, 483, 2002.

10. Broadbent, E.N., Zarin, D.J., Asner, G.P., et al., Recovery of forest structure and spectral properties after selective logging in lowland Bolivia, *Ecological Applications*, 16, 1148, 2006.

11. Aguilar-Amuchastegui, N. and Henebry, G.M., Monitoring sustainability in tropical forests: How changes in canopy spatial pattern can indicate forest stands for biodiversity surveys, *IEEE Geoscience and Remote Sensing Letters*, 3, 329, 2006.

12. Ustin, S.L. and Trabucco, A., Using hyperspectral data to assess forest structure, *Journal of Forestry*, 98, 47, 2000.

13. Clark, M.L., Roberts, D.A., and Clark, D.B., Hyperspectral discrimination of tropical rain forest tree species at leaf to crown scales, *Remote Sensing of Environment*, 96, 375, 2005.

14. Fielding, A.H., *Machine learning methods for ecological applications*. Kluwer Academic Press, Dordrecht, 1999.

15. Kalacska, M., Bohlman, S., Sanchez-Azofeifa, G.A., et al., Hyperspectral discrimination of tropical dry forest lianas and trees: Comparative data reduction approaches at the leaf and canopy levels, *Remote Sensing of Environment*, 109, 406, 2007.

16. van der Heijden, F., Duin, R.P.W., de Ridder, D., et al., *Classification, parameter estimation and state estimation: An engineering approach using Matlab*. John Wiley & Sons, West Sussex, U.K., 2004.

17. Jimenez, L.O. and Landgrebe, D., Supervised classification in high-dimensional space: Geometrical, statistical and asymptotic properties of multivariate data, *IEEE Transactions on Systems, Man and Cybernetics—Part C: Applications and Reviews*, 28, 39, 1998.

18. Holdridge, L.R., *Life zone ecology*. Tropical Science Center, San Jose, Costa Rica, 1967.

19. Defernez, M. and Kemsley, E.K., The use and misuse of chemometrics for treating classification problems, *Trac-Trends in Analytical Chemistry*, 16, 216, 1997.

20. Kruse, F.A., Lefkoff, A.B., Boardman, J.B., et al., The spectral image processing system (SIPS)—Interactive visualization and analysis of imaging spectrometer data, *Remote Sensing of Environment*, 44, 145, 1993.

21. Matricardi, E.A.T., Skole, D.L., Cochrane, M.A., et al., Multitemporal assessment of selective logging in the Brazilian Amazon using Landsat data, *International Journal of Remote Sensing*, 28, 63, 2007.

22. Pereira, R., Zweede, J.C., Asner, G.P., et al., Forest canopy damage and recovery in reduced-impact and conventional selective logging in eastern Para, Brazil, *Forest Ecology and Management*, 168, 77, 2002.

23. Dickinson, M.B., Whigham, D.F., and Herman, S.M., Tree regeneration in felling and natural treefall disturbances in a semideciduous tropical forest in Mexico, *Forest Ecology and Management*, 134, 137, 2000.

24. Kalacska, M., Sanchez-Azofeifa, G.A., Rivard, B., et al., Ecological fingerprinting of ecosystem succession: Estimating secondary tropical dry forest structure and diversity using imaging spectroscopy, *Remote Sensing of Environment*, 108, 82, 2007.

25. Asner, G.P., Biophysical and biochemical sources of variability in canopy reflectance, *Remote Sensing of Environment*, 64, 234, 1998.

26. Gamon, J.A., Kitajima, K., Mulkey, S.S., et al., Diverse optical and photosynthetic properties in a neotropical dry forest during the dry season: Implications for remote estimation of photosynthesis, *Biotropica*, 37, 547, 2005.
27. Rogge, D.M., Rivard, B., Feng, J., et al., Iterative spectral unmixing for optimizing per-pixel endmember sets, *IEEE Transactions on Geoscience and Remote Sensing*, 44, 3725, 2006.
28. Russell, S. and Norvig, P., *Artificial intelligence: A modern approach,* 2nd ed. Prentice Hall, Upper Saddle River, NJ, 2003.

10 A Technique for Reflectance Calibration of Airborne Hyperspectral Spectrometer Data Using a Broad, Multiband Radiometer

Tomoaki Miura, Alfredo R. Huete,
Laerte Guimaães Ferreira, Edson E. Sano, and
Hiroki Yoshioka

CONTENTS

10.1 INTRODUCTION

Airborne radiometers have been employed for various remote sensing applications, including optical characterizations of forest canopies [1], surface reflectance factor retrievals from satellite imagery [2], spatial extrapolation of evaporation over agricultural fields [3], and scaling up and validation of satellite data products [4,5]. These studies have found the utility of airborne radiometers because (1) they can rapidly obtain data over rather large areas, encompassing fields having considerably different surface properties [3]; and (2) they can measure reflectance factors of tall objects that cannot be reached from ground (e.g., forest canopies) [1]. Airborne or helicopter-based radiometers are typically flown at low altitudes (<150 m above ground level) "below atmosphere." On the ground, another radiometer (the same model) cross-calibrated with the ones in air is set up, continuously measuring radiances reflected off a white reference panel during the flights, to later convert airborne data to reflectance factors by ratioing (e.g., Holm et al. [5]).

Recently, several researchers deployed a field spectrometer on an airplane to acquire hyperspectral data for characterizations of top-of-canopy reflectance (e.g., Ferreira et al. [6] and Remer, Wald, and Kaufman [7]). An issue associated with these applications was the conversion of airborne hyperspectral data to reflectance factors, or reflectance calibration. Remer et al. [7] used a field spectroradiometer in a "reflectance" mode by calibrating it against a white barium sulfate plate immediately before boarding the aircraft. Ferreira et al. [6], on the other hand, used a spectrometer in an "uncalibrated, raw digital number" mode. The acquired data were calibrated to ground reflectance after the flight by taking a ratio to reference values that were linearly interpolated from the readings made over a Spectralon® white reference panel before and after the flight, based on the time stamps recorded with each of the airborne reflectance spectra. In the former case incoming solar irradiances were assumed unchanged, whereas in the latter they were assumed linearly changed. It is reasonable to presume that such a field spectrometer is too expensive and/or too valuable to have a second one dedicated solely for continuous measurements of incoming solar radiation on the ground.

In this chapter, we introduce a new technique to convert or calibrate airborne hyperspectral data to ground reflectance factors. This method takes advantage of reference panel readings continuously made with a multispectral radiometer on the ground during a flight and uses them to "adjust" the panel readings made with a spectrometer before and after the flight. In the remainder of this chapter, we first describe the theoretical background of the approach. Evaluation and validation results of the developed technique are then presented. Finally, we demonstrate the utility of reflectance data derived with this technique by presenting application results of the data to biophysical characterization in a tropical forest–savanna transitional region.

10.2 THEORY AND APPROACH

The reflectance factor is defined as the ratio of the radiant flux reflected by a surface to that reflected into the same reflected-beam geometry by an ideal (lossless), perfectly diffuse (Lambertian) standard surface irradiated under the same conditions [8].

Using a field spectrometer or radiometer, the reflectance factor for the unknown surface, R_T, is determined from

$$R_T(\theta_i) = \frac{DN_T}{DN_R} R_R(\theta_i) \tag{10.1}$$

where θ_i is the solar zenith angle and the spectrometer's optical axis is parallel to the surface normal (i.e., nadir-looking geometry); DN_T and DN_R are the digital numbers from the spectrometer (with the dark current, DC, subtracted) when the instrument is viewing the target and reference, respectively; and R_R is the reflectance factor of the reference panel with respect to a Lambertian surface of unit reflectance [9].

The issue associated with airborne field spectrometer measurements is the difficulty in obtaining DN_R that corresponds to the time when DN_T is recorded, as described briefly in the introduction section. The technique introduced in this chapter uses a multiband radiometer that continuously measures a reference surface (panel) on the ground during a flight to obtain DN_R for the reflectance factor derivations. This approach assumes that (1) there is a field spectrometer to be flown on an airplane and there is a multiband radiometer for the "continuous panel" measurements on a ground, and (2) there are two plates: one to be taken to an airport with the field spectrometer and the other to be used on the ground for the continuous measurements.

Because of the uniformity in the reflective properties of the panel surfaces, the readings made with the spectrometer and multiband radiometer over the panels can be expressed in the irradiance terms (W/m²/μm):

$$DN_{R1} = \frac{E_0 R_{R1}(\theta_i)}{c_{DN}\pi} \int_{\omega_{r1}} d\omega_{r1} \tag{10.2}$$

$$V_{R2} = \frac{E_0 R_{R2}(\theta i)}{c_v \pi} \int_{\omega_{r2}} d\omega_{r2} \tag{10.3}$$

where

  DN_{R1} is the DC subtracted digital numbers from the spectrometer over the reference panel R_1
  V_{R2} is the DC subtracted voltages from the radiometer over the continuous panel R_2
 R_{R1} and R_{R2} are the reflectance factors of the two plates
 c_{DN} and c_v are the calibration coefficients for the spectrometer and radiometer, respectively (W/m²/μm/DN or W/m²/μm/voltage)
  E_0 is the band-averaged incoming solar irradiance at ground level (W/m²/μm)
 ω_{r1} and ω_{r2} are the field of view (FOV) of the spectrometer and radiometer, respectively

Equations 10.2 and 10.3 assume that the difference in the band-averaged solar irradiances measured by these two sensors is negligible. The solar zenith angles are assumed to be the same, which is considered reasonable when the airport and the study site are in proximity.

Solving equation 10.3 for E_0 and substituting E_0 in equation 10.2 derive the following equation:

$$DN_{R1} = \frac{\int_{\omega_{r1}} d\omega_{r1}}{\int_{\omega_{r2}} d\omega_{r2}} \bullet \frac{c_v}{c_{DN}} \bullet \frac{R_{R1}(\theta_i)}{R_{R2}(\theta_i)} \bullet V_{R2} + \varepsilon \qquad (10.4)$$

or

$$DN_{R1} = C_{V \to DN} \bullet \frac{R_{R1}(\theta_i)}{R_{R2}(\theta_i)} \bullet V_{R2} + \varepsilon \qquad (10.5)$$

where

$$C_{V \to DN} = \frac{\int_{\omega_{r1}} d\omega_{r1}}{\int_{\omega_{r2}} d\omega_{r2}} \bullet \frac{c_v}{c_{DN}} \qquad (10.6)$$

The newly derived coefficient, $C_{V \to DN}$, is the cross-calibration coefficient that relates V_{R2} to DN_{R1}, which shows the overall differences in the two sensors' outputs due to FOVs and absolute calibration. This cross-calibration coefficient can be derived using the coincident measurements made with the two sensors before and after the flight.

Considering that the ground radiometer has multibands and that the estimation of DN_R involves the measurement time, the equations are expanded to explicitly include these two factors. Then, DN_{R1} for the band b at the time t is predicted from

$$\widehat{DN}_{R1}(b,t) = C_{V \to DN}(b) \bullet \frac{R_{R1}(b,\theta_i)}{R_{R2}(b,\theta_i)} \bullet V_{R2}(b,t) \qquad (10.7)$$

The predicted DN_{R1} at several discrete spectral bands can be made into a single correction factor value, $CF(t)$, by assuming that the magnitudinal changes in the incoming solar irradiance are independent of wavelength:

$$CF(t) = \frac{1}{n} \sum_{b=1}^{n} \frac{\widehat{DN}_{R1}(b,t)}{DN_{R1}^{U}(b,t)} \qquad (10.8)$$

and

$$DN_{R1}^{U} = DN_{R1}(b,t_0)\frac{t_e-t}{t_e-t_0} + DN_{R1}(b,t_e)\frac{t-t_0}{t_e-t_0} \quad (10.9)$$

where DN^U_{R1} is the linear-interpolated DN_{R1} based on those made before and after the flight at the times t_0 and t_e, respectively. Finally, the adjusted or corrected DN^C_{R1} to convert the airborne data to reflectance spectra can be derived by applying the correction factor to all the spectral bands:

$$DN_{R1}^{C}(\lambda,t) = DN_{R1}^{U}(\lambda,t) \bullet CF(t) \quad (10.10)$$

where λ is used for the bands to distinguish the narrow spectral bands of the airborne spectrometer from the broad bands used to derive CF in equations 10.7–10.9.

Theoretically, the potential source of uncertainty in the predicted DN_{R1} is the assumption used for equating (10.2) and (10.3). Relationships of DN_{R1} and V_{R2} to E_0 may be significantly different due to differences in FOV and spectral bandpasses, and spatial heterogeneity in atmospheric conditions if the airport and location where a continuous panel is set up are far apart. In the next section, we present our validation results.

10.3 PROOF-OF-CONCEPT VALIDATION IN SEMI-ARID GRASSLAND

In order to evaluate and validate this new reflectance calibration technique, a field experiment was conducted at the Jornada Experimental Range on October 5, 2002. Our design of the experiment was to simulate the airborne measurement protocols on the ground by performing transect measurements using a field spectrometer with a multiband radiometer set up for continuous panel measurements.

10.3.1 Site Description and Experimental Design

The Jornada Experimental Range is located 37 km north of Las Cruces, New Mexico, on the Jornada del Muerto Plain in the northern part of the Chihuahuan Desert. The climate is semi-arid with a mean annual precipitation of 210 mm and a mean annual temperature of 16°C. The range was once black grama (*Boutelua eriopoda*)-dominated grasslands; due to heavy grazing and fire suppression, however, the grasslands have been transformed into dunelike mesquite (*Prosopis glandulosa*) shrublands and creosote bush (*Larrea tridentate*) communities [10,11]. Black grama-dominated grasslands are found only as small patches within the range.

A 100-m transect was drawn in the north–south direction on a grassland area within the range (N 32.58914°/W 106.84277°, 1330-m elevation), on which flags were installed every 2 m as markers. On the transect, most plants were black grama, but there were also several yucca (*Yucca elata*) plants.

Near the south end of the transect, a four-band radiometer (Exotech Model 100BX) was set up on a tripod to nadir-look on the leveled surface of a Spectralon white standard panel (R_2) (Labsphere, Inc., North Sutton, New Hampshire). The bandpasses of this Exotech radiometer were adjusted to approximate the first four

TABLE 10.1

**Bandpasses (Full Width at Half Maxima) of the
Exotech Radiometer (SN 3672) and MODIS Sensor**

Band name	Exotech bandpass (nm)	MODIS bandpass (nm)
Channel A (blue)	456–475	459–479
Channel B (green)	544–564	545–565
Channel C (red)	623–670	620–670
Channel D (NIR)	838–876	841–876

Source: Huete, A. et al., *The Earth Observer*, 11, 22, 1999.

bands of a Moderate Resolution Imaging Spectroradiometer (MODIS) (table 10.1) [12]. The radiometer was fixed at 30 cm above the panel surface to ensure that no shadow was cast on the panel surface by the radiometer itself at around the local solar noon time. With 15° field-of-view (FOV) lenses attached, an 8- by 8-cm area of the panel was viewed by each of the four Exotech bands. The recording of plate-reflected radiations was made every second from 9:20 a.m. to 1:20 p.m. MST.

A full-range hyperspectral field spectrometer (Analytical Spectral Devices [ASD], Inc., Boulder, Colorado) was used to measure reflected radiation over the transect. A "fiberoptic gun" extended from the spectrometer was mounted on a back-pack "yoke," leveled, and held at 1.5 m aboveground. An 18° foreoptic was attached to the gun, and thus the spectrometer was sensing an approximately 47- by 47-cm area on the ground for each scan. The ASD spectrometer acquires spectral data in 1.4-nm intervals in the visible/near infrared (VNIR) (380–1300 nm; full width at half maximum [FWHM] = 3–4 nm) and 2.2 nm in the shortwave infrared (SWIR) region (1300–2450 nm; FWHM = 10–12 nm).

The ASD spectrometer contains three subspectrometers inside, each of which operates on different wavelength regions: the first VNIR subspectrometer measures light between 380 and 972 nm, the second SWIR1 subspectrometer covers the region between 972 and 1767 nm, and the last SWIR2 subspectrometer covers the remaining region. "Optimization" is periodically performed on the instrument to adjust sensitivity of the three subspectrometers to varying conditions of illumination. Whereas DC measurements are performed at every scan for the SWIR1 and SWIR2 subspectrometers, they are performed only at optimizations or upon a user's request for the VNIR subspectrometer. This difference in the DC correction schemes between the VNIR and the other two subspectrometers often results in a glitch or discontinuity at 972 nm in the acquired spectra.

Transect measurements were made by moving first from the south end to the north end of the transect and then from the north to south ends. Ten spectra were recorded at every flag and internally averaged to produce one spectrum per flag. Before and after the transect measurements, reference data were collected over another Spectralon white standard panel (R_1) leveled and placed near the Exotech radiometer, bracketing the ASD transect measurements. This sequence of measurements (i.e., panel, transect, and panel) took 10 minutes to complete and was repeated every 30 minutes starting at 9:30 a.m. eight times.

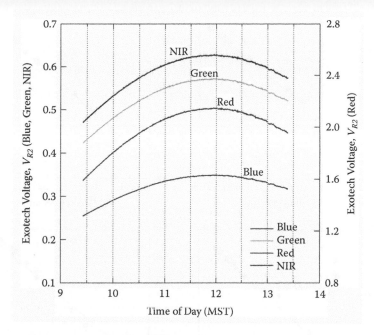

FIG. 10.1   Exotech radiometer data acquired over a Spectralon white reference panel in the Jornada Experimental Range on October 5, 2002 (local solar noon at 11:55 a.m.). See CD for color image.

10.3.2   WHITE STANDARD PANEL DATA

The uncalibrated voltage readings of the Exotech radiometer (V_{R2}) are plotted in figure 10.1. October 5, 2002, was a perfectly clear day and no cloud cover developed during the 4-hour measurement period. The Exotech voltages curve linearly changed as the solar zenith angle changed. The maximum values were obtained at the local solar noon of 11:55 a.m. MST. Based on this figure, a linear change in the voltages could reasonably be assumed if a measurement period were limited to less than 30 minutes.

In figure 10.2, the uncalibrated ASD digital numbers over the panel surface (DN_{R1}) are plotted against $(R_{R1}/R_{R2}) \cdot V_{R2}$ to evaluate validity of equation 10.5. Among the numerous number of ASD bands available, those that were positioned near the center wavelengths of the Exotech bands were chosen for this purpose. Squares of the correlation coefficients, or the coefficients of determination (R^2 values), shown in the figure were derived by fitting a simple linear model [i.e., $DN_{R1} = \beta_0 + \beta_0 \cdot (R_{R1}/R_{R2}) \cdot V_{R2} + \varepsilon$] to examine the degree of scattering between the two variables. R^2 values were larger than 0.99 for all four bands. The regression lines derived by fitting equation 10.5 to the data (i.e., no intercept model) aligned well with the data points for the blue, green, and red bands (fig. 10.2a–10.2c). There was, however, slight deviation of the regressed line from the data points for the NIR band (fig. 10.2d). This may have been attributed to uncertainties associated with the selected ASD band to correlate with V_{R2} and/or the non-Lambertian corrections of the two panels. Overall, equation 10.5 can be considered valid.

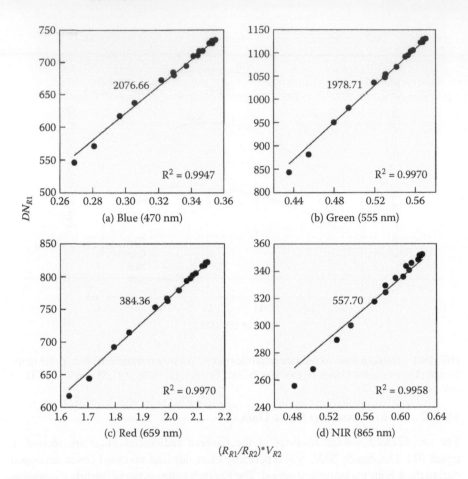

FIG. 10.2   Cross-plots of DN_{R1} and $(R_{R1}/R_{R2}) \cdot V_{R2}$ for four wavelengths corresponding to the center wavelengths of the Exotech bands. See text for descriptions of the variables.

10.3.3   CORRECTION FACTOR

The continuous panel method was applied to the whole transect data set to convert raw DN to reflectance, where the very first and last ASD panel measurements made at 9:30 a.m. and 1:10 p.m. MST, respectively, were used to obtain $C_{V \to DN}$ (b) (see equations 10.6 and 10.7), simulating a 3.5-hour airborne data acquisition scenario. The method was also applied to a subset of the transect data to simulate a 1-hour airborne data acquisition scenario using the ASD panel data measured at 10:00 a.m. and 11:10 a.m. MST. Figure 10.3 shows the correction factor values, $CF(t)$, computed for converting every ASD spectral measurement to reflectance for the two scenarios. For both scenarios, $CF(t)$ changed curve linearly; the value was nearly 1.0 at the start of the measurements, increased, and decreased down to nearly 1.0 at the end of the measurements. Except in the end of the acquisition time period, all of the values were larger than 1.0, adjusting the interpolated plate readings (DN^U_{R1}) to higher values. The maximum $CF(t)$ values were obtained at the center of the

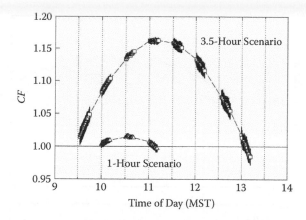

FIG. 10.3   Time series plots of the derived correction factors (*CF*). The figure shows the values for a 3.5-hour flight scenario and a ~1-hour flight scenario.

data acquisition periods: 1.17 at 11:05 a.m. MST and 1.02 at 10:35 a.m. MST for the 3.5- and 1-hour scenarios, respectively.

10.3.4   COMPARISON OF RETRIEVED REFLECTANCE

In figure 10.4, mean reflectance spectra for the 11:30 a.m. transect run derived by three different methods are compared. The 3.5-hour flight scenario was used to derive the reflectance spectra for the linear interpolation and continuous panel methods. In the linear interpolation method, the ASD panel measurements made at 9:30 a.m. and 1:10 p.m. MST were linearly interpolated to compute DN_{RI} values at the time when target measurements were made. The "true" spectral reflectance was obtained by using the ASD panel measurements made immediately before and after the transect run (i.e., a standard protocol for ground reflectance measurements). These mean values were computed by taking an arithmetic mean of all 100 spectra recorded in this transect run.

Both the linear interpolation and continuous panel methods retrieved the basic spectral signature of the target well (fig. 10.4a). The continuous panel method successfully retrieved absolute reflectance factor values as well; the derived reflectance spectrum overlapped accurately with the true reflectance spectrum (fig. 10.4a). In general, differences between the reflectance derived with the continuous panel method and the true reflectance were nearly zero throughout the wavelength range examined here (400–2400 nm) (fig. 10.4b and 10.4c). The linear interpolation method, on the other hand, resulted in a 15% overestimation; the retrieved reflectance values were 0.01–0.06 higher than the true values (fig. 10.4b and 10.4c). Since the time interval between the two ASD panel measurements was significantly long, the retrieved reflectance spectra by these two methods were subject to bias errors due to absorptions by several atmospheric constituents. These included oxygen (O_2), water vapor (H_2O), carbon dioxide (CO_2), and methane (CH_{4+}) (indicated by the arrows in fig. 10.4c).

The retrieved spectra were also subject to instrumental biases. The airborne VNIR reflectance factors (400–972 nm) were positively biased due to subtle drifts

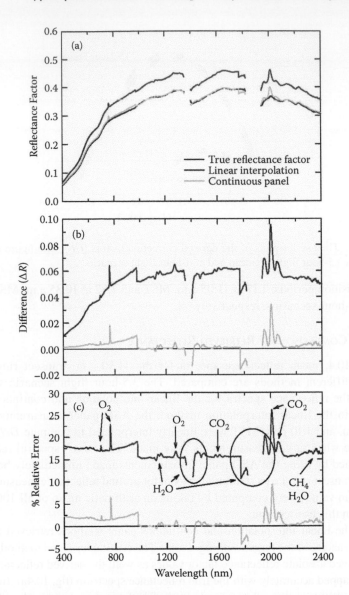

FIG. 10.4 Comparisons of mean reflectance spectra for the 11:30 a.m. MST transect run derived with three different methods: (a) reflectance factors, (b) reflectance differences, and (c) percent relative errors. For the linear interpolation and continuous panel results, the ASD panel measurements taken at 9:30 a.m. and 1:10 p.m. MST were used. See CD for color image.

(variability) in DC in this wavelength region, whereas the true reflectance factor values were seemingly slightly lower than supposed (fig. 10.4). There was another bias in the reflectance spectra at 1767 nm where data from the two SWIR subspectrometers came together (fig. 10.4). The exact reason for the glitch was unclear; however, this data set indicates that airborne ASD reflectance data at wavelengths longer than 1767 nm until the 1900-nm water absorption region can be subject to instrumental biases.

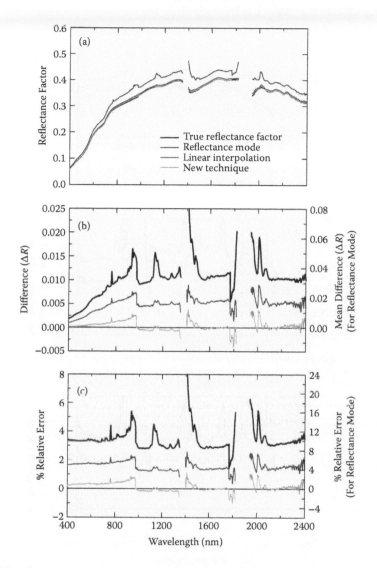

FIG. 10.5 Comparisons of mean reflectance spectra for the 10:30 a.m. MST transect run derived with four different methods. For the reflectance mode measurements, the spectrometer was calibrated at 10:00 a.m. MST. For the linear interpolation and continuous panel results, the ASD panel measurements taken at 10:00 a.m. and 11:40 a.m. MST were used. See CD for color image.

In figures 10.5 and 10.6, mean reflectance spectra for the 10:30 a.m. transect run are compared. Figures 10.5 and 10.6 show the results of 1- and 2-hour flight scenarios, respectively. In these figures, reflectance spectra measured in the "reflectance" mode are also plotted; the spectrometer was calibrated against the white reference panel before the target measurements took place.

The worst result was obtained by the reflectance mode measurements. The reflectance values were, on average, 0.03 (10%) and 0.05 (16%) higher than the true values

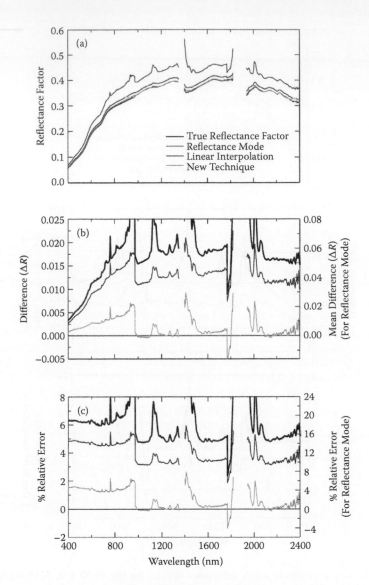

FIG. 10.6 Comparisons of mean reflectance spectra for the 10:30 a.m. MST transect run derived with four different methods. For the reflectance mode measurements, the spectrometer was calibrated at 9:40 a.m. MST. For the linear interpolation and continuous panel results, the ASD panel measurements taken at 9:40 a.m. and 11:30 a.m. MST were used. See CD for color image.

for the 1- and 2-hour flight scenarios, respectively (figs. 10.5 and 10.6). Likewise, the spectral signatures were distorted, particularly over the wavelength regions on which water vapor absorption has a strong impact. The linear-interpolation method resulted in producing reasonable reflectance spectra (figs. 10.5 and 10.6). Absolute differences in reflectance were 0.005 for the 1-hour scenario (fig. 10.5b) and increased to ~0.013 as the duration was extended to 2 hours (fig. 10.6b). The retrieved spectral signatures were subject to atmospheric absorptions (H_2O, CO_2, and O_2), but to much lesser extent than the reflectance mode measurements. The continuous panel method retrieved the most accurate reflectance spectra. Absolute differences were nearly zero for both the 1- and 2-hour flight scenarios, except for the wavelength regions affected by instrumental bias errors (figs. 10.5 and 10.6). The impact of atmospheric absorptions was of the same magnitude as for the linear-interpolation method. Regardless of these bias errors, overall accuracy of the derived spectra was less than 0.005 (1%) and 0.01 (2%) for the 1- and 2-hour flight scenarios (figs. 10.5 and 10.6).

10.3.5 UNCERTAINTY ESTIMATES

Finally, uncertainty associated with the continuous panel method was evaluated using mean differences and root mean square (RMS) errors. All the eight transect data sets were processed into reflectance values using the continuous plate method for the 3.5-hour flight scenario and the standard protocol. The data derived with the latter were used as the "true" reflectance for computing mean differences and RMS errors.

Mean differences were nearly zero except for the wavelength regions affected by either atmospheric absorptions or instrumental noise (>0.005) (fig. 10.7a). For this particular data set acquired in the semi-arid grassland, the VNIR dark current bias errors and atmospheric absorptions were both positive (except for the other instrument bias around 1800 nm). It is, however, possible that these errors are negative or larger, depending on instrument and atmospheric conditions of when and where data are acquired. RMS errors were nearly uniform at 0.005 except for the same wavelength regions affected by atmospheric and/or instrumental noise (fig. 10.7b). The RMS errors also suggest the wavelength regions where the most reliable retrievals can be expected (fig. 10.7b). Based on this data set, absolute accuracy of the reflectance derived with the continuous panel method can be considered 0.005 reflectance units with precision at the same level of 0.005 reflectance units (i.e., measurement uncertainty of 0.005 ± 0.005).

10.4 DEMONSTRATION IN TROPICAL REGION

In this section, we present application results of the continuous panel method in a tropical region in order to demonstrate the utility of spectral reflectance data obtained with this method.

10.4.1 STUDY SITE AND AIRBORNE DATA ACQUISITION

Our study area was located near Santana do Araguaia in Brazil (S 10°5′/W 50°3′) (fig. 10.8). The area represents a tropical forest–savanna transitional zone and consists of a preserved national park, the Araguaia National Park, surrounded by a complex mosaic of undisturbed forest and savanna vegetation formations, and converted pasture areas. The mean annual precipitation in this area is 1670 mm and the mean annual temperature is 26°C.

An airborne campaign was conducted on July 25 and 26, 2001 (dry season). An airborne radiometric measurement system was constructed for this campaign. The system consisted of three components: (1) the airborne radiometric package,

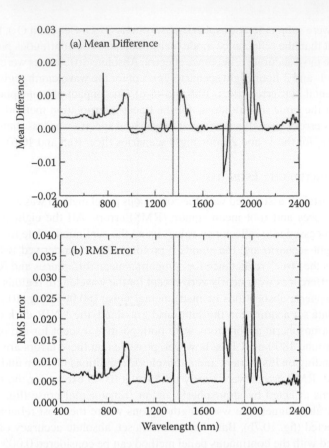

FIG. 10.7 Uncertainty estimates associated with the continuous panel method: (a) mean difference and (b) root mean square error.

including an ASD field spectrometer, GPS device (GPS III+, Garmin International, Inc., Olathe, Kansas), digital camera (C3000, Olympus Imaging America, Inc., Center Valley, Pennsylvania), and Spectralon white reference panel; (2) the ground component, including an Exotech radiometer and another Spectralon white reference panel; and (3) the data processing software package (fig. 10.9). An aluminum case was constructed to mount the fiberoptical cable of the ASD and the digital camera to the airplane. These two instruments were co-aligned carefully so as to capture approximately the same locations on the ground.

The airborne radiometric package was flown "below the atmosphere" at 150 m AGL at the speed of 30 m per second. A 5° foreoptic was attached to the spectrometer and 10 spectra were scanned and internally averaged for a single spectrum collection at every ~1 second. This resulted in the ground spatial resolution of 13 by 30 m. ASD spectrometer readings over the reference panel were made immediately before and after the flights. The Exotech radiometer was set up in the middle of an open field, soon after which the continuous panel data collection started. The flight and ground crews constantly communicated to make sure the continuous panel data period completely bracketed that of the airborne data including the ASD panel readings. The

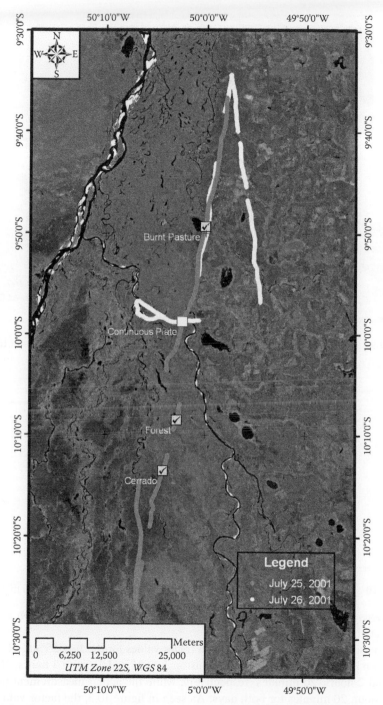

FIG. 10.8   Flight lines of airborne measurements overlaid on top of a Landsat-7 enhanced thematic mapper plus (ETM+) false color image. The image was acquired on July 21, 2001. See CD for color image.

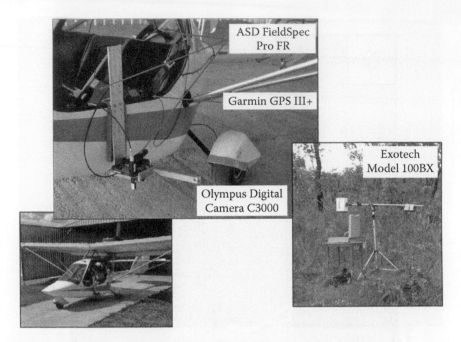

FIG. 10.9   The airborne radiometric package employed in Brazil. See CD for color image.

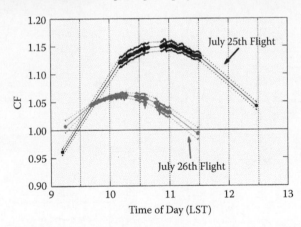

FIG. 10.10   Time series plots of the derived correction factors for the July 25 and 26 flight data.

continuous panel method was applied to the airborne radiometric data soon after the flights on the same days.

Time series of the derived correction factor values for the July 25 and 26 flights are shown in figure 10.10. Flight durations on July 25 and 26 were 3 hours, 20 minutes; and 2 hours, 20 minutes, respectively, while the actual measurement periods were 1 hour, 20 minutes for both days. As seen in figure 10.3, the factor values had cosine-curved shapes and indicate that, if reflectance had been derived using the linear-interpolation method, the retrieved values would have been 15 and 5% larger than the ones derived with the continuous panel method.

10.4.2   SPECTRAL SIGNATURES

In figure 10.11, reflectance spectra derived with the continuous panel method are plotted for three contrasting land cover types. The extraction locations of these spectra are indicated in figure 10.8. The figure demonstrates high quality of the spectral signatures obtained with the continuous panel method. These spectra were slightly subject to some atmospheric absorption features due to variability in atmospheric conditions during the flight, as observed in the preceding sections. The spectrum over a forest area had a typical spectral signature of green vegetation: the low visible reflectance with a green peak at around 550 nm, a sharp transition from the red to NIR wavelength regions followed by leaf water absorption features at around 970 and 1200 nm, and the low reflectance in the SWIR1 and SWIR2 regions (fig. 10.11a).

The spectral signature of a savanna cover type differed from that of a forest, corresponding to less abundance of green vegetation and more abundant standing litters (nonphotosynthetic vegetation [NPV]) (fig. 10.11b). It was characterized by the slightly higher visible reflectance with the green and red reflectances being nearly equal and by the much less sharp red-NIR transition in the reflectance values. The SWIR2 reflectance for this cover type was higher than that for the forest and showed strong ligno-cellulose absorptions, a typical feature of dry plant materials [13]. The spectral signature of a recently burned area was flat and lacked any absorption features, corresponding to little or no green/dry vegetation materials on the ground (fig. 10.11c).

In order to evaluate the consistency of the derived spectral signatures throughout the flight, a segment (transect) of the July 25 data was extracted. The ligno-cellulose vegetation index (LCVI), which is the reflectance difference between 2.20 and 2.33 μm and correlated with the NPV abundance [14], was computed from the reflectance data for the transect. Spatial variability in the LCVI along the transect correlated well with the changes in land cover types visually identifiable on the ETM+ false color composite (fig. 10.12). The LCVI was the lowest (<0.01) over open water bodies (lakes, rivers, etc.) and burned areas, whereas it had intermediate values of around 0.02 over forested areas, including riparian, gallery forests. For the other pasture or savanna grassland areas, the LCVI had the highest values (>0.025) and varied according to apparent brownness observed on the ETM+ image.

10.5   SUMMARY

In this study, we introduced a new technique for calibrating airborne hyperspectral spectrometer data to reflectance factors using a multiband radiometer. The theoretical justifications and approach of this "continuous panel" method were first presented and the method was then evaluated and validated using the experimental data set acquired in a semi-arid grassland. The results were compared with those obtained with the reflectance mode and linear-interpolation methods. The performance of the method was finally evaluated by applying it to an airborne campaign conducted in a tropical forest–savanna transitional area.

The reflectance mode measurements produced the worst data. As the approach does not account for the change in incoming solar radiation and calibrates only once before flights, the obtained reflectance spectra were biased and subject to large

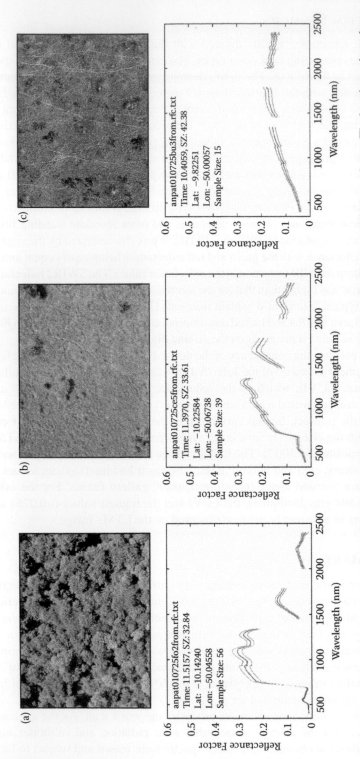

FIG. 10.11 Airborne reflectance spectra for three contrasting land cover types: (a) forest, (b) savanna, and (c) burned area. The locations where these spectra were acquired are indicated in figure 10.8. Accompanied with the spectra are aerial digital photos taken over the same locations. The photos document the surface conditions over which the airborne spectral measurements were made. See CD for color image.

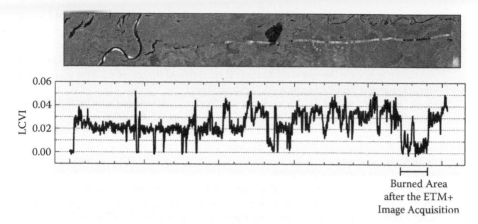

FIG. 10.12   The ligno-cellulose vegetation index along a flight segment (transect). See CD for color image.

distortions due to atmospheric absorption features, even when reflectance spectra were obtained 30 minutes after the calibration. The linear-interpolation method significantly reduced biases and distortions in the derived spectral signatures by using calibration/reference data (i.e., measurements over the reference panel) before and after the flights. In particular, if the flight duration was less than 1.5 hours, the resultant reflectance spectra were under a reasonable accuracy level (<2%). The continuous panel method outperformed these two approaches; it improved upon the linear-interpolation method by removing biases in the retrieved reflectance values. This capability was shown to be greatly advantageous as the flight durations became longer. The durations of our airborne campaigns in Brazil were longer than 2 hours, for example.

Based on the analyses of the semi-arid data set, absolute accuracy of the reflectance derived with the continuous panel method has been estimated at 0.005 reflectance units with precision at the same level of 0.005 reflectance units (i.e., measurement uncertainty of 0.005 ± 0.005). Accuracy and precision decrease if the spectrometer to be used for airborne measurements is subject to instrument noise (e.g., dark current drifts) and over the wavelength regions that are highly influenced by atmospheric absorption features (e.g., water vapor, oxygen, and carbon-dioxide absorptions). The application results of the continuous panel method have demonstrated quality of the retrieved reflectance spectra, which well captured spatial variability in the apparent amounts of NPV associated with land cover types. Airborne spectral reflectance data obtained with this method can serve as a good reference, "top-of-canopy" reflectance data set and should be found valuable for many applications.

REFERENCES

1. Hall, F.G., Huemmrich, K.F., Strebel, D.E., et al., Biophysical, morphological, canopy optical property, and productivity data from the Superior National Forest. NASA, Technical Memorandum Tm-104568, 1992.
2. Moran, M.S., Jackson, R.D., Slater, P.N., et al., Evaluation of simplified procedures for retrieval of land surface reflectance factors from satellite sensor output, *Remote Sensing of Environment*, 41, 169, 1992.

3. Jackson, R.D., Moran, M.S., Gay, L.V., et al., Evaluating evaporation from field crops using airborne radiometry and ground-based meteorological data, *Irrigation Science*, 8, 81, 1987.

4. Gao, X., Huete, A.R., and Didan, K., Multisensor comparisons and validation of MODIS vegetation indices at the semiarid Jornada Experimental Range, *IEEE Transactions on Geoscience and Remote Sensing*, 41, 2368, 2003.

5. Holm, R.G., Jackson, R.D., Yuan, B., et al., Surface reflectance factor retrieval from thematic mapper data, *Remote Sensing of Environment*, 27, 47, 1989.

6. Ferreira, L.G., Yoshioka, H., Huete, Y., et al., Optical characterization of the Brazilian savanna physiognomies for improved land cover monitoring of the cerrado biome; preliminary assessments from an airborne campaign over an LBA core site, *Journal of Arid Environments*, 56, 425, 2004.

7. Remer, L.A., Wald, A.E., and Kaufman, Y.J., Angular and seasonal variation of spectral surface reflectance ratios; implications for the remote sensing of aerosol over land, *IEEE Transactions on Geoscience and Remote Sensing*, 39, 275, 2001.

8. Nicodemus, F.E., Richmond, J.C., Hsia, J.J., et al., *Geometrical considerations and nomenclature for reflectance.* U.S. Department of Commerce, National Bureau of Standards. Washington, D.C., 1977.

9. Slater, P.N., Radiometric considerations in remote sensing, *Proceedings of the IEEE*, 73, 997, 1985.

10. Buffington, L.C. and Herbel, C.H., Vegetational changes on a semidesert grassland range from 1858 to 1963, *Ecological Monographs*, 35, 139, 1965.

11. Schlesinger, W.H., Reynolds, J.F., Cunningham, G.L., et al., Biological feedbacks in global desertification, *Science*, 247, 1043, 1990.

12. Huete, A., Keita, F., Thome, K., et al., A light aircraft radiometric package for Modland Quick Airborne Looks (Mquals), *The Earth Observer*, 11, 22, 1999.

13. Elvidge, C.D., Visible and near-infrared reflectance characteristics of dry plant materials, *International Journal of Remote Sensing*, 11, 1775, 1990.

14. Elvidge, C.D., Examination of the spectral features of vegetation in 1987 AVIRIS data, in *Proceedings of the First AVIRIS Performance Evaluation Workshop*, Pasadena, CA, 1988.

11 Assessment of Phenologic Variability in Amazon Tropical Rainforests Using Hyperspectral Hyperion and MODIS Satellite Data

Alfredo R. Huete, Youngwook Kim,
Piyachat Ratana, Kamel Didan,
Yosio E. Shimabukuro, and Tomoaki Miura

CONTENTS

11.1 INTRODUCTION

Phenology represents the seasonal timing and annual repetition of biologic life cycle events and is a characteristic property of ecosystem functioning and predictor of ecosystem processes. Shifts in phenology depict a canopy's integrated response to environmental change and influence local biogeochemical processes, including nutrient dynamics, photosynthesis, water cycling, soil moisture depletion, and canopy physiology [1]. An understanding of vegetation phenology is prerequisite to interannual studies and predictive modeling of land surface responses to climate change [2–4].

Despite its importance, the phenology of tropical rainforest ecosystems and the environmental conditions controlling its variability are not well understood. This is in part due to the complexity of tropical forest canopies, where a highly diverse tree species population can result in a wide variety of phenologic responses to the same or common environmental factors, such as rainfall, temperature, and photoperiod [5–8]. There are many challenges in both field and satellite assessments of phenology, with field methods generally time consuming, limited to fairly narrow spatial and temporal scales, and difficult to extend over large areas [9,10]. Satellite measurements integrate numerous trees and species; however, they are constrained by the persistence of cloud cover and saturation of reflectances in high biomass tropical forest canopies, both of which hamper the accurate assessment of seasonal variability in forest canopy properties.

In this chapter, we examine patterns of optical-phenologic variability in an evergreen broadleaf tropical forest in the Amazon, using space-borne hyperspectral and moderate resolution satellite measurements. Moderate resolution satellite data provide high frequency but poor spatial resolution data of limited spectral content, while hyperspectral data offer finer resolution and spectral detail but at infrequent time intervals. We focus here on the potential contributions of hyperspectral information in phenology studies and their scaling with high temporal frequency, moderate resolution satellite data for improved characterization of tropical rainforest phenology.

11.2 PHENOLOGY IN TROPICAL FORESTS

The phenology and interactions of evergreen broadleaf tropical rainforests with environmental, climate, and anthropogenic factors are poorly understood. At the landscape level, many climate and growth models characterize tropical rainforests as having no seasonal variation in biophysical plant properties such as greenness, leaf area index (LAI), fraction of absorbed photosynthetically active radiation (FPAR), and albedo [11,12]. Yet, in many local field and flux tower studies one observes consistent seasonal changes in tropical forest canopy characteristics, including synchronized flushing and exchange of new leaves, periods of decreased foliage density, leaf aging, senescence, and litter fall [13]. Wright and Schaik [5] found leaf and flower production at sites across eight different rainforests, including the Ducke Reserve near Manaus in the central Amazon, to coincide with seasonal peaks in solar irradiance in the dry season. In the more seasonally dry eastern Amazon rainforests, Asner et al. [14] and Carswell et al. [15] also measured increases in LAI in the dry season period of rapid leaf turnover at the Flona Tapajós and Caxiuanã rainforest sites. Tower flux measurements have shown increased photosynthesis activity in the dry season at both sites [15,16].

In contrast, forest phenology and photosynthetic activity are much more closely associated with moisture limitations in the transitional and drier tropical forests of the southern Amazon. Randow et al. [17] found seasonality in photosynthesis to be small at the Jaru Biological Reserve tower flux site in Rondonia, and Vourlitis et al. [18] found strong positive correlations between photosynthesis and water availability in a transitional forest (cerradão) flux tower site near Sinop, Mato Grosso. Herrerias-Diego, Quesada, and Stoner [19] similarly found more synchronous flowering and fruiting events in the early wet season in tropical dry forests of Mexico, noting that these forests may directly brown-down and have lower photosynthetic capacity due to reduced water availability (see also Singh and Kushwaha [20]).

There is increasing concern on how land surface phenology in tropical regions could change in response to global warming and shifts in land cover change and land use activities [21,22]. Land cover changes resulting from forest disturbance and conversion alter landscape phenology and influence seasonal ecosystem functioning and its role in atmospheric CO_2 concentrations [23]. Fires and forest disturbance result in large amounts of carbon released to the atmosphere, while regenerating secondary forests can rapidly sequester carbon through rapid regrowth [24–26]. Thus, at the regional scale, a more complex mosaic of tropical forest phenologies can be expected in response to variations in land use activities, light and water controlling factors, forest structural variations, and associated ecological conditions (soils, topography, nutrients) [14,27].

11.3 SATELLITE PHENOLOGY OF TROPICAL FORESTS

Optical remote sensing data have been widely used to monitor and map tropical forests and their temporal and spatial dynamics of deforestation, regrowth, and fire [28–30]. Most of these studies utilize finer resolution satellite data (e.g., Landsat), which offer more accurate discrimination of forest classes and disturbance events (e.g., Vieira et al. [31] and Kalacska et al. [32]). However, it is difficult to obtain cloud-free images at the frequencies needed to define accurate phenologic trends. Much of what is known about seasonal vegetation dynamics in the tropics comes from moderate and coarse resolution satellite measurements, which are commonly used for large-scale vegetation monitoring and vegetation-climate studies [33,34]. For example, Defries et al. [29] utilized multitemporal phenologic metrics to derive land cover classifications from the NOAA-advanced very high resolution radiometer (AVHRR) time series data at 8-km spatial resolution.

These sensors provide high temporal frequency measurements, but with poor spatial resolution and limited spectral content, resulting in low spectral sensitivity for tracking temporal and spatial variability in tropical forest characteristics, including phenology. This is partly attributed to a low optical depth of penetration through densely vegetated forest canopies, whereby spectral vegetation indices, such as the normalized difference vegetation index (NDVI), become "saturated" and insensitive to forest properties, including more subtle phenologic characteristics [35]. Often the phenology of moist, tropical, evergreen broadleaf forests is characterized as "flat" and seasonally constant [35–37]. In a study of tropical successional forests, Sader et al. [38] found the NDVI to be a poor predictor of biomass in these rapidly growing regenerating forests, which quickly maximize the absorption of red radiation and NDVI sensitivity to biophysical properties.

In addition to saturation of the chlorophyll signal, the usefulness of AVHRR-NDVI time series data records in tropical forests is also constrained by canopy reflectance contamination from clouds and cloud shadows and difficulties in atmosphere correction of seasonally variable atmosphere water vapor and aerosols [39]. Kobayashi and Dye [40] found strong seasonal signals from clouds and aerosols in the AVHRR-NDVI data sets over the Amazon that dominated the relatively weak "apparent" seasonal signal from the tropical forests themselves.

Some of the more recent advanced moderate resolution satellite sensors, such as SPOT-VEGETATION (VGT) and moderate resolution imaging spectroradiometer (MODIS) [41], have better sensor capabilities for phenologic characterization of tropical forest ecosystems. These sensors offer improved calibration and atmosphere correction, narrower spectral bands, and finer resolution observations (250–1000 m) that facilitate cloud filtering and noise removal [42,43]. In contrast to the AVHRR-NDVI data, strong local and region-wide phenologic patterns were found in Amazon forests with MODIS and SPOT-VGT enhanced vegetation index (EVI) data [44–48].

Xiao et al. [45] also found SPOT-VGT and MODIS-derived land surface water index (LSWI) sensitivity to seasonal fluctuations in canopy/leaf water contents in Amazon tropical forest. Myneni et al. [49] similarly found large seasonal swings in MODIS-derived LAI across the Amazon region that were correlated strongly with sunlight and inversely related to precipitation. Ratana et al. [48] looked at the tropical forest phenology across the humid–dry forest transition zone in the southern Amazon and found mixed light, moisture, and anthropogenic controls on observed MODIS seasonal patterns. These studies suggest tropical forests are highly dynamic, with strong phenologic responses to light and moisture as well as land use and land cover change.

11.3.1 HYPERSPECTRAL REMOTE SENSING

Hyperspectral remote sensing measurements add spectral fidelity to the extraction of optical-phenologic information from tropical forests. Hyperspectral imagery can potentially be useful in characterizing forest tree diversity and their phenology by exploiting spectral variability in leaves and crowns associated with leaf age, leaf morphology, chlorophyll content, senescence, epiphyll colonization, water stress, and canopy structure [50–52].

The launch of the space-borne imaging spectrometer, EO-1 Hyperion, provides new opportunities to study tropical canopies in the spectral domain [53,54]. Hyperion measures reflected radiances in 224 contiguous spectral bands from 400 to 2500 nm, across a narrow 11-km-wide swath, at 16-day frequencies and with a nominal spatial resolution of 30 m. Miura et al. [55] found Hyperion measurements clearly depicted such diagnostic absorption features of vegetation as the red edge, red near-infrared

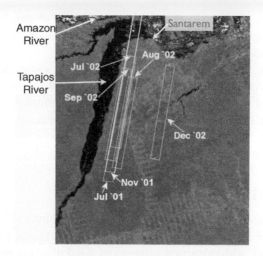

FIG. 11.1   Location of 6 EO-1 Hyperion scene acquisitions within the Floresta Nacional do Tapajós and surrounding areas, south of Santarém in the state of Pará, Brazil; each Hyperion image is 11 km wide. See CD for color image.

TABLE 11.1
Hyperion Scenes Analyzed over the Study Area
(Path 227 Row 062) in 2001 and 2002

Solar zenith/sensor view angles	Acquisition date (day of year)	Aerosol optical depth at 675 nm/water
38.0/–4.6	July 30, 2001 (211)	0.098/4.15 cm
28.9/–4.9	November 3, 2001 (307)	0.125/4.45 cm
39.0/+8.9	July 8, 2002 (189)	0.048/3.89 cm
35.7/+9.0	August 9, 2002 (221)	0.135/3.67 cm
27.1/+8.7	September 26, 2002 (269)	0.128/3.24 cm
33.9/–14.5	December 1, 2002 (335)	0.490/4.18 cm

(NIR) transition, and ligno-cellulose absorptions, enabling more effective biochemical characterization and discrimination of Brazilian cerrado physiognomies (grass, shrub, tree). Asner et al. [14] used hyperspectral metrics from the Hyperion sensor to assess variations in canopy water content and light-use efficiency in an Amazon drought study, and reported improved accuracies in estimating ecosystem productivity.

In this chapter we examine the optical-phenologic characteristics of the protected Floresta Nacional do Tapajós and surrounding areas south of Santarém in the state of Pará, Brazil, with a temporal sequence of six EO-1 Hyperion images acquired during the 2001 and 2002 dry seasons (July to December) (fig. 11.1, table 11.1). This region consists mostly of undisturbed, dense tropical evergreen broadleaf rainforest with smaller patches of disturbed and converted forest areas, and has received considerable study as part of the large-scale biosphere-atmosphere experiment in Amazonia (LBA) [27]. The primary tropical forest canopies are 40 m tall with high LAIs of 5–7 [56]. Mean annual precipitation ranges from 1900 to 2300 mm with a dry season (defined as monthly rainfall less than 100 mm) period of ~5 months from early July through early December. The soils are primarily nutrient-poor clay Oxisols with some sandy Ultisols [57].

The radiometrically calibrated Hyperion data were atmospherically corrected using ACORN4 (Analytical Imaging and Geophysics LLC, Boulder, Colorado), a MODTRAN-based radiative transfer model, to convert from radiance to apparent surface reflectance (Miura et al. [55]). Aerosol and optical depth parameters were derived from the AERONET sun photometer network (http://aeronet.gsfc.nasa.gov/data_frame.html) at the Belterra site (Lat. 2.647 S, Long. 54. 952 W) [58].

11.4   SPECTRAL SIGNATURES

Spectral reflectance signatures within primary tropical rainforest, regenerating successional forests of varying age classes, and forest conversion to pasture/agriculture areas were sampled, with extraction windows of 5 × 5 pixels in the smaller agriculture patches to 25 × 25 pixels in larger pasture and forest areas. Areas of regenerating successional forests of varying age classes as well as agriculture/pasture sites were selected from sites identified in Shimabukuro et al. [54] and Espirito-Santo, Shimabukuro, and Kuplich [59].

11.4.1   SPECTRAL SIGNATURES AT BEGINNING OF DRY SEASON

The Hyperion-measured spectral signatures of the main vegetation types at the start of the dry season are shown in figure 11.2 (July 8, 2002). Most of the spectral variations among tropical rainforest, regenerating successional forest, and pasture/agriculture occur in the NIR region between 800 and 1300 nm, but there is also high spectral variability in the shortwave infrared (SWIR) regions, while the least spectral variation occurs in the visible region. There are no significant variations between

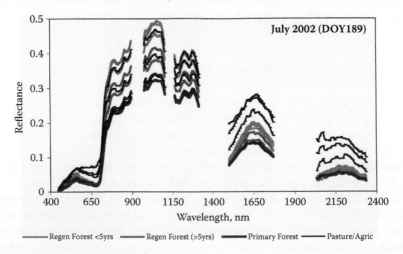

FIG. 11.2 Hyperion-measured spectral signatures of primary forest, regenerating forest, and pasture/agriculture at the start of the dry season (July 8, 2002).

primary and regenerating tropical forest types in the visible and the 2100-nm SWIR regions. The pasture and agriculture areas are easily discriminated by their relatively high visible and SWIR reflectances; however, the pasture/agriculture areas are highly variable in NIR reflectances, a result of the more heterogeneous surface cover conditions resulting from land use activities.

On the other hand, primary and regenerating forest spectral signatures are clearly distinct in the NIR and 1600-nm SWIR regions, with the youngest age class of regenerating forests (<5 years) exhibiting the highest NIR and SWIR reflectances and the mature primary forests showing the lowest reflectances. The older regenerating forests (>5 years) had spectral signatures between those of primary forests and younger regenerating forests (fig. 11.2). This optical gradient of decreasing NIR spectral reflectances with forest maturity has been noted in previous studies in which variations in the structural properties of the canopy were shown to have a major influence in the NIR [26].

Mature, primary tropical forests are generally darkest in the NIR due to a more complex stand structure with different tree strata at different heights and more self- and interspecies shading influences. Early successional forest canopies that follow forest conversion consist of fast-growing pioneer trees with an understory of small shrubs and herbaceous vegetation, which appear bright in the NIR [26,60]. Older regenerating forests are characterized with increasing species richness, a multi-layered tree canopy, and an understory vegetation increasingly limited by low light levels [61]. Hence, in progressing from younger to older successional forests and mature forests, many structural variables vary, including number of canopy layers, height and continuity of the upper canopy, and the presence and number of emergent trees [62]. Spectrally, the successional forests are a transitional class between pasture and primary tropical forest, with the initial stages of forest regeneration spectrally similar to pasture, while the final stages of forest regeneration are more spectrally related to primary tropical forest.

11.4.2 Spectral Signatures in Mid Dry Season

When we plot the spectral signatures of the same vegetation types midway into the dry-season period (September, DOY 269), we find appreciable changes revealing significant phenologic changes and associated biophysical canopy properties (fig. 11.3). The most dramatic optical changes are the greatly diminished variations in NIR reflectances among all vegetation canopy types and the much greater variations found in visible and SWIR reflectances, especially near 1600 nm. Optical contrasts between the primary and regenerating forests diminished significantly between July and September, as the NIR reflectances from primary forests increased by 20% (~0.23–0.28, at 800 nm), while the younger and older regenerating forests decreased in NIR reflectances by 20% (0.36–0.32, at 800 nm) and 10% (0.30–0.27, at 800 nm), respectively, resulting in very little remaining contrast among the forest canopy types (fig. 11.3). The pasture/agriculture sites also showed strong decreases in NIR reflectances, resulting in the lowest NIR reflectances among all vegetation types. Despite the strong observed optical changes in the NIR, there remained only slight differences between primary and regenerating forests in the visible region, with slightly

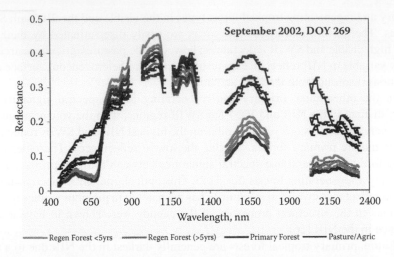

FIG. 11.3 Hyperion-measured spectral signatures of primary forest, regenerating forest, and pasture/agriculture at the middle of the dry season (September 26, 2002).

higher SWIR reflectances in the younger regenerating forests relative to the primary forests. There are dramatic reflectance increases in visible and SWIR reflectances in the pasture/agriculture areas, a result of the drying and senescing vegetation.

We can better visualize the trends, magnitude, and synchrony in phenologic patterns across the various vegetation canopies by geometrically registering and subtracting the September from July 2002 scenes (fig. 11.4). Negative values indicate decreasing reflectances, while positive values represent reflectance increases through the first half of the dry season. The primary tropical forests increased in NIR reflectances while all other vegetation types showed decreasing NIR reflectances over the same time period. The smallest negative change in NIR reflectances occurred in the older regenerating forests, followed by stronger negative NIR changes in the younger regenerating forests and very pronounced negative changes in the rapidly drying converted pasture/agriculture sites. In contrast, dry-season changes in visible and SWIR reflectances are small and slightly positive for the primary and regenerating forest sites, and strongly positive for the pasture/agriculture sites. Overall, there are strong optical-phenologic differences among the dominant vegetation types, related to vegetation type, disturbance, and land use history, with a wide gradient in reflectance shifts from positive (0.08 and 0.10, in NIR and SWIR, respectively) to very negative (–0.12 in the NIR) (fig. 11.4a).

The increase in NIR reflectances in the primary forests is related to leaf flushing and leaf exchange that occur with seasonal peaks in sunlight during the dry season [5,46,49,51]. In the Tapajós region, seasonal changes in cloud cover are the main controller of incident solar radiation and photosynthetic active radiation (PAR), with solar zenith angle variations playing a secondary role. Temperatures are slightly warmer and evapotranspiration is higher in the dry season as trees maintain access to soil water throughout the dry season [63,64]. Thus, while soil moisture near the surface (5 cm) may vary markedly during the dry season in response to rain events,

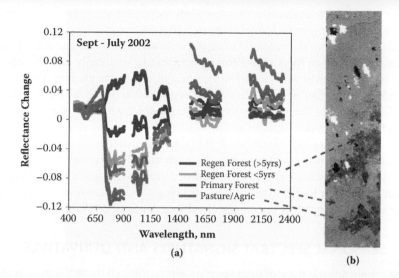

FIG. 11.4 Subtraction of the September from July 2002 Hyperion images: (a) spectral "reflectance change" signatures for the different vegetation canopies, with negative values indicating decreasing reflectances, while positive values represent reflectance increases through the first half of the dry season; (b) medium gray (640 nm), NIR (854 nm), and SWIR (2133 nm) composite "reflectance change" image with light gray denoting positive NIR change; dark gray colors are negative NIR and positive medium gray and SWIR changes.

deep soil moisture (>2.5 m) levels are comparatively constant year round with only gradual declines in the dry season [63].

The decreasing NIR reflectances and increasing visible and SWIR reflectances in the pasture and agriculture areas are direct optical responses to moisture limitations due to the very shallow rooted herbaceous vegetation and the drying of the upper soil profile. The regenerating forests are also vulnerable to moisture limitations, primarily in the shallow-rooted, understory herbaceous vegetation layer, and possibly the trees themselves, depending on age class and root development. Thus, the decline in NIR reflectance seen in regenerating forests is associated with the drying of the herbaceous layer as well as its greater vulnerability to drought relative to the more mature primary forests, which are more drought tolerant. In the older regenerating forests, the understory vegetation is increasingly limited by low light levels, and the decrease in NIR reflectances is much smaller. The drying of the herbaceous understory is more evident in the NIR spectral region due to the greater optical depth penetration of this spectral region through the overlying forest canopy. In contrast, the more absorptive visible and SWIR regions have low optical depth penetration through these forests, resulting in very little variation.

The color composite image of reflectance differences using red (640 nm), NIR (854 nm), and SWIR (2133 nm) wavelengths shows the spatial variability and unique phenologic behavior of primary, regenerating, and pasture/agriculture areas over the study area (fig. 11.4b). Spatially, there is an extensive pattern of positive NIR reflectance changes across the intact primary forests (depicted as light gray). The pasture/agriculture areas, with strong negative NIR reflectance changes and strong positive medium gray and SWIR changes show up while the regenerating forests are depicted with various dark gray colors, depending on age class.

This demonstrates the potential to map spatial variations and landscape patterns in phenology based on their asynchronous response to seasonally variable environmental forcings, such as water availability, sunlight, and nutrients. This also provides a methodology to map and quantify forest regeneration areas, which is generally quite difficult due to the large variety of age classes from early growth to nearly mature forests [65–67]. Carreiras et al. [67] attempted to map regenerating forests with SPOT-VGT data and concluded that phenologic differences may be more evident and useful for mapping such areas. In more water-limited, dry tropical forests, attempts to map forests, pasture, and disturbed areas with satellite data have been difficult due to their highly synchronous phenologic patterns, which exhibit similar drying and leaf-off patterns during the cloud-free dry season [19,50].

11.5   SEASONAL SPECTRAL SIGNATURES AND DERIVATIVES

An optical-phenologic trace of each vegetation type through the dry season is shown by adding the November 2001 Hyperion data to the July, August, and September 2002 sequence (fig. 11.5). The spectral reflectance signatures of the primary tropical rainforest sites show continued increases in NIR reflectances from early-dry-season (July, August) through mid-dry-season (September) and late-dry-season (November) periods (fig. 11.5a). Fewer variations are evident in the highly absorbing visible and SWIR regions where both total chlorophyll content and canopy water content are very high and possibly saturating the reflectances, although a notable increase in SWIR reflectances at 1600 nm can be seen in November.

Sun angle variations through the dry season may influence the seasonal canopy spectral signatures through shading and illumination effects. Cloud seasonality and sun angle can also affect the levels of diffuse and direct incoming PAR and influence photosynthesis within the canopy. The smallest solar zenith angle at Tapajós (27°) occurs in the mid-dry-season (September) Hyperion image with sun zenith angles increasing to 39° in the early dry season (July) and 34° toward the late dry season (December) (table 11.1). Thus, during the first part of the dry season, the sun angle is becoming smaller and canopy shading may decrease, which may contribute to the observed increases in reflectances, particularly in the NIR. However, in the latter half of the dry season, sun angles and shading become larger again and the reflectances should decrease, which was not observed here.

First- and higher-order derivative analysis of spectral signatures, band ratioing, and continuum removal have been noted to be useful in assessing spectral features while controlling variations in illumination (sun angle, topography) and background [68–71]. The first derivatives of the NIR–red transition (721 nm) for the seasonal spectral signatures show a pattern of increasing of red-edge slopes from early dry season to mid dry season and a much stronger increase in slope in the latter part of the dry season (fig. 11.5b).

The seasonal dynamics of the pasture/agriculture spectral signatures exhibit a general decreasing trend in NIR reflectances and pronounced increases in the visible and SWIR reflectances (fig. 11.6a). These sites are more variable due to the range of

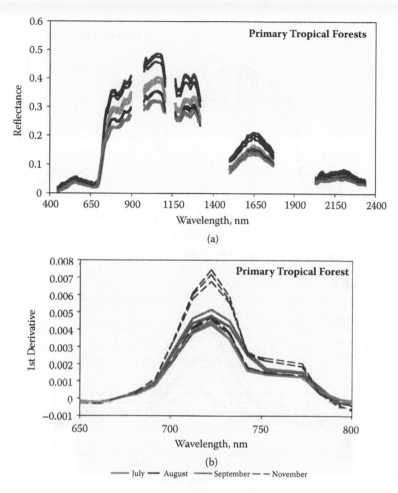

FIG. 11.5   Hyperion-measured seasonal sequence of (a) spectral reflectance signatures and (b) first derivative signatures for primary tropical forest. See CD for color image.

land use practices. The large SWIR reflectance increases are associated with drying of the vegetation and the visible reflectance increases are related to browning (loss of chlorophyll) of the vegetation. The first derivative spectral curves at 721 nm show decreasing slopes from start to end of the dry season, with each monthly period resulting in significant slope decreases (fig. 11.6b).

The regenerating forests exhibit more complex optical-phenologic patterns through the dry season (figs. 11.7 and 11.8). The younger regenerating forest sites (<5 years old) show a decreasing trend in NIR reflectances in the first half of the dry season from July and August through September, followed by an increase in NIR reflectances in the latter part of the dry season in November (fig. 11.7a). In the SWIR region, there are slight decreases in reflectances in the early dry season followed by a strong increase in November. The first derivative signatures also show decreasing slopes in the early dry season followed by an increase in the late dry season (fig. 11.7b). The older regenerating forest sites (>5 years) show more subtle decreases

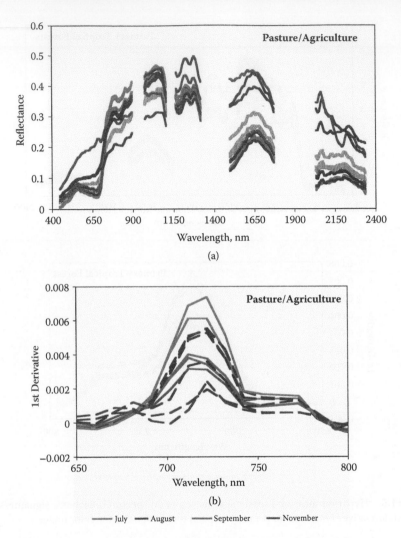

FIG. 11.6 Hyperion-measured seasonal sequence of (a) spectral reflectance signatures and (b) first derivative signatures for pasture/agriculture areas. See CD for color image.

in NIR reflectances in the early dry season, followed by strong NIR increases in the latter part of the dry season (fig. 11.8a). SWIR reflectances at 1600 nm are also higher in November and there are much smaller variations in the visible and SWIR regions at 2100 nm. The first derivative spectral signature follows the same trend as in the younger regenerating forests, except that the early-dry-season slope decreases are very weak, while the slope increase in November is much stronger.

The more complex phenologic patterns of the regenerating forests are partly a result of the contribution of understory vegetation reflectances that mix with those of the forest canopy. The NIR reflectance decreases in the first half of the dry season are due to the drying of the understory herbaceous vegetation layer, with this drying

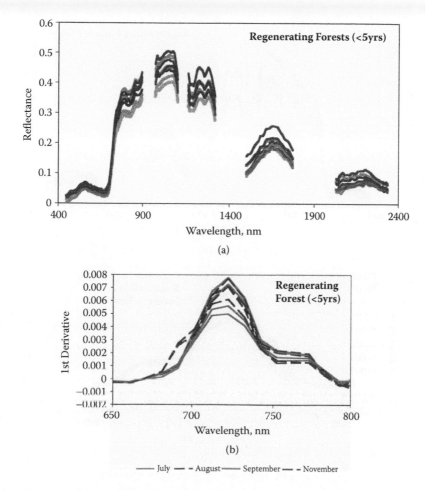

FIG. 11.7 Hyperion-measured seasonal sequence of (a) spectral reflectance signatures and (b) first derivative signatures for young regenerating forests (<5 years). See CD for color image.

effect more pronounced in the younger and more open forest canopies, relative to the older, closed, and more developed regenerating forests. The overall increase in NIR reflectances in the second half of the dry season, during which the understory vegetation should remain dry, suggests that these forests also have a greening phenology (leaf flushing and exchange) in the dry season. Greening is very apparent in the older regenerating forests where NIR reflectances in November are significantly higher than in July, while greening is more subtle in the younger forests where NIR reflectances in November were similar to those in July (figs. 11.7a and 11.8a).

However, the exact timing of the "greening" phase in the regenerating forests is difficult to determine due to the unknown extent of mixing of the herbaceous understory vegetation (with its asynchronous phenology) with the tree canopy phenology. Greening in the latter half of the dry season has also been observed in transitional and drier tropical forests [17], and regenerating forests may behave

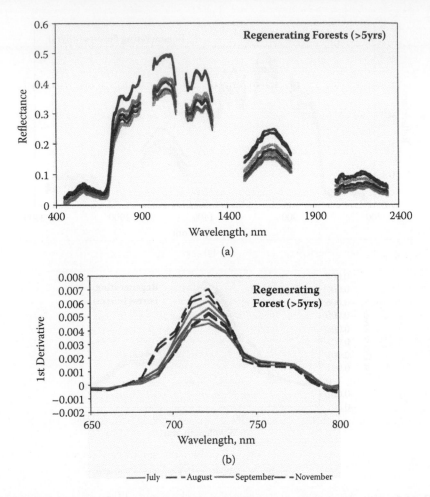

FIG. 11.8 Hyperion-measured seasonal sequence of (a) spectral reflectance signatures and (b) first derivative signatures for older regenerating forests (>5 years). See CD for color image.

more similarly to these forests, given the drier environment created by less well-developed rooting systems.

11.5.1 Subtraction of End from Start of Dry-Season Periods (2001)

A depiction of start to end of the dry-season phenology for 2001 is shown in figure 11.9 using geometrically registered November minus July Hyperion images. This is generally similar to the 2002 dry season with positive increases in NIR reflectances over the primary forests and older regenerating forests and negative shifts in NIR reflectances over the younger regenerating forests. Pasture/agriculture areas show pronounced negative NIR shifts and strong positive SWIR reflectance changes (fig. 11.9a). In the reflectance differences composite image (fig. 11.9b; 640, 845, and 2133 nm), the different vegetation classes are discriminable based on their unique asynchronous phenologic responses in the dry season. It is worth noting that the subtraction of the end

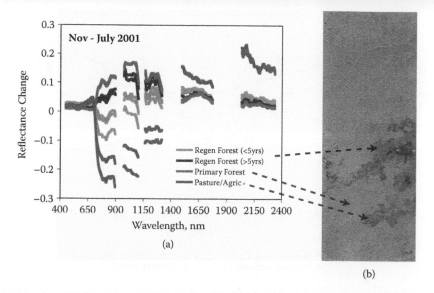

(b)

FIG. 11.9 Subtraction of the November from July 2001 Hyperion images: (a) spectral "reflectance change" signatures with negative values indicating decreasing reflectances, while positive values represent reflectance increases through the entire dry season period; (b) red (640 nm), NIR (854 nm), and SWIR (2133 nm) composite "reflectance change" image with green denoting positive NIR change; pink and brown colors are negative NIR and positive red and SWIR changes. See CD for color image.

of dry season from start of dry-season periods fails to capture the initial drying of secondary forests, as observed in the 2002 data.

11.6 SEASONALITY IN SPECTRAL BIOPHYSICAL INDICES

In this section, we look at the seasonal traces of various optical measures of greenness (chlorophyll) and wetness for their potential use to characterize landscape phenology in tropical forests, including first derivative peaks, vegetation indices, and vegetation water indices. We computed these biophysical indices using MODIS bandwidths, which were convolved from the Hyperion spectral signatures. In general, vegetation indices are sensitive to canopy greenness, which is a composite property of canopy cover, leaf area, chlorophyll content, and canopy architecture. The vegetation water indices are more sensitive to the water content of vegetation, which is also a composite property of leaf water content, leaf area, specific leaf area, and canopy architecture.

11.6.1 SEASONALITY OF THE FIRST DERIVATIVE VEGETATION INDEX

The phenologic profiles depicted by the first derivative peak values at 721 nm are shown in figure 11.10 for the main vegetation types across all 2001 and 2002 Hyperion acquisition dates. As the December 1, 2002, Hyperion image is outside our study area (fig. 11.1), areas that resembled primary forest, regenerating forest, and

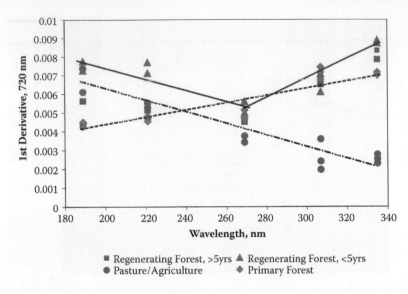

FIG. 11.10 The phenology profiles depicted by the first derivative peak values, at 720 nm for all vegetation types across the 2001 and 2002 Hyperion acquisition dates. See CD for color image.

pasture/agriculture were used, but without ground verification. This metric of greenness, based on the slope of the NIR–red transition, is indicative of the abundance and activity of pigment absorbers—namely, chlorophyll—in the leaves and is also known as the first derivative vegetation index (FDVI) [70]. Three main phenologic patterns are apparent: (1) the primary rainforests that generally show increasing FDVI values from beginning to end of the dry season, (2) the pasture/agriculture areas that continuously decrease or brown down during this same period, and (3) the regenerating forests, with values that decline in the early dry season followed by increasing FDVI values in the late dry season. The younger regenerating forests also have higher FDVI values than older regenerating forests at the beginning of the dry season.

11.6.2 SEASONALITY OF VEGETATION INDICES (NDVI, EVI)

There were major differences in the phenologic profiles derived using the NDVI to represent greenness dynamics (fig. 11.11a). In contrast to the FDVI, peak NDVI values occur at the beginning of the dry season (July) in all vegetation types and there is only small variability in NDVI values across the dry season, with slight NDVI value decreases from July to August and small increase in September and November. In contrast, there are significant NDVI decreases in pasture/agriculture areas through the dry season. The saturation of NDVI values not only results in weak seasonal variability, but also inhibits discrimination of primary and regenerating forest canopies. At the leaf level, Roberts et al. [51] noted that NDVI may not detect seasonal variations in greenness in dense tropical forest canopies since NDVI is primarily responsive to the red band, which is not as impacted by leaf aging and epiphyll activity compared with other spectral regions, such as the NIR.

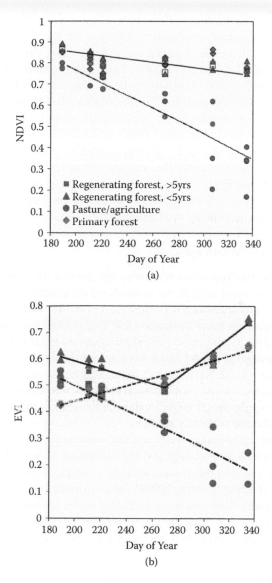

FIG. 11.11   Dry-season phenologic profiles depicted by the (a) NDVI and (b) EVI for all vegetation types across the 2001 and 2002 Hyperion acquisition dates. See CD for color image.

On the other hand, the phenologic profiles obtained with the enhanced vegetation index (EVI) strongly resemble those of the FDVI greenness measure (figs. 11.10 and 11.11b). EVI values of primary tropical forest increase continuously from start to end of the dry season, while decreasing continuously over pasture/agriculture sites. The EVI profiles of regenerating forest show browning in the first half of the dry season followed by a reversal with greening in the second half of the dry season, in agreement with the FDVI patterns. Previous studies with the MODIS instrument also found EVI values to peak in the late dry season and not saturate in Amazon tropical rainforests [44–46]. This is also confirmed by seasonal tower flux measurements of photosynthesis (gross primary productivity, GPP) at the Tapajós primary forest flux tower site (67 km), with a more physiologically active canopy in the dry season and strong correlations with MODIS EVI [16,46]. The EVI has greater penetration through dense canopies by maintaining sensitivity to the nonabsorbing NIR region. Xiao et al. [45] and Zhang et al. [43] reported that EVI responds to the "green" chlorophyll component of the fraction of absorbed photosynthetically active radiation (FPAR), while the NDVI is more correlated with "total" FPAR at the canopy level (i.e., absorption by both photosynthetically active and nonphotosynthetically active vegetation).

Xiao et al. [47] also noted that the seasonal dynamics of NDVI did not reflect changes in leaf age, chlorophyll contents, nor canopy structure in seasonally moist tropical forests, and questioned the use of NDVI as inputs to production efficiency models that estimate GPP and net primary production (NPP) of forest ecosystems [72,73].

11.6.3 Vegetation Water Indices

Water-sensitive spectral vegetation measures are derived through combinations of NIR and SWIR bands and include the normalized difference water index (NDWI) and land surface water index (LSWI) (table 11.2) [74,75]. These indices are sensitive to canopy equivalent water thickness (EWT, g H_2O/m^2), and in open canopies the LSWI is sensitive to surface soil moisture status as well. The seasonal canopy wetness profiles using the NDWI and LSWI across the dry-season drought period are depicted in figure 11.12. NDWI values are mostly below zero and show a slightly positive canopy wetness trend in the primary forests through the dry season. The primary forests have the lowest NDWI values at the start of the dry season and the largest values by the end of the dry season.

In contrast, the regenerating forests and the pasture/agriculture areas show small and large drying trends, respectively, and the separation among all vegetation types increases as the dry season progresses, indicating unique drying patterns across all canopy types (fig. 11.12a). Increasing NDWI values in the dry season may be attributed to the higher greenness of the canopy, as indicated by the seasonal dynamics of FDVI and EVI (figs. 11.10 and 11.11) and the greater proportion of young leaves, which have higher leaf water content than older leaves [51]. In the case of regenerating forests,

---

TABLE 11.2

MODIS Bandwidths and Biophysical Indices Computed from Them

MODIS bands	Biophysical indices	Ref.
1. (red), 620–670 nm	NDVI = (B2 – B1)/(B2 + B1)	
2. (NIR), 841–876 nm	EVI = 2.5 * (B2 – B1)/(1 + B2 + 6 * B1 – 7.5 * B3)	42
3. (blue), 459–479 nm		
4. (green), 545–565 nm		
5. (NIR), 1230–1250 nm	NDWI = (B2 – B5)/(B2 + B5)	74
6. (SWIR), 1628–1652 nm	LSWI = (B2 – B6)/(B2 + B6)	45
7. (SWIR), 2105–2155 nm		

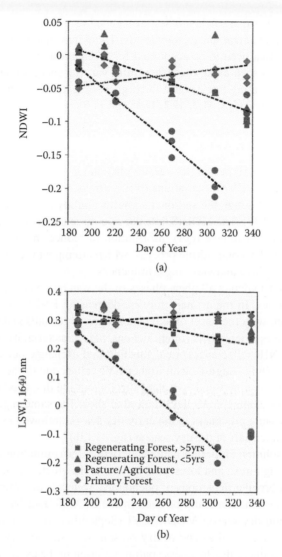

FIG. 11.12   Dry-season phenologic profiles depicted by the (a) NDWI and (b) LSWI for all vegetation types across the 2001 and 2002 Hyperion acquisition dates. See CD for color image.

NDWI tracks canopy greenness in the first half of the dry season; in the latter half, the canopy continues to dry, although canopy greenness reverses and starts to increase (figs. 11.10, 11.11b, and 11.12a). Thus, the regenerating forest canopies may be experiencing some water stress or there may be some interplay of influences among the understory layer, canopy structure, and leaf water content.

The LSWI canopy wetness profiles show weaker seasonal variability and lower sensitivity as LSWI values appear to saturate across primary and regenerating forest canopies, similar to NDVI, and there is only a slight drying trend in the regenerating forests (fig. 11.12b). Xiao et al. [45] also reported no indication of water stress in the Tapajós forest over the 1998–2002 dry seasons using SPOT-VGT measures of

LSWI and noted that more field studies are needed to measure seasonal variations in leaf water content at leaf and canopy levels. There may also be important canopy structural effects that need to be accounted for when extracting canopy water content information, as seen along the structurally varying successional forests where the youngest regenerating forests have the highest reflectances in both the NIR and SWIR spectral regions (figs. 11.2, 11.3, 11.4, and 11.9).

11.7   MODIS COMPARISONS

When we overlay the Hyperion dry-season phenology with an entire year of coincident MODIS 250-m EVI observations (3 × 3 arrays, $n = 9$), we obtain consistent seasonal patterns, although the seasonal signal is slightly dampened in the coarser MODIS data (fig. 11.13). The MODIS data use operational and globally applied algorithms with quality assurance (QA) information to reduce the uncertainty associated with the multiday compositing process, which attempts to remove clouds while minimizing atmosphere and view angle influences.

The MODIS data depict all phenophases of the vegetation canopies, across both wet and dry seasons. In the primary forests, decreasing EVI begins in the middle of the wet season, with the onset of leaf senescence (litter fall) and colonization of older, mature leaves by epiphylls (fungi, lichens, algae, bacteria, liverworts), both of which diminish NIR reflectances. Leaf flushing and exchange occur near the start of the dry season (July–August), with higher EVI values due to the younger foliage (fig. 11.13a). The greening trend continues throughout the dry season with new leaf expansion and development. MODIS data also show the same regenerating forest patterns of decreasing greenness in the early dry season followed by a reversal with greening in the latter half of the dry season (fig. 11.13b).

The finer resolution Hyperion data show better discrimination among primary forest, regenerating forest, and forest conversion areas, which can be combined with MODIS data for broader depictions of phenology. To test if the phenologic patterns found in the Hyperion data could be seen in the MODIS data on a regional scale, we subtracted mid-dry-season MODIS EVI (September) from start of dry-season MODIS EVI (July), as well as end of dry season from start of dry-season MODIS EVI (November minus July), to assess spatial variation in the extent of greening and browning (fig. 11.14). In the September–July difference image, we expect "browning" in all forest conversion and regenerating forest areas with simultaneous "greening" in the undisturbed primary forests (fig. 11.14a). However, in the November–July difference image, only the pasture and agriculture areas brown-down, as the primary and regenerating forests are both greening (fig. 11.14b), in accordance with the Hyperion results presented in figure 11.11(b).

It is not possible to differentiate regenerating forests in any single image, as all three dates are needed since regenerating forests brown-down with pasture/agriculture in the first dry-season phase, and they green-up along with primary forests in the second dry-season phase. Hence, with proper training, MODIS-based phenology is useful in distinguishing and tracking tropical forest canopies and forest disturbance classes.

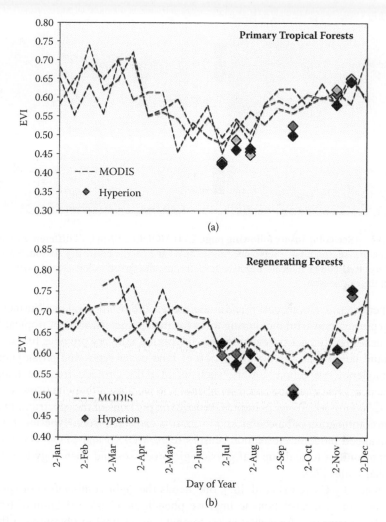

FIG. 11.13 Hyperion dry-season phenologic profiles plotted with annual MODIS 250-m EVI observations for (a) primary tropical forests and (b) regenerating forests.

11.8 DISCUSSION

In this chapter we evaluated the spectral and temporal variability in landscape phenologic patterns of different physiognomic vegetation types in an evergreen broadleaf primary tropical forest, including regenerating successional forests and pasture and agricultural areas. Our goal was to investigate landscape phenologic patterns in complex tropical rainforests and assess the extent, magnitude, and synchrony of phenologic patterns. We found wide variability and large seasonality in spectral signatures in the tropical forests analyzed here, with much of the variation occurring in forest conversion and regenerating areas of varying age classes and type of secondary forest regrowth. Unique phenologic responses to seasonal drought periods were observed across all vegetation types, both spectrally and temporally.

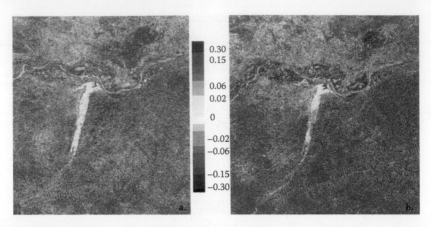

FIG. 11.14   (See color insert following page 134) MODIS 250-m EVI difference image for (a) the first dry-season phase (September minus July) and (b) the entire dry season, November minus July. Red colors indicate negative EVI trends and green colors depict positive EVI trends in the dry season.

Hyperspectral remote sensing and moderate resolution temporal satellite measurements represent powerful monitoring tools for the characterization of tropical landscape phenology and ecosystem processes. Hyperion imagery provides fine spectral and spatial detail that improves detection of land cover types and physiognomies, while moderate resolution sensors, such as MODIS, provide frequent temporal measurements for obtaining complete phenologic profiles, although at coarser resolutions (250 m to 8 km) and in limited, broadband portions of the spectrum. In addition, the combination of biochemical information extracted from Hyperion, together with advanced moderate resolution sensors (e.g., MODIS and SPOT-VGT), offers new opportunities to scale-up leaf physiologic processes and phenology to regional and global scales [72].

As noted by Carreiras et al. [67], one needs the right combination of spectral, spatial, or temporal resolutions to improve phenologic characterization of tropical forests. High temporal frequency measurements are critical in obtaining sufficient acquisitions of cloud-free data and achieve greater sensitivity to phenologic variations. In the spatial domain, disturbance and land-use classes and forest fragmentation patterns are best defined at fine resolutions < 30 m, and spectral fidelity and sensitivity are required to extract important biochemical canopy features that contribute to tropical forest phenologic patterns.

Many of the unique phenologic patterns found in the tropical forests presented here required adequate spectral, spatial, and temporal information. For example, attention to phenology was vital to the characterization of regenerating forests from space-borne sensors—however, only after the right spatial detail and correct spectral information were used. The finer spatial and spectral detail of Hyperion provided training data for MODIS and enabled extension and scaling of the more detailed and optically rich hyperspectral data. Only in realizing the spectral-spatial detail present in fine resolution hyperspectral data, can one effectively interpret and characterize phenologic variations in coarser-scale satellite imagery.

We found large phenologic variability in moist tropical forests, largely driven by the availability of solar radiation. Large spectral changes occurred throughout the dry season in response to leaf flushing and leaf exchange and associated canopy structure, leaf age, and leaf physiological changes, all of which enhanced the NIR signal. This enhanced dry-season greening signal became weaker with less developed regenerating forests, and was completely reversed in pasture and agriculture areas, where moisture limitations became the dominant control on phenology. Regenerating forests also appear to exhibit leaf flushing and green-up prior to the start of the new rainy season. Thus, optical-phenologic variations were observed in response to increases in physiologic activity as well as canopy structural changes.

Further studies are needed to more fully understand the biophysical/biochemical information present in the optical-phenologic signatures. Field work involving seasonal measurements of leaf water content, chlorophyll, dry matter, leaf phenology (leaf age), and canopy structure (LAI, shadows) is needed in order to better understand tropical forest phenology at the stand level. Such measurements become even more crucial in decoupling the asynchronous phenologies found in mixed, understory/overstory, tropical forest canopies. Attention to phenology will also be useful in distinguishing forest fragmented patches of tropical forest from the neighboring vegetation disturbance classes and in quantifying varying stages of forest degradation [26].

There are many optical measures and techniques that can also be used to more fully exploit hyperspectral remote sensing, including higher order derivative analysis of spectral signatures [70], band ratioing and continuum removal, ligno-cellulose absorption index [76], mixture modeling [77–81] (chapter 8), and spectral physiologic indices such as the photochemical reflectance index [82] and the "green" index of chlorophyll [83]. Mixture modeling of hyperspectral imagery offers much potential for structural and biochemical characterization of intact and disturbed tropical forests through subpixel mapping of green vegetation, soil, shadows, and nonphotosynthetic vegetation (NPV) quantities.

ACKNOWLEDGMENTS

We thank Dr. Brent Holben and the Aeronet staff for establishing and maintaining the 2001 and 2002 aerosol information from the Belterra site used in this investigation.

REFERENCES

1. Reich, P.B. and Borchert, R., Water stress and tree phenology in a tropical dry forest in the lowlands of Costa Rica, *Journal of Ecology*, 72, 61, 1984.
2. Myneni, R.B., Keeling, C.D., Tucker, C., et al., Increased plant growth in the northern high latitudes from 1981 to 1991, *Nature*, 386, 698, 1997.
3. Vina, A. and Henebry, G.M., Spatio-temporal change analysis to identify anomalous variation on the vegetated land surface: ENSO effects in tropical South America, *Geophysical Research Letters*, 32, 1, 2005.
4. White, M.A., Hoffman, F., Hargrove, W.W., et al., A global framework for monitoring phenological responses to climate change, *Geophysical Research Letters*, 32, doi:10.1029/2004GL021961, 2005.

5. Wright, S.J. and Schaik, C.P., Light and the phenology of tropical trees, *American Naturalist*, 143, 192, 1994.

6. Reich, P.B., Uhl, C., Walters, M.B., et al., Leaf demography and phenology in Amazonian rain forest: A census of 40,000 leaves of 23 tree species, *Ecological Monographs*, 74, 3, 2004.

7. Prior, L.D., Bowman, D.M.J.S., and Eamus, D., Seasonal differences in leaf attributes in Australian tropical tree species: Family and habitat comparisons, *Functional Ecology*, 18, 707, 2004.

8. Kushwaha, C.P. and Singh, K.P., Diversity of leaf phenology in a tropical deciduous forest in India, *Journal of Tropical Ecology*, 21, 47, 2005.

9. Bohlman, S.A., Adams, J.B., Smith, M.O., et al., Seasonal foliage changes in the Eastern Amazon basin detected from Landsat thematic mapper satellite images, *Biotropica*, 30, 376, 1998.

10. Curran, P.J. and Williamson, H.D., Sample size for ground and remotely sensed data, *Remote Sensing of Environment*, 20, 31, 1986.

11. Lean, J. and Rowntree, P.R., A GCM simulation of the impact of Amazonian deforestation on climate using an improved canopy representation, *Quarterly Journal of the Royal Meteorological Society*, 119, 509, 1993.

12. Sellers, P.J., Meeson, B.W., Hall, F.G., et al., Remote sensing of the land surface for studies of global change: Models–algorithms–experiments, *Remote Sensing of Environment*, 51, 3, 1995.

13. Frankie, G.W., Baker, H.G., and Opler, P.A., Comparative phenological studies of trees in tropical wet and dry forests in the lowlands of Costa Rica, *Journal of Ecology*, 62, 881, 1974.

14. Asner, G.P., Nepstad, D., Cardinot, G., et al., Drought stress and carbon uptake in an Amazon forest measured with space-borne imaging spectroscopy, *Proceedings of the National Academy of Sciences*, 101, 6039, 2004.

15. Carswell, F.E., Costa, A.L., Palheta, M., et al., Seasonality in CO_2 and H_2O flux at an eastern Amazonian rain forest, *Journal of Geophysical Research*, 107, 8076, 2002.

16. Saleska, S.R., Miller, D.M., Matross, M.L., et al., Carbon in Amazon forests: Unexpected seasonal fluxes and disturbance-induced losses, *Science*, 302, 1554, 2003.

17. Randow, C.V., Manzi, A.O., Kruijt, B., et al., Comparative measurements and seasonal variations in energy and carbon exchange over forest and pasture in southwest Amazonia, *Theoretical and Applied Climatology*, 78, 5, 2004.

18. Vourlitis, G.L., Filho, N.P., Hayashi, M.M.S., et al., Seasonal variations in the net ecosystem CO_2 exchange of a mature Amazonian transitional tropical forest (cerradao), *Functional Ecology*, 15, 388, 2001.

19. Herrerias-Diego, Y., Quesada, M., and Stoner, K.E., Fragmentation on phenological patterns and reproductive success of the tropical dry forest tree *Ceiba aesculifolia*, *Conservation Biology*, 20, 1111, 2006.

20. Singh, K.P. and Kushwaha, C.P., Emerging paradigms of tree phenology in dry tropics, *Current Science*, 89, 964, 2005.

21. Cochrane, M.A., Using vegetation reflectance variability for species level classification of hyperspectral data, *International Journal of Remote Sensing*, 21, 2075, 2000.

22. Houghton, R.A., Skole, D.L., Nobre, C.A., et al., Annual fluxes of carbon from deforestation and regrowth in the Brazilian Amazon, *Nature*, 403, 301, 2000.

23. Lambin, E.F., Geist, H.J., and Lepers, E., Dynamics of land-use and landcover change in tropical regions, *Annual Review of Environment and Resources*, 28, 205, 2003.

24. Brown, S. and Gaston, G., *Estimates of biomass density for tropical forest*. MIT Press, Cambridge, MA, 1996.

25. Foody, G., Palubinskas, G., Lucas, R., et al., Identifying terrestrial carbon sinks: Classification of successional stages in regenerating tropical forest from Landsat TM data, *Remote Sensing of Environment*, 55, 205, 1996.
26. Castro-Esau, K., Sanchez-Azofeifa, G.A., and Rivard, B., Monitoring secondary tropical forests using space-borne data: Implications for Central America, *International Journal of Remote Sensing*, 24, 1853, 2003.
27. Keller, M., Alencar, A., Asner, G.P., et al., Ecological research in the large-scale biosphere-atmosphere experiment in Amazon: Early results, *Ecological Applications*, 14, S3, 2004.
28. Cihlar, J., Ly, H., Li, J., et al., Multitemporal, multichannel AVHRR data sets for land biosphere studies artifacts and corrections, *Remote Sensing of Environment*, 60, 35, 1997.
29. Defries, R.S., Field, C.B., Fung, I., et al., Mapping the land surface for global atmosphere-biosphere models—Toward continuous distributions of vegetation functional properties, *Journal of Geophysical Research-Atmospheres*, 100, 20867, 1995.
30. Mayaux, P. and Lambin, E.F., Tropical forest area measured from global land-cover classifications: Inverse calibration models based on spatial textures, *Remote Sensing of Environment*, 59, 29, 1997.
31. Vieira, I.C., Almeida, A.S., Davidson, E.A., et al., Classifying successional forest stages using Landsat spectral properties and ecological characteristics in eastern Amazonia, *Remote Sensing of Environment*, 87, 470, 2003.
32. Kalacska, M.E.R., Sanchez-Azofeifa, G.A., Calvo-Alvarado, J.C., et al., Effects of season and successional stage on leaf area index and spectral vegetation indices in three mesoamerican tropical dry forests, *Biotropica*, 37, 486, 2005.
33. Running, S.W., Loveland, T.R., Pierce, L.L., et al., A remote sensing based vegetation classification logic for global land cover analysis, *Remote Sensing of Environment*, 51, 39, 1995.
34. Running, S.W., Justice, C., Salomonson, V.V., et al., Terrestrial remote sensing science and algorithms planned for EOS/MODIS, *International Journal of Remote Sensing*, 15, 3587, 1994.
35. Skole, D.L. and Qi, J., *Optical remote sensing for monitoring forest and biomass change in the context of the Kyoto Protocol*, *Cgceo/Ra01-01/W*, Michigan State University, East Lansing, 2001.
36. Justice, C.O., Townshend, J.R.G., Holben, B.N., et al., Analysis of the phenology of global vegetation using meteorological satellite data, *International Journal of Remote Sensing*, 6, 1271, 1985.
37. Eva, H.D., Belward, A.S., De Miranda, E.E., et al., A land cover map of South America, *Global Change Biology*, 10, 731, 2004.
38. Sader, S.A., Waide, R.B., Lawrence, W.T., et al., Tropical forest biomass and successional age class relationships to a vegetation index derived from Landsat TM data, *Remote Sensing of Environment*, 28, 143, 1989.
39. Goward, S.N., Markham, B., Dye, D.G., et al., Normalized difference vegetation index measurements from the advanced very high resolution radiometer, *Remote Sensing of Environment*, 35, 257, 1991.
40. Kobayashi, H. and Dye, D.G., Atmospheric conditions for monitoring the long-term vegetation dynamics in the Amazon using normalized difference vegetation index, *Remote Sensing of Environment*, 97, 519, 2005.
41. Justice, C., Hall, D., Salomonson, V.V., et al., The moderate resolution imaging spectroradiometer (MODIS): Land remote sensing for global change research, *IEEE Transactions on Geoscience and Remote Sensing*, 36, 1228, 1998.
42. Huete, A., Didan, K., Miura, T., et al., Overview of the radiometric and biophysical performance of the MODIS vegetation indices, *Remote Sensing of Environment*, 83, 195, 2002.

43. Zhang, Q., Xiao, X., Braswell, B., et al., Estimating light absorption by chlorophyll, leaf and canopy in a deciduous broadleaf forest using MODIS data and a radiative transfer model, *Remote Sensing of Environment*, 99, 357, 2005.
44. Huete, A., Ratana, P., Didan, K., et al., Seasonal biophysical dynamics along an Amazon eco-climatic gradient using MODIS vegetation indices, in *Anais XI SBSR*, Belo Horizonte, Brazil, 2003.
45. Xiao, X., Zhang, Q., Saleska, S.R., et al., Satellite-based modeling of gross primary production in a seasonally moist tropical evergreen forest, *Remote Sensing of Environment*, 94, 105, 2005.
46. Huete, A.R., Didan, K., Shimabukuro, Y.E., et al., Amazon rainforests green-up with sunlight in the dry season, *Geophysical Research Letters*, 13, doi:10.1029/2005GL025583, 2006.
47. Xiao, X., Hagen, S., Zhang, Q., et al., Detecting leaf phenology of seasonally moist tropical forests in South America with multitemporal MODIS images, *Remote Sensing of Environment*, 103, 465, 2006.
48. Ratana, P., Spatial and temporal amazon vegetation dynamics and phenology using time series satellite data, University of Arizona Dissertation, 157 pp., 2006. http://etd.library.arizona.edu/etd/SearchServlet.
49. Myneni, R.B., Yang, W., Nemani, R.R., et al., Large seasonal swings in leaf area of Amazon rainforests, *Proceedings of the National Academy of Sciences*, 104, 4820, 2007.
50. Sanchez-Azofeifa, G.A., Castro, K., Rivard, B., et al., Remote sensing research priorities in tropical dry forest environments, *Biotropica*, 35, 134, 2003.
51. Roberts, D.A., Nelson, B.W., Adams, J.B., et al., Spectral changes with leaf aging in Amazon Caatinga, *Trees-Structure and Function*, 12, 315, 1998.
52. Sanchez-Azofeifa, G.A. and Castro, K.L., Canopy observations on the hyperspectral properties of a community of tropical dry forest lianas and their host trees, *International Journal of Remote Sensing*, 27, 2101, 2006.
53. Ungar, S., Pearlman, J., Mendenhall, J., et al., Overview of the Earth observing one (EO-1) mission, *IEEE Transactions on Geoscience and Remote Sensing*, 41, 1149, 2003.
54. Shimabukuro, Y.E., Miura, T., Huete, A., et al., Analise dos dados hyperespectrais Do EO-1 obtidos sobre a Floresta Nacional de Tapajos no sstado do para, in *Anais XI SBSR*, 05-10 April, Belo Horizonte, Brazil, 2003.
55. Miura, T., Huete, A.R., Ferreira, L., et al., Discrimination and biophysical characterization of Brazilian Cerrado physiognomies with EO-1 hyperspectral Hyperion, in *12th JPL Airborne Earth Science Workshop*, R.E. Green, ed. NASA, Jet Propulsion Laboratory, California Institute of Technology, 2003, 207.
56. Goulden, M.L., Miller, S.D., da Rocha, H.R., et al., Diel and seasonal patterns of tropical forest CO_2 exchange, *Ecological Applications*, 14, S42, 2004.
57. Silver, W.L., Neff, J., McGroddy, M., et al., Effects of soil texture on belowground carbon and nutrient storage in a lowland Amazonian forest ecosysytem, *Ecosystems*, 3, 193, 2000.
58. Holben, B.N., Eck, T.F., Slutsker, I., et al., Aeronet—A federated instrument network and data archive for aerosol characterization, *Remote Sensing of Environment*, 66, 1, 1998.
59. Espirito-Santo, F.D.B., Shimabukuro, Y.E., and Kuplich, T.M., Mapping forest successional stages following deforestation in Brazilian Amazonia using multitemporal Landsat images, *International Journal of Remote Sensing*, 26, 635, 2005.
60. Ewel, J.J., Tropical succession: Manifold routes to maturity, *Biotropica*, 12(suppl.), 1, 1980.
61. Brown, S.A. and Lugo, A.E., Tropical secondary forests, *Journal of Tropical Ecology*, 6, 1, 1990.

62. Steininger, M., Satellite estimation of tropical secondary forest aboveground biomass: Data from Brazil and Bolivia, *International Journal of Remote Sensing*, 21, 1139, 2000.
63. da Rocha, H.R., Goulden, M.L., Miller, S.D., et al., Seasonality of water and heat fluxes over a tropical forest in Eastern Amazonia, *Ecological Applications*, 14, S22, 2004.
64. Nepstad, D.C., de Carvalho, C.R., Davidson, E., et al., The role of deep roots in the hydrological and carbon cycles of Amazonian forests and pastures, *Nature*, 372, 666, 1994.
65. Lucas, R.M., Honzak, M., Foody, G.M., et al., Characterizing tropical secondary forests using multitemporal Landsat sensor imagery, *International Journal of Remote Sensing*, 14, 3016, 1993.
66. Lucas, R.M., Honzak, M., Curran, P.J., et al., Mapping the regional extent of tropical forest regeneration stages in the Brazilian legal Amazon using NOAA AVHRR data, *International Journal of Remote Sensing*, 21, 2855, 2000.
67. Carreiras, J.M.B., Pereira, J.M.C., Campagnolo, Y.E., et al., Assessing the extent of agriculture pasture and secondary successsion forest in the Brazilian legal Amazon using spot vegetation data, *Remote Sensing of Environment*, 101, 283, 2006.
68. Demetriades-Shah, T.H., Steven, M.D., and Clark, J.A., High resolution derivative spectra in remote sensing, *Remote Sensing of Environment*, 33, 55, 1990.
69. Penuelas, J., Gamon, J.A., Fredeen, A.L., et al., Reflectance indices associated with physiological changes in nitrogen- and water-limited sunflower leaves, *Remote Sensing of Environment*, 48, 135, 1994.
70. Elvidge, C.D. and Chen, Z., Comparison of broadband and narrowband red and near-infrared vegetation indices, *Remote Sensing of Environment*, 54, 38, 1995.
71. Tsai, F. and Philpot, W., Derivative analysis of hyperspectral data, *Remote Sensing of Environment*, 66, 41, 1998.
72. Field, C.B., Behrenfeld, M.J., Randerson, J.T., et al., Primary production of the biosphere: Integrating terrestrial and oceanic components, *Science*, 281, 237, 1998.
73. Nemani, R.R., Keeling, C.D., Hashimoto, H., et al., Climate-driven increases in global terrestrial net primary production from 1982 to 1999, *Science*, 300, 1560, 2003.
74. Gao, B.C., NDWI: A normalized difference water index for remote sensing of vegetation liquid water from space, *Remote Sensing of Environment*, 58, 257, 1996.
75. Ceccato, P.S., Flasse, S., Tarantola, S., et al., Detecting vegetation leaf water content using reflectance in the optical domain, *Remote Sensing of Environment*, 77, 22, 2001.
76. Elvidge, C.D., Visible and near-infrared reflectance characteristics of dry plant materials, *International Journal of Remote Sensing*, 11, 1775, 1990.
77. Adams, J.B., Sabol, D., Kapos, V., et al., Classification of multispectral images based on fractions of endmembers; application to land-cover change in the Brazilian Amazon, *Remote Sensing of Environment*, 52, 137, 1995.
78. Cochrane, M.A. and Souza, C.M., Linear mixture model classification of burned forests in the Eastern Amazon, *International Journal of Remote Sensing*, 19, 3433, 1998.
79. Asner, G.P. and Lobell, D.B., A biogeophysical approach for automated SWIR unmixing of soils and vegetation, *Remote Sensing of Environment*, 74, 99, 2000.
80. Foody, G.M., Palubinskas, G., Lucas, R.M., et al., Identifying terrestrial carbon sinks: Classification of successional stages in regenerating tropical forest from Landsat TM data, *Remote Sensing of Environment*, 55, 205, 1996.
81. Shimabukuro, Y.E., Holben, B.N., and Tucker, C.J., Fraction images derived from NOAA AVHRR data for studying the deforestation in the Brazilian Amazon, *International Journal of Remote Sensing*, 15, 517, 1994.
82. Gamon, J.A., Serrano, L., and Surfus, J.S., The photochemical reflectance index: An optical indicator of photosynthetic radiation use efficiency across species, functional types, and nutrient levels, *Oecologia*, 112, 4921, 1997.
83. Gitelson, A.A., Vina, A., Ciganda, V., et al., Remote estimation of canopy chlorophyll content in crops, *Geophysical Research Letters*, 32, LO8403, 2005.

62. Saleska, S.R., Didan, K., Huete, A.R., et al., Amazon forests green-up during 2005 drought, *Science*, 318, 612, 2007.

63. Huete, A.R., Didan, K., Shimabukuro, Y.E., et al., Amazon rainforests green-up with sunlight in dry season, *Geophysical Research Letters*, 33, L06405, 2006.

64. Samanta, A., Ganguly, S., Hashimoto, H., et al., Amazon forests did not green-up during the 2005 drought, *Geophysical Research Letters*, 37, L05401, 2010.

65. Brando, P.M., Goetz, S.J., Baccini, A., et al., Seasonal and interannual variability of climate and vegetation indices across the Amazon, *Proceedings of the National Academy of Sciences*, 107, 14685, 2010.

66. Atkinson, P.M., Dash, J., and Jeganathan, C., Amazon vegetation greenness as measured by satellite sensors over the last decade, *Geophysical Research Letters*, 38, L19105, 2011.

67. Morton, D.C., Nagol, J., Carabajal, C.C., et al., Amazon forests maintain consistent canopy structure and greenness during the dry season, *Nature*, 506, 221, 2014.

12 Hyperspectral Remote Sensing of Canopy Chemistry, Physiology, and Biodiversity in Tropical Rainforests

Gregory P. Asner

CONTENTS

12.1   INTRODUCTION

Whether one is observing from the ground or from above the canopy, humid tropical forests are among the most complex of biological constructs found in nature. Species diversity is usually high, and the vertical layering of canopies is often pronounced. Tree mortality and regrowth occur throughout a rainforest, resulting in a mosaic of canopy gaps and tree ages, with concomitant variability of local species dominance and canopy structural properties [1]. Overlain on these natural sources of variation are community- and ecosystem-level changes that come at the hand of the human enterprise. We clear-cut, selectively log, occupy, and/or abandon thousands of square kilometers of tropical forests each year [2,3], causing regional variations of taxonomic composition and canopy structure. Together, these natural and anthropogenic sources of variation result in a complexity of ecosystem form and function that challenges ground-based studies.

Many fundamental questions in tropical ecology revolve around the biochemistry, physiology, and biodiversity of forest canopies. Because canopies are a locus of biogeochemical processes in an ecosystem, canopy chemistry is core to understanding the spatial and temporal variability of carbon, nutrient, and hydrological cycling. Canopy physiology, which controls gross and net primary production, is also mediated by canopy biochemistry. Tree biodiversity largely determines the biochemical and physiological properties of a tropical forest ecosystem [4], but with functionally important control by substrate age and fertility [5,6]. In all, the biochemistry, physiology, and biodiversity of the tree canopies are intimately linked and cannot be easily studied without acknowledging these linkages.

Until recently, remote sensing has provided a very limited treatment of tropical forest function, partly due to the inadequacy of available sensor technologies, but also because we have lacked an understanding of how canopy biochemical and physiological properties vary across tropical landscapes. Not long ago, tropical forests were treated as "dense dark vegetation" targets for calibrating global satellite sensors such as the Advanced Very High Resolution Radiometer (AVHRR) [7]. In contrast today, the global remote sensing community acknowledges the seasonal and spatial dynamics of tropical forest function using multispectral sensors like Terra MODIS [8]. Moderate resolution sensors, such as Landsat, have also been used to assess phenological changes in tropical forests [9], but the far more common use of these sensors has been natural [10] and human-caused disturbances [11]. Meanwhile, very high spatial resolution (<2 m) satellite monitoring has rapidly expanded to allow estimation of tree crown properties and mortality in tropical forests [12,13]. For the most part, these studies highlight the structural complexity of tropical forests, but biochemical and physiological functioning cannot be probed with the limited spectral information contained in high spatial resolution satellite sensors (e.g., IKONOS, Quickbird).

With the advent of hyperspectral imaging (also called imaging spectroscopy), there has been an expanding potential for landscape and regional studies of canopy biochemistry and physiology. Most of this work has focused on temperate forest ecosystems [14–20], showing that canopy water, leaf pigments, and nitrogen can be estimated using hyperspectral imaging. However, many of these studies have also grappled with the covariance between the intended measurement—leaf chemistry—and other factors such as vegetation type (e.g., coniferous vs. broadleaf), leaf area index (LAI), and phenology [21]. To date, we still do not know the precise relative importance of leaf and canopy properties determining the hyperspectral signatures of forest canopies, although we recognize that vegetation type and LAI are the principal drivers of reflectance variability in many forests [22].

In this chapter, I discuss issues and methods relating to the remote exploration of canopy biochemistry, physiology, and biodiversity in humid tropical forests using airborne and space-based hyperspectral imaging. I begin with the spectral properties of tropical rainforests, building from the leaf to canopy to stand scales. A key issue rests in the sources of spectral variation as one ascends in scale, which ultimately determines what can be measured remotely from airborne and space-based imaging spectrometers. Throughout the scaling process, I highlight three remote sensing approaches for exploring the biochemistry and physiology of tropical forests: partial least squares (PLS) regression analysis, radiative transfer modeling, and hyperspectral reflectance indices. Some methods work well with high-performance field and airborne hyperspectral sensors, while others are needed for the current low-fidelity space-borne sensors. I then offer a conceptual approach to mapping canopy diversity using biochemical signatures accessible with high-fidelity hyperspectral imaging. In covering these issues and methods, I focus specifically on examples—some published and some newly reported in this chapter—from my work in humid tropical forests of the Brazilian Amazon and Hawaii.

12.2   TROPICAL FOREST BIOPHYSICS AND BIOCHEMISTRY

Closed-canopy tropical forests are often characterized by an intermingling of relatively large tree crowns, many of which have high LAI and mostly random or spherical foliage distributions (fig. 12.1). This basic biophysical setting creates radiative conditions that allow for a range of biochemical and physiological studies of tropical canopies. To understand how these biophysical conditions translate to remote biochemical determinations, it is important to first summarize the way that leaf chemistry, and then canopy structure, is expressed in the optical spectrum (400–2500 nm).

FIG. 12.1   Airborne panchromatic image of lowland tropical rainforest in Hawaii. (G. Asner, unpublished data.)

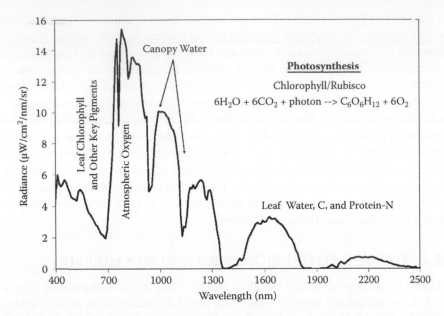

FIG. 12.2 The radiance spectrum of a leaf or canopy expresses the presence and abundance of the building blocks and products of photosynthesis. Elemental and molecular contributions to the spectrum are labeled. This spectrum was acquired over lowland Hawaiian rainforest using the Airborne Visible and Infrared Imaging Spectrometer (AVIRIS).

12.2.1 LEAF SPECTRAL-BIOCHEMICAL PROPERTIES

The spectral properties of live foliage set up the radiation field in a tropical forest canopy, and these spectral properties express the presence and abundance of both the inputs to and products of photosynthesis (fig. 12.2). Leaf pigments such as chlorophylls, carotenoids, and anthocyanins are expressed in the 400- to 700-nm range, matching the wavelength region of maximum solar input to the biosphere. The relative contribution of pigments to the reflectance and transmittance properties of foliage varies by wavelength in this region, and all pigments have overlapping absorption features (fig. 12.3). Chlorophyll a (chl-a) displays maximum absorptions in the 410- to 430-nm and 600- to 690-nm regions, whereas chlorophyll b (chl-b) shows maximum absorptions in the 450- to 470-nm range. Carotenoids absorb most efficiently between 440 and 480 nm. In the foliage of many canopy species, chl-a dominates the overall absorption spectrum at shorter and longer wavelengths in the visible spectrum, whereas carotenoids can be a major contributor at slightly longer wavelengths (fig. 12.3).

In the near infrared (700–1300 nm) and shortwave infrared (1300–2500 nm), the leaf spectrum is dominated by water content, thickness, and, to a lesser degree, protein-nitrogen (N), cellulose, and lignin content (fig. 12.2) [23]. In particular, live foliage is very efficient at scattering radiation in the 750- to 1300-nm range; this is caused by internal scattering at the air–cell–water interfaces within the leaves [24,25]. At longer wavelengths (>1300 nm), relatively small amounts of water (and resulting air–water interfaces) effectively trap radiation, resulting in absorption that

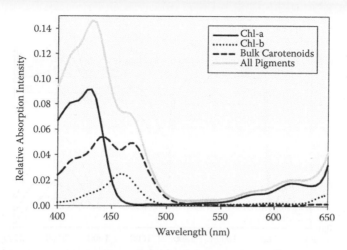

FIG. 12.3   Relative absorption intensity of chlorophyll a and b, as well as bulk carotenoids, in *Metrosideros polymorpha* collected in a lowland Hawaiian rainforest. (G. Asner, unpublished data.)

exceeds scattering processes in this portion of the spectrum. Meanwhile, leaf structural properties, especially area per mass (specific leaf area or SLA), play an underlying but integral role in containing the biochemicals in a functional form adapted for leaf carbon fixation and related photosynthetic processes [26,27].

What is the contribution of leaf N to reflectance signatures of tropical forests? Proteins have absorptions spread throughout the shortwave-infrared (IR) spectrum [23] that are partially obscured by water absorptions. In addition, chlorophyll is ~6% nitrogen by mass, and thus chlorophyll and N tend to be broadly correlated with one another. Remote estimation of chlorophyll tends to scale with N, with a range of correlation coefficients (r) from about 0.7 to 0.9 [28]. Moreover, spectral analyses between leaf or canopy spectra and nitrogen often show spectral correlations in the wavelength regions associated with both chlorophyll and protein-N [9,16,17]. But are correlations between spectra and N concentration direct, or are they indirect by way of leaf chlorophyll and other leaf constituents? Leaf construction involves a stoichiometric balance among chlorophyll, nitrogen, water, and other biochemicals, all convolved with leaf structure, especially SLA [29,30]. Leaf protein is dominated by the enzyme rubisco, which mediates carbon fixation in concert with light capture by chlorophyll [31,32]. Leaf water concentrations scale with SLA, and SLA is correlated with leaf N per unit area [33]. These leaf parameters are therefore not independent, biophysically or ecologically, and thus the retrieval of leaf N tends to be wrapped up in the retrieval of the other parameters.

I can demonstrate these spectral-biochemical linkages using partial least squares (PLS) regression analysis to relate the entire leaf reflectance or transmittance spectrum to any given biochemical constituent. I took the fresh foliage spectra of 150 canopy species collected in a 50-ha forest stand in the eastern Brazilian Amazon (fig. 12.4; appendix 12.1), and ran a PLS regression against their respective leaf N concentrations. Spectral weighting vectors generated by the PLS calculation correspond to features in the spectra that predict the chemical constituent analyzed. The spectral features that

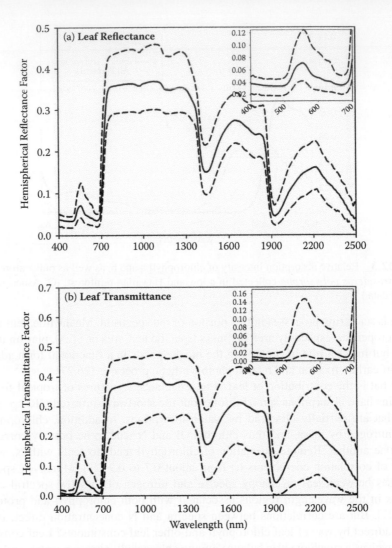

FIG. 12.4 Mean, minimum, and maximum reflectance and transmittance spectra of 150 canopy tree species from the eastern Amazon basin (G. Asner, unpublished data). Inset shows zoom of visible (400–700 nm) portion of the spectrum.

contribute most to the regression are given the largest positive and negative weightings. Figure 12.5 shows that spectral reflectance (and transmittance) features centered near 500, 530, 705, 730, 1580, 1695, 1790, 2050, 2125, 2180, 2240, and 2315 nm all strongly contributed to the prediction of leaf N concentration of the Amazon forest canopy species. The 500- to 730-nm features are associated with pigments (fig. 12.3); the 1580- to 1790-nm features result from carbon constituents related to SLA—namely, lignin, cellulose, and starches [17]. The 2050- to 2240-nm features are directly associated with protein N; and the 2315-nm feature is usually linked to leaf oils [23]. A constellation of stoichiometrically balanced leaf biochemicals forms the spectrum, and thus a regression based on only one of those chemicals—nitrogen—expresses indirect

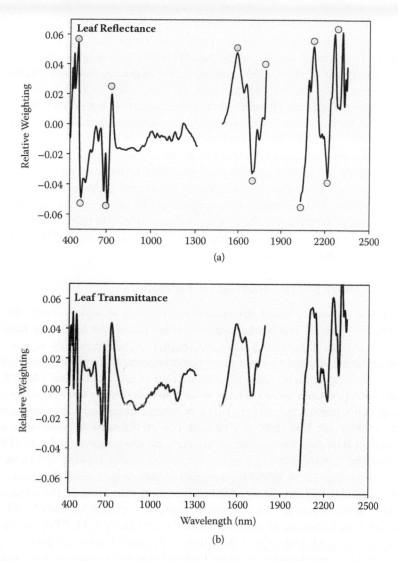

FIG. 12.5 Relative spectral weightings for predicting leaf nitrogen concentration from the reflectance and transmittance spectra of 150 tropical forest species collected in the eastern Brazilian Amazon (fig. 12.4). Gray dots in upper panel show peaks of maximum importance in the leaf N predictions; these peaks align well with published studies that highlight spectral sensitivities to pigment, N, and leaf water.

contributions from other leaf constituents. The take-home message is that any single-parameter estimate of leaf (or canopy) chemistry must take into account the potential indirect contributions made by other chemicals.

Aside from the causal linkages between leaf spectral and biochemical properties in tropical forests, another major issue rests in the variability among species. Do the leaves of species look the same spectrally or biochemically? No, but the degree of spectral and biochemical variability dictates whether we might be able to remotely sense differences among species [34,35]. The 150 species collected in the eastern Amazon (fig. 12.4) demonstrate a common observation that leaf reflectance and transmittance are most variable in the near-IR (NIR) wavelengths (750–1300 nm), with the shortwave IR (1500–2500 nm) appearing as the second most variable. The visible region (400–700 nm) has the least variable reflectance and transmittance among rainforest species. However, these observations do not account for the fact that spectral variation can both under- and overestimate biochemical variation. In both the visible and near-IR spectral ranges, small variations in leaf reflectance/transmittance are very strongly expressed at canopy scales, but for very different reasons [36]. In the visible, scattering is low, and thus hyperspectral remote sensing observations of canopies are sensitive to top-of-canopy reflectance variability that is driven by leaf pigments. In contrast, near-IR observations are subject to large amounts of scattering in the canopy, and thus small changes in leaf water are expressed as an integral change in canopy volumetric water content. These concepts are presented in the next section.

12.2.2 Canopy Spectral-Biochemical Properties

For airborne and space-based remote sensing, it is not enough to know the linkages between spectral and biochemical properties at the leaf level. The basic leaf biochemical-spectral properties described earlier are differentially expressed at the canopy scale, depending upon canopy architecture, leaf volume, and chemical composition. For example, when vegetation cover is less than about 70–80%, gap fraction is the first-order driver of reflectance signatures in any terrestrial ecosystem [37], especially tropical forests [38]. However, when canopy cover approaches 100%, canopy volume and leaf chemistry become co-determinants of reflectance signatures. Here, I introduce the concept of "effective photon penetration depth" (EPPD), defined as the canopy depth to which a downward (nadir) viewing spectrometer is most sensitive (fig. 12.5). EPPD is analogous to path length and light extinction in classical radiative transfer theory [39]; however, I use it to estimate the actual depth in number of leaf layers to which we might measure biochemicals and physiological properties. It is important to note that EPPD is not simply LAI. EPPD is the convolved effect of LAI, leaf angle distribution, and foliar biochemical properties, the latter of which is wavelength dependent. Estimation of EPPD requires a combination of detailed field measurements and a canopy radiative transfer model (*sensu* [36]). Utilizing these methods, one can account for the specific absorption and scattering properties of the foliage in three dimensions, thereby backing out LAI as an index of photon penetration into the canopy, and the subsequent foliar depth expressed in a hyperspectral signature measured from above the canopy.

Since live foliage strongly absorbs in the visible (400–700 nm) region, EPPD averages 1–1.5 LAI units in these wavelengths (fig. 12.6). Therefore, when the canopy is closed, only upper-canopy pigments can be directly measured with downward-looking spectrometers. In contrast, foliage efficiently scatters photons in the near IR (700–1300 nm); the overall reflectance of this region (the so-called "near-IR plateau") increases nonlinearly as LAI approaches values of about 3–4 (fig. 12.6). Thereafter, the plateau saturates, but the liquid water features centered at 980

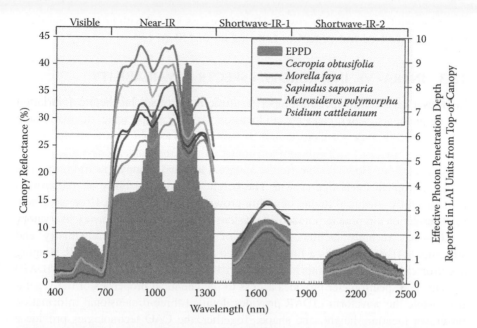

FIG. 12.6 Effective photon penetration depth (EPPD) of a tropical forest canopy in Hawaii. This EPPD estimate was derived by combining field measurements of leaf hyperspectral-optical properties, canopy shape, and architectural data, and LAI in a three-dimensional canopy radiative transfer model. EPPD was calculated as the depth from the top of the canopy to which photons no longer contribute to the top-of-canopy reflectance.

and 1190 nm continue to deepen, widen, and shift up to LAI values of 8 or more [15,40,41]. EPPD is therefore high in the near-IR wavelength region, so this spectral range can provide biochemical information deep into a foliage canopy. However, the only major biochemical contributor to the near IR is water; thus, canopy water content can be measured in the near IR. The shortwave IR (1500–2500 nm) is variably sensitive to both leaf and canopy water content. The 1500- to 1800-nm range saturates at LAI of about 2, whereas the 2000- to 2500-nm range saturates at LAI values of only 1.5 or less (fig. 12.6). Therefore, the EPPD is moderate to very low in the shortwave IR, and so a detailed knowledge of this spectral region can be used to maximize sensitivity to upper-canopy or leaf-level water content [42].

The LAI of humid tropical forest canopies averages 4.8 (s.d. = 1.7), but with maximum reported values exceeding 10 LAI units [43]. From the preceding description of the contributing biochemicals to the optical spectrum and the subsequent EPPD, it is clear that pigments can only be sensed from uppermost-canopy leaves, in units of either concentration or concentration per area. In contrast, both leaf and canopy water content can be estimated using a combination of near-IR and shortwave-IR regions. Where scattering dominates (750–1300 nm), hyperspectral reflectance measurements are uniquely sensitive to canopy water content. Where absorption dominates (>1300 nm), the reflectance measurements are most sensitive to leaf water concentration in the upper canopy. Nitrogen is again expressed at the upper canopy via chlorophyll and subtle shortwave-IR features associated with proteins, but these are often convolved with other leaf constituents and SLA.

12.3   INTRA- VS. INTERCROWN SPECTRAL VARIABILITY

Leaf-level variation in spectral properties directly expresses biochemical variation among pigments, water, nutrients, and carbon. At the canopy scale, the expression of leaf spectral properties is mediated by factors such as LAI, leaf orientation, and canopy gap fraction (with concomitant contributions from underlying soils, surface litter, and shade). How variable are the apparent leaf and canopy spectral properties within a tropical forest tree crown? The height and inaccessibility of tree crowns have generally precluded studies of within-crown spectral variability. However, a new high-resolution airborne remote sensing system—the Carnegie Airborne Observatory (CAO)—has recently facilitated studies of tropical forest tree chemistry and structure [44]. The CAO is a fully integrated 288-channel visible/near-IR imaging spectrometer and small-footprint waveform light detection and ranging (LiDAR) system. The spectrometer provides access to the chemical properties of the vegetation, while the waveform LiDAR provides detailed three-dimensional information on crown location, height, and shape. Together, the CAO technologies provide a highly calibrated measurement of vegetation spectral and structural properties at up to 0.3-m spatial resolution.

I used the CAO to assess within-crown spectral reflectance variation of a rainforest stand on the island of Hawaii (mean annual precipitation = 2500 mm) (fig. 12.7). The site is located in the Laupahoehoe Forest Reserve, which contains trees that are 20–30 m tall, with the largest crowns averaging 8–12 m in diameter, on some of the most fertile substrates in the Hawaiian Archipelago [6,45]. The image area is dominated by a single native Hawaiian forest species, *Metrosideros polymorpha* (ohia), providing a means to consider within-species variability in canopy spectroscopy. The CAO imagery was collected at 0.5-m spatial resolution, resulting in 150–200 pixels per tree crown. The image was atmospherically corrected using the ACORN-5 code (ImSpec LLC, Palmdale, California), but no other smoothing functions were applied. Within a 50-ha area, I selected 10 trees from which to extract the spectra from fully sunlit portions of each crown. Shade pixels were avoided to limit confounding effects of illumination angle. I then calculated the mean and standard deviation spectra throughout the 400- to 1050-nm wavelength range.

Intercrown reflectance variability was highest in the 750- to 1050-nm range, with a maximum 48% variation in reflectance values at 828 nm (fig. 12.8a). The lowest intercrown reflectance variability was found in the 500- to 510-nm and 695- to 705-nm regions, with a minimum 12% reflectance variation at 501 nm. These results follow the earlier discussion that the EPPD of canopies is maximal in the near-IR spectral region dominated by scattering related to leaf and canopy water; the EPPD in the visible spectrum is low due to efficient absorption caused primarily by leaf chlorophyll and carotenoid pigments. These within-crown spectral signatures suggest that canopy water content is highly variable across individual tree crowns, likely due to localized variations in canopy thickness and LAI. In contrast, pigment

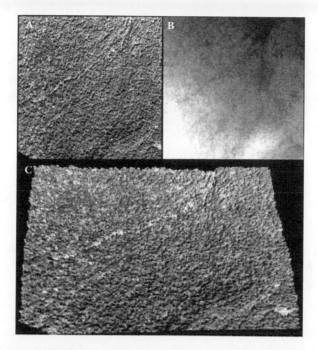

FIG. 12.7 Fusion of high spatial resolution imaging spectrometer and waveform-LiDAR data from the Carnegie Airborne Observatory (http://cao.stanford.edu). The imagery covers a portion of the state of Hawaii's Laupahoehoe Forest Reserve. The three-dimensional imagery provides a convenient means to select only sunlit portions of *Metrosideros polymorpha* tree crowns for analysis of intra- and intercrown variability in canopy reflectance. See CD for color image.

variation is highly conservative across these same tree crowns, at the aggregate scale of 0.5-m sampling resolution.

Within-crown variation in canopy reflectance (fig. 12.8b) was, on average, only 10% of the magnitude of the reflectance variation between tree canopies (fig. 12.8a). This is a critically important result demonstrating that intercrown reflectance variability far exceeds that of within-crown variation, and hence the spectral signatures of a species—at least, *M. polymorpha*—are relatively conservative. This suggests that within-crown biochemical variability, as expressed, for example, in the visible wavelength region by top-of-canopy pigment absorptions, is relatively low in comparison to intercrown differences. Strong intercrown variation thus suggests that microsite conditions, tree age, and canopy development exert some control over the canopy biochemistry of species in a localized area. The next level of analysis considers the broader impacts of environmental conditions on canopy spectral properties within a species.

12.4 CLIMATE-SUBSTRATE EFFECTS ON HYPERSPECTRAL REFLECTANCE INDICES

At the aggregated scale of moderate-resolution satellite pixels (e.g., 30 × 30 m), closed rainforest canopies often appear homogeneously green, with topography being one

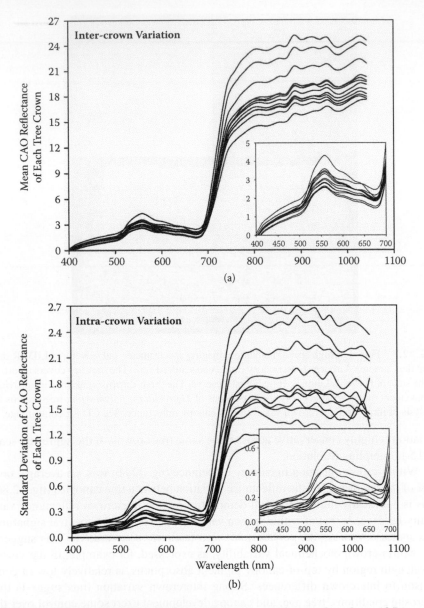

FIG. 12.8 (a) Mean reflectance of 10 *Metrosideros polymorpha* trees from figure 12.7 ($n = 100$ pixels per tree); (b) standard deviation of reflectance of the same trees. In comparing ranges of mean and standard deviation spectra, note the major difference in magnitudes of x-axes.

of the primary sources of perceived spatial variation in an image [46]. However, two of the most important factors affecting forest productivity are soil fertility and climate [6,47], even though few Landsat-type studies have been able to assess their relative importance in determining rainforest dynamics. Space-borne hyperspectral remote sensing has opened a window through which we can assess biochemical and physiological responses of rainforests to soil fertility gradients and climate change.

The Earth Observing-1 (EO-1) Hyperion sensor now provides global accessibility of imaging spectroscopy to tropical forests, although with much lower signal performance than that of most airborne sensors [48]. Nonetheless, Hyperion can provide hyperspectral reflectance indices with some sensitivity to forest pigment chemistry, which is directly related to canopy physiology, especially light-use efficiency (LUE), or the amount of carbon fixed per unit energy absorption.

We recently analyzed the expressed chemistry and physiology of moist and wet forests in Hawaii using Hyperion reflectance indices (fig. 12.9) [49]. We used vegetation, soil, and climate maps to isolate forests with known combinations of substrate age (fertility) and mean annual precipitation. We also controlled for forest composition by selecting stands dominated by the native keystone tree *M. polymorpha* (similar to fig. 12.7). This resulted in 28 substrate age/climate combinations for Hyperion analysis of *M. polymorpha* stands (fig. 12.9). We then calculated four common hyperspectral reflectance indices associated with leaf pigment, greenness, and canopy water properties (table 12.1). The normalized difference vegetation index (NDVI) is most sensitive to canopy fractional absorption of photosynthetically active radiation (fPAR) and LAI [22,50]. Two pigment indices—the photochemical reflectance index (PRI) [51] and carotenoid reflectance index (CRI) [52]—are sensitive to relative proportions of carotenoid and chlorophyll pigments [53–55]. The normalized difference water index (NDWI) detects relative changes in canopy water content [56].

We found that all indices of *M. polymorpha* forests increased in value with both precipitation and substrate age, but that there were differences among indices. The NDVI peaked in rainforest precipitation zones of about 4075 mm yr^{-1} and then declined; it also increased to maximum substrate ages of 140,000 years old (fig. 12.9a). In contrast, the water index NDWI never saturated, reaching a maximum at 6000 mm yr^{-1} and 140,000-year-old substrate (fig. 12.8b). This result supports the growing consensus that water indices, such as the NDWI, track canopy leaf area and structural dynamics with far more sensitivity than that of the traditional NDVI [20,57].

The upper-canopy pigment indices behaved differently from the canopy greenness/water metrics (figs. 12.9c and 12.9d). The PRI peaked at mean annual precipitation values of 4200 mm yr^{-1} and 130,000-year-old substrate. The CRI reached a maximum at about 3800 mm yr^{-1} on the older substrates. High PRI and CRI values generally indicate low carotenoid/chlorophyll ratios and high LUE [52–54,58–60], so these Hyperion results strongly suggest that forest canopy growth rates (gross and/or net primary production) are highest in moderately wet conditions on substrates with well-developed soils. Interestingly, these satellite observations concur with a number of field-based studies that have looked at *M. polymorpha* production: Herbert and Fownes [6] showed that the LUE and production of these forests peak on 20,000- to 150,000-year-old substrates that have high N and P availability; Schuur and Matson [61] demonstrated that *M. polymorpha* forests on Haleakala volcano (Maui) reach peak production in precipitation zones of about 3500 mm yr^{-1}; Raich, Russell, and Vitousek [62] showed that *M. polymorpha* forests on Mauna Loa volcano (Hawaii Island) have maximum growth rates at about 3000–3500 mm yr^{-1} rainfall. Martin and Asner [63] provide the most direct evidence to support the Hyperion observations, showing that the PRI, CRI, and light-use efficiency of *M. polymorpha* leaves peak at ~3500 mm yr^{-1} on the most fertile substrates. This test of space-borne hyperspectral

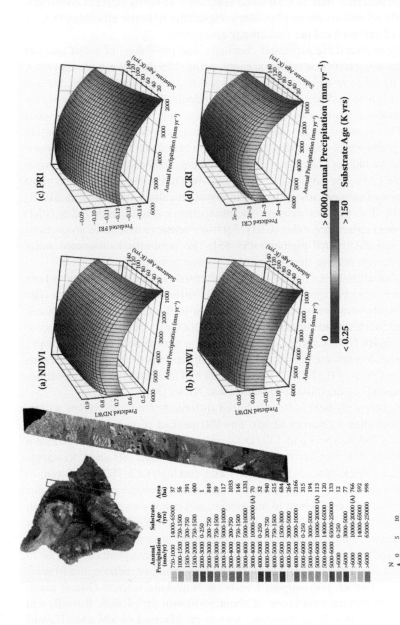

FIG. 12.9 (See color insert following page 134) Effects of substrate age (fertility) and precipitation on four narrowband spectral reflectance indices of *Metrosideros polymorpha* forests in Hawaii. Combinations of substrate age and precipitation were mapped, and the (a) NDVI, (b) NDWI, (c) PRI, and (d) CRI indices (table 12.1) were measured and interpolated as shown. (Reprinted from Asner, G.P. et al., *Remote Sensing of Environment*, 96, 497, 2005. With permission of Elsevier.)

TABLE 12.1

Four Narrowband, Hyperspectral Indices Available from the EO-1 Hyperion Space-borne Imaging Spectrometer

Index	Name	Equation	Ref.
NDVI	Normalized difference vegetation index	$(\rho_{800} - \rho_{680})/(\rho_{800} + \rho_{680})$	90
NDWI	Normalized difference water index	$(\rho_{857} - \rho_{1241})/(\rho_{857} + \rho_{1241})$	56
PRI	Photochemical reflectance index	$(\rho_{531} - \rho_{570})/(\rho_{531} + \rho_{570})$	51
CRI	Carotenoid reflectance index	$(1/\rho_{510}) - (1/\rho_{550})$	52

Note: Subscripts on reflectance (ρ) denote wavelength values in nanometers.

reflectance indices demonstrates that canopy biochemical and physiological properties can be measured to track forest responses to soil fertility and climate change.

The focus on rainforest–climate interactions has increased in recent years because predictions call for drier conditions in many humid tropical regions of the world, either by way of altered circulation patterns or from changes in El Nino southern oscillation (ENSO) frequency [64]. Multispectral satellite observations of both seasonal [8] and interannual [65] variations in greenness suggest that Amazon tropical forests are highly dynamic. Current hyperspectral observations complement these broad-scale satellite studies by exploring the detailed biochemical and physiological changes that may be occurring in forest canopies in Amazonia. For example, we used EO-1 Hyperion to collect hyperspectral reflectance indices of paired 1-ha forest stands in the eastern Amazon at the beginning and end of the dry season (July, November) [66]. One forest stand was artificially dried by removing 80% of incoming throughfall off-site [67]; the other stand was not manipulated. Field measurements of both stands included monthly LAI, leaf water potential, and soil water availability, showing that plant-available water decreased by more than 50% throughout the dry season, whereas dry-season effects on LAI were minimal [66]. Leaf water potential decreased, tracking changes in plant-available water in the soil. A Hyperion water index (SWAM), which is similar to the NDWI, was highly sensitive to seasonal drying of the canopy, even though the NDVI showed no sensitivity (fig. 12.10). In addition, the PRI and another pigment index, the anthocyanin reflectance index (ARI) [68], were highly sensitive to the artificial dry-down, capturing pigment changes directly related to decreases in light-use efficiency and primary production.

Interestingly, the control stand expressed changes in pigments that suggested increased light-use efficiency and photosynthetic rates during the dry season, a result that concurs with the Terra MODIS enhanced vegetation index (EVI) observations of the larger Amazon region by Huete et al. [8]. These results demonstrate that rainforest biochemical and physiological responses to climate variability can be observed and linked to canopy processes measured in the field. They also quantify both intraspecific (the Hawaii case) and temporal (the Amazon case) variation in expressed biochemical and physiological processes at stand scales.

Despite decades of advancement in remote sensing, the basic variability of tropical forest reflectance is still largely unexplored. Multispectral studies have provided insight to spatial variation in greenness, but moderate resolution satellite sensors

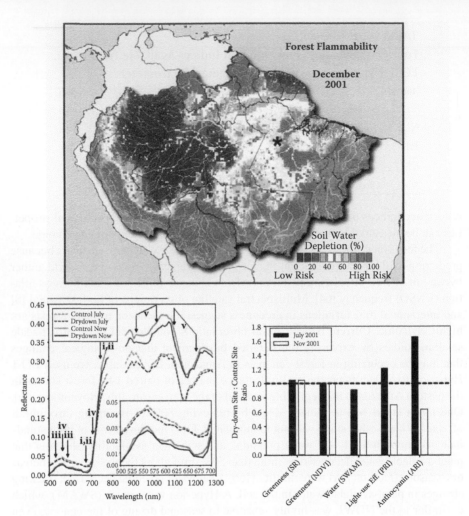

FIG. 12.10 Drought stress caused by land-use and climate change is a major concern for humid tropical forest productivity and diversity. The top panel shows a fire flammability prediction map generated by the Risque model run under drought conditions. The asterisk indicates the location of a forest drought experiment site in the eastern Amazon. Lower left panel shows EO-1 Hyperion spectra of the forest monitoring sites, including a control stand and an artificially dried forest site ("drydown"); lower right panel shows the sensitivity of Hyperion reflectance indices to forest drought conditions at the beginning (July) and end (November) of the dry season. (Reprinted from Asner, G.P. et al., *Proceedings of the National Academy of Sciences*, 101, 6039, 2004. With permission of the National Academy of Science, USA.) See CD for color image.

(e.g., Landsat) cannot easily resolve variation in closed-canopy forest. Hyperion represents the first chance to study hyperspectral reflectance variation in any tropical forest. In an exploratory study for this chapter, I collected 12 Hyperion images distributed across nine forest types [69] in the Brazilian Amazon basin from 2002 to 2003 (fig. 12.11). The images were acquired near the beginning of the dry season

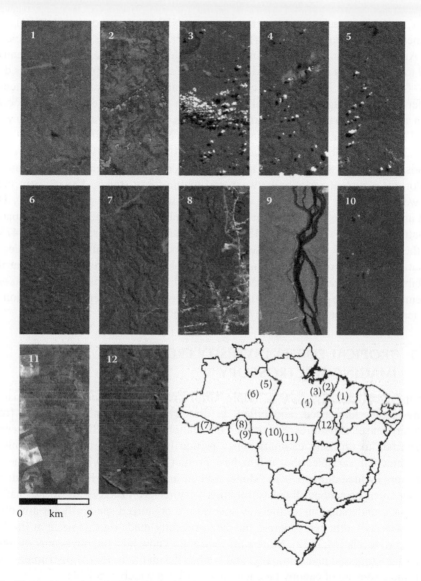

FIG. 12.11 A series of EO-1 Hyperion images of humid tropical forests throughout the Brazilian Amazon. Intact, closed-canopy forest areas were selected for analysis of spatial variations in canopy water and pigment hyperspectral reflectance indices (table 12.2). See CD for color image.

(June–August), and all were taken from nadir-viewing orbits (Hyperion is capable of pointing from adjacent orbits). I ran the ACORN-5 atmospheric correction algorithm on each image using a tropical atmosphere setting and 50-km visibility/aerosol setting. I then subset each image into 1000-ha forest stands with no sign of disturbance, and I avoided any obvious topographic shade. For each image subset, I calculated the four hyperspectral reflectance indices listed in table 12.1, and compared the spatial statistics of the indices. My primary goal was to understand the basic variability of each index, to see how each expresses the greenness (NDVI), water content (NDWI), and pigment concentration/physiology (PRI, CRI) of the forest stands of mixed species.

Among the 12 humid tropical forests, I found that the NDVI varied from a mean of 0.79 to 0.93, but that the spatial variability within a stand was extremely small (s.d. = 0.01–0.02) (table 12.2). The spatial coefficient of variation (CV) was also calculated to buffer the analysis against potential differences in image brightness that might occur from atmospheric artifacts (e.g., aerosol). The CVs of the NDVI values were also very small (1–2%). In contrast, the mean (±s.d.) of the NDWI ranged from 0.008 ± 0.021 to 0.033 ± 0.028, a fourfold range between forest stands, and the spatial CV for these stands varied from 82 to 462%. The NDWI was thus highly sensitive to structural and/or water content variation between and within forest stands. The pigment indices were also highly variable, both within and between forest stands (table 12.2). The PRI and CRI indices extended over a 4- and 17-fold range between sites. The spatial variation within sites ranged from 15 to 37% for the PRI, and a surprising 167 to 475% for the CRI. Clearly, additional research should focus on explaining these sources of variation; yet this collection of images and simple analyses already shows that the variability among canopy water and pigment indices far exceeds that of traditional greenness indices (e.g., NDVI). This, in turn, suggests that the remote exploration of tropical forest biochemistry and physiology is feasible today.

12.5   TROPICAL FOREST DIVERSITY FROM IMAGING SPECTROSCOPY

Canopy biodiversity is truly a "Holy Grail" in tropical forest remote sensing. Many studies have tried, and failed, to directly estimate canopy diversity from remotely sensed signatures. Successful application of multispectral and multisensor approaches to diversity estimation has primarily come via indirect mapping of environmental variables that might best predict diversity [70,71]. Environmental indicators include climate, soils, slope, aspect, and greenness. However, the direct measurement of canopy diversity remains very elusive because we do not have a systematic understanding of the way species are organized spectrally (or any other way remotely), although progress has been recently made on this issue at the leaf and crown levels [34,35,72]. Moreover, we do not know how the phylogeny of forest species is expressed biochemically, and neither the methods nor the instrumentation for remote sensing of canopy biochemistry has been widely available.

Advances in high-fidelity imaging spectroscopy, canopy radiative transfer modeling, and taxonomically organized biochemistry are just starting to address these issues for tropical forest mapping. For example, using the Jet Propulsion Laboratory's (JPL) airborne visible and infrared imaging spectrometer (AVIRIS), along with a new radiative transfer inverse modeling technique, we mapped canopy fractional cover, leaf N concentration, and canopy water content across a 1360-ha montane rainforest area in Hawaii Volcanoes National Park on the island of Hawaii (fig. 12.12) [73]. The upper-canopy leaf N and whole-canopy water content indicated the spatial distribution of unique plant functional types and invasive species. Canopies with high leaf N concentrations and water content were dominated by the invasive N-fixing tree *Myrica faya*; canopies with low leaf N and water content were

TABLE 12.2

Mean, Standard Deviation, and Coefficient of Variation (CV; %) of Four Hyperspectral Reflectance Indices[a] Collected by the EO-1 Hyperion Satellite Sensor over 12 Rainforest Regions in the Brazilian Amazon[b]

Image	NDVI			NDWI			PRI			CRI		
	Mean	SD	CV	Mean	SD	CV	Mean	SD	CV	Mean	SD	CV
1	0.88	0.01	1.3	0.021	0.017	82.8	0.110	0.028	26.0	0.042	0.089	213.9
2	0.90	0.01	1.5	0.033	0.028	85.5	0.087	0.032	37.1	0.025	0.069	279.8
3	0.90	0.02	1.7	0.010	0.044	435.2	0.132	0.034	25.9	0.051	0.138	271.4
4	0.86	0.01	1.7	0.004	0.020	462.2	0.105	0.025	23.3	0.030	0.056	185.7
5	0.89	0.01	1.3	0.016	0.018	113.9	0.097	0.024	25.0	0.025	0.043	167.7
6	0.91	0.01	1.3	0.008	0.021	262.7	0.150	0.032	20.9	0.071	0.336	475.3
7	0.79	0.02	2.0	0.029	0.014	49.8	0.076	0.018	23.3	0.012	0.023	189.1
8	0.91	0.01	1.2	0.020	0.016	82.2	0.139	0.022	15.5	0.043	0.077	180.9
9	0.86	0.01	1.7	0.030	0.025	82.0	0.107	0.025	23.2	0.027	0.056	208.0
10	0.89	0.02	2.1	0.029	0.028	93.4	0.121	0.030	25.1	0.043	0.151	349.9
11	0.91	0.02	1.7	0.019	0.027	142.9	0.173	0.040	23.2	0.166	0.492	295.7
12	0.93	0.01	1.0	0.021	0.020	92.8	0.243	0.040	16.6	0.445	0.904	202.9

[a] Table 12.1.
[b] Figure 12.11.

FIG. 12.12 **(See color insert following page 134.)** High-fidelity airborne imaging spectroscopy and three-dimensional canopy radiative transfer modeling were used to estimate leaf N concentration, canopy water content, and vegetation cover in a montane rainforest in Hawaii [73]. The areas of high leaf N and canopy water content correlated with the N-fixing invasive tree species *Myrica faya*, whereas areas of low N and high water content were dominated by the invasive forest understory herb *Hedychium gardnerianum*. These results highlight the potential for mapping plant functional types and invasive species in tropical rainforest.

dominated by the native tree *Metrosideros polymorpha*. Surprisingly, the areas displaying low leaf N and high canopy water were dominated by the understory herb *Hedychium gardnerianum*. This study highlighted the potential use of biochemical signatures for mapping unique plant functional types and/or invasive species.

How can hyperspectral imaging take us beyond plant functional types, and instead directly to taxonomic diversity among tree canopies? Radiative transfer and other hyperspectral approaches could yield accurate canopy diversity maps if the following requirements could be met: (1) Biochemical variability tracks species diversity, (2) biochemical "signatures" relate to species at some level of taxonomic aggregation, and (3) biochemicals are systematically expressed in hyperspectral reflectance measurements. Many studies show that the latter requirement is possible to meet; reflectance spectra contain biochemical information [23,57]. To address the first requirement related to biochemical diversity, we have been collecting foliage from canopy trees in humid tropical forests around the world and have been looking at the biochemical variability among these species on a spatial basis. In general, we find that an ensemble of biochemicals (chl-a, chl-b, carotenoids, anthocyanins, water,

N, P, lignin, cellulose), combined with SLA, tends to represent a species at a site, and that these biochemical signatures scale with canopy diversity to the landscape level.

How well does biochemical diversity express taxonomic diversity? To explore this issue, I developed a landscape canopy chemistry model that randomly populates a virtual forest with species and their measured biochemical signatures. First, a single species is randomly selected from the total community of n species collected in the field. The mean biochemical values of this species are taken from the database, along with the field-measured variance of each biochemical property to accommodate natural intraspecific variability. Other species are then randomly selected and added to the landscape in the same way. As the species richness of the virtual forest increases, I track the change in the total range (maximum–minimum) of biochemical values until the entire community is populated, or until a prescribed richness level is obtained. My model uses a Monte Carlo approach to calculate an average change in the biochemical variability of 1000 virtual forests as biodiversity is randomly increased. The model can be run for a single biochemical (e.g., chl-a) or for a biochemical signature index (Λ) that combines any number of leaf chemical properties per species (i). This index is calculated as the square root of the sum of squared normalized values of any number of biochemical parameters measured in the field. For example, a three-chemical signature incorporating leaf water, carotenoids, and N would be calculated:

$$\Lambda_i = \text{sqrt}[(H_2O_i/H_2O_{min})^2 + (Car_i/Car_{min})^2 + (N_i/N_{min})^2] \qquad (12.1)$$

To demonstrate this approach, I used the 150 species from the eastern Amazon (appendix 12.1) to calculate stand-level diversity of N and P. As canopy species richness increases, the variability (range) of N and P increases nonlinearly and gradually reaches an asymptote after which adding species has a diminishing effect on biochemical variability (fig. 12.13). Surprisingly, neither the N nor the P variability fully saturates. This then calls into question the spatial sampling of the trees. For this data set, I collected the upper-canopy leaves from randomly selected trees within a 50-ha area. Other smaller stands, and/or more tightly packed tree crowns, would change the shape of these curves. Nonetheless, the increasing biochemical variability does track increasing species richness, even with only one chemical element. Combining multiple leaf properties into a single signature Λ (equation 12.1) serves to diversify the biochemical portfolio of a species, which translates to an improved dynamic range as expressed in a forest canopy landscape (fig. 12.13). The increased information content of Λ, or another multidimensional biochemical metric, may be key to developing the most flexible and applicable biochemical signature for a given species or assemblage of species.

As a side note, this modeling technique also allows us to consider length scales of biochemical variation with respect to taxonomic diversity across a landscape. The initial slope of the biochemical range (e.g., for the first 10 "neighboring" tree species) indicates the biochemical diversity at a very local scale (fig. 12.13). Among trees in the eastern Amazon site, leaf N is more variable locally (steeper slope in fig. 12.13) than is P, and thus N is a better indicator of localized canopy diversity in

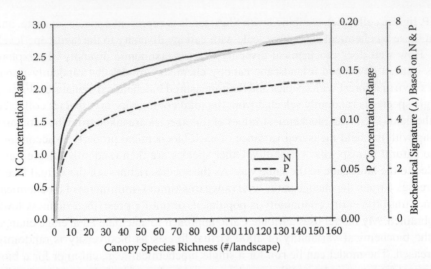

FIG. 12.13   Leaf nitrogen (N) and phosphorus (P) variability (range of values) simulated for a community of canopy tree species measured in the eastern Brazilian Amazon (appendix 12.1). The biochemical signature index Λ combines N and P using equation 12.1.

this particular system. Other sites, with differing environmental controls, as well as other biochemicals, likely tell a different story.

Can this conceptual framework for mapping diversity from biochemistry actually work? The answer depends on both the dynamic range of biochemical signatures in a forest, and the expression of the biochemistry in the reflectance spectrum. We have considered this latter issue in some detail in Hawaiian forest ecosystems. Using AVIRIS in a high resolution mode with 3- × 3-m pixels, we collected canopy spectra of sunlit portions of the crowns and crown clusters of trees throughout a range of forested landscapes on the island of Hawaii (appendix 12.2). We quantified the spectral separability of 31 species and four major groups of rainforest trees, finding that nearly all were statistically different in some combination of the reflectance spectrum (fig. 12.14) [74]. We also found that introduced and invasive tree species were spectrally and biochemically unique from native trees. Most importantly, PLS regression analyses of AVIRIS spectra with field-measured leaf and canopy biochemical properties showed that each chemical constituent, along with SLA and LAI, defined the canopy spectral signatures, thereby accounting for the separability of species in Hawaiian forests (fig. 12.15) [74].

Combining the theoretical and methodological approaches described before in a field test of biochemical-diversity remote sensing, Carlson et al. [75] show how ranges of leaf water, chlorophyll, and N content change with species richness across lowland rainforest ecosystems in Hawaii. First, we collected the top-of-canopy foliage from 13 tree species, and repeated the effort for just *M. polymorpha*, the native tree that is found throughout this study region. This allowed us to quantify the intraspecific variation in leaf biochemical properties of *M. polymorpha* in comparison to variation across species. We found that, in these relatively low-diversity forests, variation among biochemical signatures increases with species richness

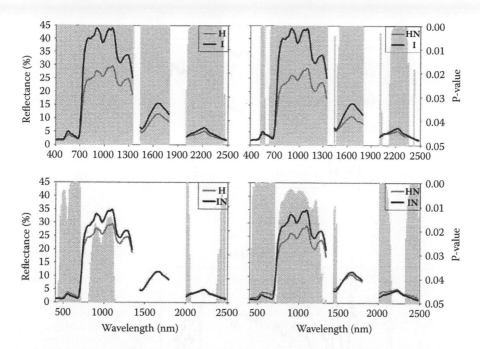

FIG. 12.14   Mean reflectance of Hawaiian (H), Hawaiian nitrogen-fixing (HN), introduced (I), and introduced nitrogen-fixing (IN) species, with band-by-band *t*-tests (gray bars; *p*-values < .05) showing the spectral separability of species. (Reprinted from Asner, G.P. et al., *Remote Sensing of Environment*, in press. With permission of Elsevier.)

(fig. 12.16a), and the variability is much higher between than within species. We then used AVIRIS spectral signatures to link the biochemical variation to species richness, finding that spatial variation (range) in the shape of the AVIRIS spectra in wavelength regions associated with upper-canopy pigments, water, and nitrogen content was well correlated with species richness across field sites (figs. 12.16b and 12.16c). This approach allowed for a detailed mapping of forest canopy richness throughout lowland Hawaiian rainforest ecosystems on the island of Hawaii [75].

This biochemical-diversity mapping concept may be appropriate for other forests when high-fidelity imaging spectroscopy is available, and where the spatial resolution of the imagery is appropriate to the scale of diversity being mapped. However, the particular species richness prediction algorithm that we are developing may not be directly applicable to other areas or different vegetation regimes. Lowland Hawaiian rainforests are relatively unique because of their young basalt-derived soils and relatively low species richness (≤17 canopy species per 0.1 ha), and our early algorithms may depend on these distinctive regional attributes to calculate diversity. In other forests, different spectral regions or wavelengths may be better correlated with species richness via their biochemical properties. Hyperspectral data are therefore a key component to this type of diversity mapping, because the fine spectral resolution allows greater flexibility in the wavelength choice, which is not available from multispectral data.

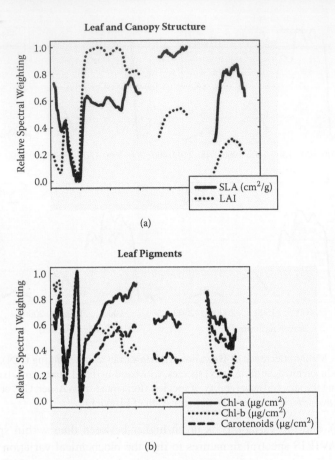

(a)

(b)

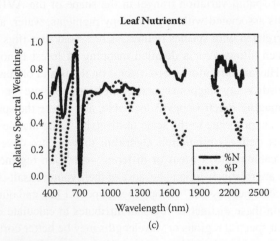

(c)

FIG. 12.15 Relative spectral weightings of (a) leaf SLA and canopy LAI; (b) leaf chl-a, chl-b, and carotenoids; and (c) leaf N and P, for AVIRIS reflectance spectra of Hawaiian forest species listed in appendix 12.2. (Reprinted from Asner, G.P. et al., *Remote Sensing of Environment*, in press. With permission of Elsevier.)

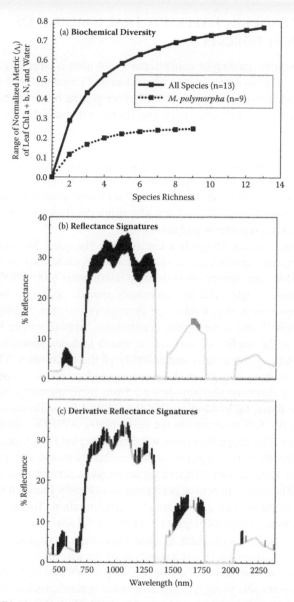

FIG. 12.16 (a) Biochemical variability of lowland tropical forests on the island of Hawaii using the biochemical signature index Λ (equation 12.1) with leaf chlorophyll a+b, nitrogen (N), and water concentrations. A comparison of 13 neighboring species is shown in comparison to the spatial variability of a dominant native Hawaiian species *Metrosideros polymorpha*. (b–c) Pearson correlation coefficients between tree species richness and AVIRIS spectral bands for reflectance and derivative reflectance signatures. Significance of the relationships varied across the spectrum; black marks indicate $p < .01$, gray marks indicate $.01 \leq p < .05$, and no mark indicates $.05 \leq p$. The height of the mark indicates the magnitude of the Pearson correlation value (r), with taller marks having higher values. The spectrum over which these coefficients are graphed is a mean spectrum derived from all trees in the study. (Adapted from Carlson, K.M. et al., *Ecosystems*. With permission from Springer.)

12.6   CHALLENGES IN HYPERSPECTRAL IMAGING OF TROPICAL FORESTS

Hyperspectral remote sensing of ecosystems is not new; a large number of studies have focused on a range of ecosystem properties and processes over the years [57]. On the other hand, airborne and space-based hyperspectral remote sensing of tropical forests is still very new, with only a few studies recently published [49,60,66,76,77]. More than anything, these efforts highlight the potential contributions of imaging spectroscopy to studies of tropical forest biochemistry, physiology, and diversity. I believe that hyperspectral remote sensing can serve as an interactive, exploratory tool for probing tropical forests, allowing for both new and improved knowledge about their structure and function. However, numerous challenges remain, including issues of sensor design and accuracy, algorithm development, and even the very basics of tropical forest ecology and taxonomy.

Imaging spectrometer design is a key issue in the quest for more quantitative approaches to remote sensing. Some spectrometers operate only in the visible and near IR (400–1100 nm); others are full-range instruments (400–2500 nm). Applied to tropical forests, visible–NIR spectrometers provide access to pigment, water, and some nitrogen chemistry. Full-range systems extend this capability to include leaf water, protein-N, and some carbon constituents. Arguably more important than spectral range is the fidelity of the sensor, which includes signal-to-noise performance, uniformity of the image, and stability of the electronics. These engineering issues have a direct impact on the precision and accuracy of any biochemical, physiological, or diversity product derived from hyperspectral imagery. A good example can be found by looking outside the tropical forest literature at the evolution of the JPL AVIRIS program. In the early 1990s, AVIRIS signal-to-noise ratio was in the 10s to 100s depending upon wavelength region [78]. Since then, AVIRIS has evolved, through major upgrades and constant hardware and software adjustment, to provide spectra with effective signal-to-noise performances of many 100s to 1000s [48]. The payoff in sensor performance is clearly found in the spectra and in the derived products such as expressed N concentration. Today, researchers are able to apply consistent methodologies to hyperspectral analysis of canopy N [19,79], whereas previous studies with older AVIRIS technology struggled to do so [14,80].

The sensor technology also bears squarely upon the state of hyperspectral algorithm development, an issue demonstrated within this chapter. With the high fidelity of AVIRIS and the CAO, we are able to apply physically based methods, such as canopy radiative transfer models, to explore the spectra in some detail. With low-fidelity systems such as EO-1 Hyperion, we are often relegated to narrowband vegetation indices, which, although revolutionary in their application to tropical forest exploration, fall short in providing the true quantitative spectroscopy available from the best airborne sensors.

The basic know-how for hyperspectral analysis of ecosystems is also evolving, but much progress is yet to be made to reach the full potential of high-fidelity imaging spectroscopy. A good example of this problem can be found by further considering tropical forest physiognomy with respect to hyperspectral reflectance signatures. As discussed, the particular molecular interactions with radiation determine the wavelength regions to exploit as well as the EPPD that dictates whether leaf concentration or canopy content can be estimated for any given biochemical. However, this framework is limited without an accounting of shade on the upper surface of the canopy. Shade occurs at two distinct geometric scales: intercrown shadows and leaf self-shading. Both depend on sun angle and viewing geometry. Intercrown shadowing is driven by the orientation and proximity of individual tree crowns or branches [81]. Leaf self-shading is driven by foliage orientation [82]. Although leaf self-shading can be largely accounted for by using canopy radiative transfer models or band normalization techniques, intercrown shadows are more difficult to predict.

How pronounced is the intercrown shadowing issue in humid tropical forests, and what can be done about it? Using 44 1-m IKONOS images collected across the Brazilian Amazon, we showed that intercrown shadows averaged 25% (±12%), with large local-scale variations in shadow that would hamper hyperspectral estimates of canopy biochemistry using sensors with Landsat-like resolutions (~30 m) [83]. Therefore, the most direct measurements of canopy chemistry should be those made at sufficiently high spatial resolution to account for (or mask out) shaded portions of tree crowns [18]. Shade results from canopy three-dimensional structure, so hyperspectral studies can also benefit from a simultaneous measurement of structure via technologies such as LiDAR, interferometric synthetic aperture radar (InSAR), or multiview angle (MVA) optical sensing [84–86]. The frontier in tropical forest remote sensing may lie in the fusion of hyperspectral and structural measurement approaches [87].

12.7 CONCLUSIONS

Hyperspectral remote sensing offers to revolutionize studies of tropical forest biochemistry, physiology, and biodiversity that could bear on conservation, management, and policy development efforts. However, the efficacy of imaging spectroscopy for rainforest research and monitoring rests in understanding the sources of variance in spectral signatures at different biophysical, taxonomic, community, and ecosystem levels of organization. In addition, the methods and sensors play a major role in determining what can be done today, and what might be possible in the future.

From our ongoing studies in Hawaii and the continental tropics, it is clear that leaf spectral variability is driven by a constellation of biochemical and structural properties of foliage. No single wavelength band or region is uniquely sensitive to a particular biochemical. Thus, spectroscopic studies of leaf biochemistry and physiology must account for relative and/or ensemble contributions of leaf constituents between taxa and across environmental gradients. At the crown or canopy level, hyperspectral reflectance properties are driven by leaf properties, but are mediated by canopy structural and architectural characteristics. This causes the biochemicals to be differentially expressed at leaf (e.g., visible wavelengths) to upper-canopy (shortwave IR) to fully integrated-canopy (e.g., near IR) scales in airborne and space-based hyperspectral images. In turn, this affects the types of canopy chemical and physiological properties that can be estimated and monitored in tropical forests.

Intra- vs. interspecific variation in canopy reflectance has been poorly addressed in tropical forest remote sensing, yet it is key to understanding the relative importance of environmental and taxonomic variation in hyperspectral (or any other) imagery.

I have shown that the native Hawaiian tree *Metrosideros polymorpha*, which occurs across an enormous range of environmental conditions, can be used to explore the contributions of microsite, substrate, and climate controls on canopy reflectance. Other studies should seek to analyze local and regional sources of spectral variation within a species, wherever that may be possible. Our *M. polymorpha* studies suggest that intracrown variability of sunlit pixels is low in comparison to microsite-, substrate-, or climate-mediated variability in intercrown reflectance. If this is broadly true for tropical forest ecosystems, then environmental impacts, especially climate change, can be monitored with hyperspectral data collected at the right spatial scale. Very little is currently known regarding the effects of climate change on rainforest dynamics, and thus the quantitative monitoring approaches offered by imaging spectroscopy are likely to become central to large-scale studies of vegetation–climate interactions in remote tropical forest regions.

The spectral-biochemical properties of tropical canopies also play out taxonomically. Species exert major control over leaf and canopy biochemistry, which, at some taxonomic level of organization, translates to spectral signatures and spatial variability in hyperspectral images. We do not know the degree of biochemical expression in the spectroscopy, nor do we understand how spectral signatures play out at the species, genus, or family levels among rainforest plants. Here, I presented a conceptual framework for analysis of biochemical and spectral variability in tropical forest landscapes, and I showed how this framework can be used to estimate canopy species richness in lowland Hawaiian rainforests. However, the general applicability and scalability of the biochemical-diversity approach remains unknown. We are now focusing on this issue across a wide range of tropical forest sites.

Whether one is interested in the biochemistry, physiology, or diversity of tropical forest canopies, the importance of sensor fidelity, measurement resolution, and analytical technique cannot be overemphasized. High-fidelity, high spatial resolution sensors from airborne platforms allow for nearly any analytical technique to be applied. Low-fidelity systems, whether from aircraft or EO-1 Hyperion (the only space-borne system widely available), limit the types of analyses that can be performed, which directly impacts the accuracy, breadth, and usefulness of the data products derived from the hyperspectral imagery. With the developing high-fidelity satellite missions such as Flora [88] and ENMAP [89], exploration of tropical forest biochemistry, physiology, and diversity may become possible anywhere in the world. Until then, effort should continue to focus on quantifying and understanding sources of spectral variation at differing scales of biological and geographic aggregation.

ACKNOWLEDGMENTS

I thank J. Buckley, K. Carlson, F. Hughes, M. Jones, M. Keller, D. Knapp, R. Martin, A. Townsend, and P. Vitousek for many spirited discussions on the topic of tropical forest ecology and remote sensing. I thank the Jet Propulsion Laboratory's AVIRIS Team, especially R. Green and M. Eastwood, and the NASA Goddard Space Flight Center's Earth Observing-1 office (S. Ungar, T. Brakke, L. Ong) for data acquisition support. This work was supported by NASA Terrestrial Ecology Program Biodiversity Research Grant NNG-06-GI-87G and The Carnegie Institution.

REFERENCES

1. Denslow, J.S., Tropical rainforest gaps and tree species diversity, *Annual Review of Ecology and Systematics*, 18, 431, 1987.
2. Nepstad, D.C., Versissimo, A., Alencar, A., et al., Large-scale impoverishment of Amazonian forests by logging and fire, *Nature*, 398, 505, 1999.
3. Achard, F., Eva, H.D., Stibig, H.J., et al., Determination of deforestation rates of the world's humid tropical forests, *Science*, 297, 999, 2002.
4. Townsend, A.R., Cleveland, C.C., Asner, G.P., et al., Controls over foliar N:P ratios in tropical rain forests, *Ecology*, 88, 107, 2007.
5. Vitousek, P.M. and Farrington, H., Nutrient limitation and soil development: Experimental test of a biogeochemical theory, *Biogeochemistry*, 37, 63, 1997.
6. Herbert, D.A. and Fownes, J.H., Forest productivity and efficiency of resource use across a chronosequence of tropical montane soils, *Ecosystems*, 2, 242, 1999.
7. Holben, B.N., Vermote, E., Kaufman, Y.J., et al., Aerosol retrieval over land from AVHRR data-application for atmospheric correction, *IEEE Transactions on Geoscience and Remote Sensing*, 30, 212, 1992.
8. Huete, A.R., Didan, K., Shimabukuro, Y.E., et al., Amazon rainforests green-up with sunlight in the dry season, *Geophysical Research Letters*, 13, doi:10.1029/2005GL025583, 2006.
9. Bohlman, S.A., Adams, J.B., Smith, M.O., et al., Seasonal foliage changes in the eastern Amazon basin detected from Landsat thematic mapper satellite images, *Biotropica*, 30, 376, 1998.
10. Nelson, B.W., Kapos, V., Adams, J.B., et al., Forest disturbance by large blowdowns in the Brazilian Amazon, *Ecology*, 75, 853, 1994.
11. Skole, D. and Tucker, C., Tropical deforestation and habitat fragmentation in the Amazon: Satellite data from 1978 to 1988, *Science*, 260, 1905, 1993.
12. Asner, G.P., Palace, M., Keller, M., et al., Estimating canopy structure in an Amazon forest from laser range finder in IKONOS satellite observations, *Biotropica*, 34, 483, 2002.
13. Clark, D.B., Read, J.M., Clark, M.L., et al., Application of 1-m and 4-m resolution satellite data to ecological studies of tropical rain forests, *Ecological Applications*, 14, 61, 2004.
14. Wessman, C.A., Aber, J.D., Peterson, D.L., et al., Remote sensing of canopy chemistry and nitrogen cycling in temperate forest ecosystems, *Nature*, 335, 154, 1988.
15. Gao, B.C. and Goetz, A.F.H., Retrieval of equivalent water thickness and information related to biochemical components of vegetation canopies from AVIRIS data, *Remote Sensing of Environment*, 52, 155, 1995.
16. Martin, M.E. and Aber, J.D., High spectral resolution remote sensing of forest canopy lignin, nitrogen and ecosystem processes, *Ecological Applications*, 7, 431, 1997.
17. Kokaly, R.F., Investigating a physical basis for spectroscopic estimates of leaf nitrogen concentration, *Remote Sensing of Environment*, 75, 153, 2001.
18. Zarco-Tejada, P.J., Miller, J.R., Mohammed, G.H., et al., Vegetation stress detection through chlorophyll a+b estimation and fluorescence effects on hyperspectral imagery, *Journal of Environmental Quality*, 31, 1433, 2002.
19. Smith, M.L., Martin, M.E., Plourde, L., et al., Analysis of hyperspectral data for estimation of temperate forest canopy nitrogen concentration: Comparison between an airborne (AVIRIS) and space-borne (Hyperion) sensor, *IEEE Transactions on Geoscience and Remote Sensing*, 41, 1332, 2003.
20. Roberts, D.A., Ustin, S.L., Ogunjemiyo, S., et al., Spectral and structural measures of northwest forest vegetation at leaf to landscape scales, *Ecosystems*, 7, 545, 2004.
21. Kupiec, J.A. and Curran, P.J., Decoupling effects of the canopy and foliar biochemicals in AVIRIS spectra, *International Journal of Remote Sensing*, 16, 1731, 1995.

22. Myneni, R.B., Ramakrishna, R.N., and Running, S.W., Estimation of global leaf area index and absorbed par using radiative transfer models, *IEEE Transactions on Geoscience and Remote Sensing*, 35, 1380, 1997.
23. Curran, P.J., Remote sensing of foliar chemistry, *Remote Sensing of Environment*, 30, 271, 1989.
24. Thomas, J.R., Namken, L.N., Oerther, G.F., et al., Estimating leaf water content by reflectance measurements, *Agronomy Journal*, 63, 845, 1971.
25. Hunt, J.E.R., Rock, B.N., and Nobel, P.S., Measurement of leaf relative water content by infrared reflectance, *Remote Sensing of Environment*, 22, 249, 1987.
26. Jacquemoud, S., Ustin, S.L., Verdebout, J., et al., Estimating leaf biochemistry using the prospect leaf optical properties model, *Remote Sensing of Environment*, 56, 194, 1996.
27. Jacquemoud, S., Bacour, C., Poilive, H., et al., Comparison of four radiative transfer models to simulate plant canopies reflectance: Direct and inverse mode, *Remote Sensing of Environment*, 74, 471, 2000.
28. Yoder, B.J. and Pettigrew-Crosby, R.E., Predicting nitrogen and chlorophyll content and concentrations from reflectance spectra (400–2500 nm) at leaf and canopy scales, *Remote Sensing of Environment*, 53, 199, 1995.
29. Reich, P.B., Walters, M.B., and Ellsworth, D.S., From topics to tundra: Global convergence in plant functioning, *Proceedings of the National Academy of Sciences*, 94, 13730, 1997.
30. Evans, J.R. and Poorter, H., Photosynthetic acclimation of plants to growth irradiance: The relative importance of specific leaf area and nitrogen partitioning in maximizing carbon gain, *Plant Cell and Environment*, 24, 755, 2001.
31. Niinemets, U. and Tenhunen, J.D., A model separating leaf structural and physiological effects on carbon gain along light gradients for the shade-tolerant species *Acer saccharum*, *Plant Cell and Environment*, 20, 845, 1997.
32. Stylinski, C.D., Oechel, W.C., Gamon, J.A., et al., Effects of lifelong (CO_2) enrichment on carboxylation and light utilization of *Quercus pubescens* Willd. Examined with gas exchange, biochemistry and optical techniques, *Plant Cell and Environment*, 23, 1353, 2000.
33. Wright, I.J., Reich, P.B., Westoby, M., et al., The worldwide leaf economics spectrum, *Nature*, 428, 821, 2004.
34. Gamon, J.A., Kitajima, K., Mulkey, S.S., et al., Diverse optical and photosynthetic properties in a neotropical dry forest during the dry season: Implications for remote estimation of photosynthesis, *Biotropica*, 37, 547, 2005.
35. Castro-Esau, K., Sanchez-Azofeifa, G.A., Rivard, B., et al., Variability in leaf optical properties of mesoamerican trees and the potential for species classification, *American Journal of Botany*, 93, 517, 2006.
36. Asner, G.P., Biophysical and biochemical sources of variability in canopy reflectance, *Remote Sensing of Environment*, 64, 234, 1998.
37. Baret, F., Clevers, J.G.P.W., and Steven, M.D., The robustness of canopy gap fraction estimates from red and near-infrared reflectances: A comparison of approaches, *Remote Sensing of Environment*, 54, 141, 1995.
38. Asner, G.P., Knapp, D.E., Cooper, A.N., et al., Ecosystem structure throughout the Brazilian Amazon from Landsat observations and automated spectral unmixing, *Earth Interactions*, 9, 1, 2005.
39. Chandrasekar, S., *Radiative transfer*. Dover Publications, New York, 1960.
40. Roberts, D.A., Green, R.O., and Adams, J.B., Temporal and spatial patterns in vegetation and atmospheric properties from AVIRIS, *Remote Sensing of Environment*, 62, 223, 1997.

41. Green, R.O., Dozier, J., Roberts, D., et al., Spectral snow-reflectance models for grain size and liquid water fraction in melting snow for the solar-reflected spectrum, *Annals of Glaciology*, 34, 71, 2002.
42. Ceccato, P.S., Flasse, S., Tarantola, S., et al., Detecting vegetation leaf water content using reflectance in the optical domain, *Remote Sensing of Environment*, 77, 22, 2001.
43. Asner, G.P., Scurlock, J.M., and Hicke, J.A., Global synthesis of leaf area index observations: Implications for ecological and remote sensing studies, *Global Ecological Biogeography*, 12, 191, 2003.
44. Asner, G.P., Jones, M.O., Knapp, D.E., et al., The Carnegie airborne observatory: In-flight fusion of hyperspectral and waveform-LiDAR data for 3-D studies of carbon, water and biodiversity in terrestrial ecosystems, *Remote Sensing of Environment*, online journal. Vol. 1 DOI:10.1117/1.279408.
45. Crews, T.E., Kitayama, K., Fownes, J.H., et al., Changes in soil phosphorus fractions and ecosystem dynamics across a long chronosequence in Hawaii, *Ecology*, 76, 1407, 1995.
46. Foody, G.M. and Hill, R.A., Classification of tropical forest classes from Landsat TM data, *International Journal of Remote Sensing*, 17, 2353, 1996.
47. Tian, H., Melillo, J.M., Kicklighter, D.W., et al., Climatic and biotic controls on annual carbon storage in Amazonian ecosystems, *Global Ecology and Biogeography*, 9, 315, 2000.
48. Green, R.O., Pavri, B.E., and Chrien, T.G., On-orbit radiometric and spectral calibration characteristics of EO-1 Hyperion derived with an underflight of AVIRIS and in situ measurements at Salar De Arizaro, Argentina, *IEEE Transactions on Geoscience and Remote Sensing*, 6, 1194, 2003.
49. Asner, G.P., Carlson, K.M., and Martin, R.E., Substrate and precipitation effects in Hawaiian forest canopies from space-borne imaging spectroscopy, *Remote Sensing of Environment*, 96, 497, 2005.
50. Verstraete, M.M., Retrieving canopy properties from remote sensing measurements, in *Imaging spectrometry: A tool for environmental observations*, J. Hill and J. Megier, eds. ECSC, EEC, EAEC, 1994, 109.
51. Gamon, J.A., Field, C.B., Bilger, W., et al., Remote sensing of the xanthophyll cycle and chlorophyll fluorescence in sunflower leaves and canopies, *Oecologia*, 85, 1, 1990.
52. Gitelson, A.A., Zur, Y., Chivkunova, O.B., et al., Assessing carotenoid content in plant leaves with reflectance spectroscopy, *Photochemistry and Photobiology*, 75, 272, 2002.
53. Sims, D.A. and Gamon, J.A., Relationships between leaf pigment content and spectral reflectance across a wide range of species, leaf structures and developmental stages, *Remote Sensing of Environment*, 81, 337, 2002.
54. Guo, J. and Trotter, C.M., Estimating photosynthetic light-use efficiency using the photochemical reflectance index; variation among species, *Functional Plant Biology*, 31, 255, 2004.
55. Martin, R.E., Asner, G.P., and Sack, L., Genetic variation in leaf pigment, optical and photosynthetic function among diverse phenotypes of *Metrosideros polymorpha* grown in a common garden, *Oecologia*, 151(3), 387, 2006.
56. Gao, B.C., NDWI: A normalized difference water index for remote sensing of vegetation liquid water from space, *Remote Sensing of Environment*, 58, 257, 1996.
57. Ustin, S.L., Roberts, D.A., Gamon, J.A., et al., Using imaging spectroscopy to study ecosystem processes and properties, *Bioscience*, 54, 523, 2004.
58. Gamon, J.A., Serrano, L., and Surfus, J.S., The photochemical reflectance index: An optical indicator of photosynthetic radiation use efficiency across species, functional types, and nutrient levels, *Oecologia*, 112, 4921, 1997.
59. Stylinski, C.D., Gmon, J.A., and Oechel, W.C., Seasonal patterns of reflectance indices, carotenoid pigments and photosynthesis of evergreen chaparral species, *Oecologia*, 131, 366, 2002.

60. Asner, G.P., Martin, R.E., Carlson, K.M., et al., Vegetation climate interaction among native and invasive species in Hawaiian rainforest, *Ecosystems*, 9, 1106, 2006.

61. Schuur, E.A.G. and Matson, P.A., Net primary productivity and nutrient cycling across a Mesic to wet precipitation gradient in Hawaiian montane forest, *Oecologia*, 128, 431, 2001.

62. Raich, J.W., Russell, A.E., and Vitousek, P.M., Primary productivity and ecosystem development along an elevational gradient on Mauna Loa, Hawaii, *Ecology*, 78, 707, 1997.

63. Martin, M. and Asner, G.P., Leaf biochemical and optical variation in *Metosideros polymorpha* across substrate and climate gradients in Hawaii, *Oecologia*, in press.

64. Clark, D.A., Tropical Forests and global warming: slowing it down or speeding it up, *Frontiers in Ecology and the Environment*, 2, 73, 2004.

65. Asner, G.P., Townsend, A.R., and Braswell, B.H., Satellite observations of El Nino effects on Amazon forest phenology and productivity, *Geophysical Research Letters*, 27, 981, 2000.

66. Asner, G.P., Nepstad, D., Cardinot, G., et al., Drought stress and carbon uptake in an Amazon forest measured with space-borne imaging spectroscopy, *Proceedings of the National Academy of Sciences*, 101, 6039, 2004.

67. Nepstad, D.C., Moutinho, P., Dias, M.B., et al., The effects of partial throughfall exclusion on canopy processes, aboveground production and biogeochemistry of an Amazon forest, *Journal of Geophysical Research*, 107, 1, 2002.

68. Gitelson, A.A., Merzlyak, M.N., and Chivkunova, O.B., Optical properties and non-destructive estimation of anthocyanin content in plant leaves, *Photochemistry and Photobiology*, 74, 38, 2001.

69. Instituto Socioambiental, Amazonia Brasileira 2000. Pages map of forest types, land-use change and protected areas in the Amazon, in *Instituto Socioambiental* 2000.

70. Nagendra, H., Using remote sensing to assess biodiversity, *International Journal of Remote Sensing*, 22, 2377, 2001.

71. Turner, W., Spector, S., Gardiner, N., et al., Remote sensing for biodiversity and conservation, *Trends in Ecology and Evolution*, 18, 306, 2003.

72. Cochrane, M.A., Using vegetation reflectance variability for species level classification of hyperspectral data, *International Journal of Remote Sensing*, 21, 2075, 2000.

73. Asner, G.P. and Vitousek, P.M., Remote analysis of biological invasion and biogeochemical change, *Proceedings of the National Academy of Sciences*, 102, 4383, 2005.

74. Asner, G.P., Jones, M.O., Martin, M.E., et al., Remote sensing of native and invasive species in Hawaiian forests, *Remote Sensing of Environment*, in press.

75. Carlson, K.M., Asner, G.P., Hughes, F.R., et al., Hyperspectral remote sensing of canopy biodiversity in Hawaiian lowland rainforests, *Ecosystems*, 10, 536, 2007.

76. Held, A., Ticehurst, C., Lymburner, L., et al., High resolution mapping of tropical mangrove ecosystems using hyperspectral and radar remote sensing, *International Journal of Remote Sensing*, 24, 2739, 2003.

77. Clark, D.A., Roberts, D.A., and Clark, D.A., Hyperspectral discrimination of tropical rain forest tree species at leaf to crown scales, *Remote Sensing of Environment*, 96, 375, 2005.

78. Vane, G., Green, R.E., Chrien, T.G., et al., The airborne visible infrared imaging spectrometer (AVIRIS), *Remote Sensing of Environment*, 44, 127, 1993.

79. Ollinger, S.V. and Smith, M.L., Net primary production and canopy nitrogen in a temperate forest landscape: An analysis using imaging spectroscopy, modeling and field data, *Ecosystems*, 8, 760, 2005.

80. Curran, P.J., Dungan, J.L., Macler, B.A., et al., Reflectance spectroscopy of fresh whole leaves for the estimation of chemical concentration, *Remote Sensing of Environment*, 39, 153, 1992.

81. Li, X. and Strahler, A.H., Geometric-optical bidirectional reflectance modeling of the discrete crown vegetation canopy: Effect of crown shape and mutual shadowing, *IEEE Transactions on Geoscience and Remote Sensing*, 30, 276, 1992.
82. Hapke, B.D., DiMucci, R., Nelson, R., et al., The cause of the hot spot in vegetation canopies and soils: Shadow-hiding versus coherent backscatter, *Remote Sensing of Environment*, 58, 63, 1996.
83. Asner, G.P. and Warner, A.S., Canopy shadow in IKONOS satellite observations of tropical forests and savannas, *Remote Sensing of Environment*, 87, 521, 2003.
84. Barnsley, M.J., Allison, D., and Lewis, P., On the information content of multiple view angle (MVA) images, *International Journal of Remote Sensing*, 18, 1937, 1997.
85. Treuhaft, R.N. and Siqueira, P.R., Vertical structure of vegetated land surfaces from interferometric and polarimetric radar, *Radio Science*, 35, 141, 2000.
86. Lefsky, M.A., Cohen, W.B., Parker, G.G., et al., LiDAR remote sensing for ecosystem studies, *Bioscience*, 52, 19, 2002.
87. Gillespie, T.W., Brock, J., and Wright, C.W., Prospects for quantifying structure, floristic composition and species richness of tropical forests, *International Journal of Remote Sensing*, 25, 707, 2004.
88. Asner, G.P., Knox, R.G., Green, R.O., and Ungar, S.G., *The Flora mission for ecosystem composition disturbance, and productivity,* National Academy of Sciences of the United States of America, Washington, D.C., 2004.
89. Stuffler, T., Kaufmann, C., Hofer, S., et al., The ENMAP hyperspectral imager—An advanced optical payload for future applications in Earth observation programs, in *Proceedings of the 57th IAC (International Astronautical Congress)*, Valencia, Spain, 2006.
90. Tucker, C.J., Red and photographic infrared linear combinations for monitoring vegetation, *Remote Sensing of Environment*, 8, 127, 1979.

APPENDIX 12.1: TROPICAL FOREST SPECIES COLLECTED IN THE EASTERN AMAZON FOR LEAF SPECTRAL-BIOCHEMICAL ANALYSES PRESENTED IN FIGURES 12.2 AND 12.4

Species	N (%)	Species	N (%)	Species	N (%)
Alexa grandiflora	3.05	*Goupia glabra*	2.07	*Pourouma paraensis*	1.94
Annona paludosa	2.40	*Guatteria poeppigiana*	1.74	*Pouteria guianensis*	1.37
Apeiba buschella	2.45	*Haptodendron* spp.	2.26	*Pouteria guianensis*	2.60
Apuleia moralis	1.87	*Himatanthus sucuuba*	1.51	*Pouteria* sp.	3.04
Aspidosperma desmanthum	2.32	*Hymenaea courbaril*	2.37	*Pouteria* spp.	1.55
Aspidosperma nitidum	2.03	*Hymenaea courbaril*	2.02	*Pouteria* spp.	1.66
Astronium lecointei	2.29	*Hymenaea parviflora*	1.57	*Pouteria* spp.	1.33
Bagassa guianensis	2.72	*Hymenolobium excelsum*	2.16	*Pouteria* spp.	2.52
Bellucia guianensis	2.06	*Inga alba*	2.10	*Pouteria velutina*	1.80
Brosimum guianense	2.50	*Inga edulis*	2.55	*Proteum decandrum*	1.37
Brosimum latifolia	1.26	*Inga heterophylla*	3.35	*Protium trifoliatum*	1.44
Brosimum rubescens	1.88	*Inga* spp.	2.22	*Pseudopiptadenia psilostachya*	2.34

Species	N (%)	Species	N (%)	Species	N (%)
Byrsonima spp.	1.95	*Inga stipularis*	2.34	*Pterocarpus amazonia*	2.70
Caryocar glabrum	1.69	*Isolernia paraense*	1.89	*Rinorea guianensis*	1.77
Caryocar glabrum	1.84	*Jacaranda copaia*	3.26	*Rollinia* spp.	1.86
Caryocar villosum	1.92	*Jacaranda copaia*	3.37	*Rwollfia* spp.	2.53
Casearia javitensis	1.86	*Joannaesia hevioides*	2.79	*Sacoglottis* spp.	1.37
Cecropia palmata	2.89	*Lacunaria guianensis*	1.83	*Sagotia racemosa*	1.56
Cecropia sciadophylla	2.56	*Laetia procera*	2.93	*Sapium marmieri*	2.91
Ceiba pentandra	1.52	*Laetia procera*	2.40	*Schefflera morototoni*	2.21
Chamaecrista spp.	2.33	*Lecithys pisonis*	2.40	*Schefflera morototoni*	2.61
Chamaecrista spp.	3.37	*Lecythis idatimon*	1.51	*Sclerolobium* spp.	1.83
Chamaecrista spp.	2.57	*Lecythis lurida*	2.51	*Sclerolobium* spp.	2.87
Chimarrhis guianensis	2.22	*Lecythis lurida*	2.44	*Simaba cedron*	2.17
Chrysophyllum guianensis	2.56	*Lecythis pisonis*	2.89	*Simarouba amara*	1.75
Clarisia racemosa	1.71	*Licania canescens*	2.94	*Simarouba amara*	2.13
Conceveiba guianensis	2.24	*Licania* spp.	1.18	*Sloanea nitida*	1.31
Copaifera multijuga	2.62	*Luehea speciosa*	1.92	*Solanum* spp.	2.60
Cordia bicolor	2.62	*Macrolobium campestre*	2.07	*Sterculia* spp.	1.89
Cordia goeldiana	2.50	*Manilkara amazonica*	1.24	*Sterculia* spp.	1.60
Couepia spp.	1.54	*Manilkara huberi*	1.46	*Sterculia xixa*	1.79
Courataria guianensis	1.57	*Manilkara huberi*	1.38	*Stryphnodendron paniculatum*	4.18
Croton maturensis	3.10	*Mapuria guianensis*	2.68	*Swietenia macrophylla*	2.09
Cynometa guianensis	1.65	*Marmaroxylon racemosum*	2.90	*Tabebuia impetiginosa*	2.63
Dialium guianensis	2.60	*Mezilaurus lindaviana*	3.30	*Tabernaemontana* spp.	2.45
Dinizia excelsa	1.95	*Miconia guianensis*	2.45	*Tachigalia guianensis*	1.80
Diplotropis purpurea	2.41	*Mistiluma guianensis*	1.54	*Terminalia amazonicum*	1.70
Dipteryx odorata	1.58	*Mouriri brevipes*	2.91	*Terminalia dichotoma*	2.20
Endopleura uchi	1.57	*Nectandra* sp.	2.08	*Tetragastris altissima*	1.65
Enterolobium schomburgkii	3.08	*Ocotea baturitensis*	2.37	*Tetragastris trifoliata*	1.48
Eriodendron samauma	2.07	*Ocotea* spp.	2.92	*Theobroma speciosa*	1.77
Eschweileira curiosa	1.95	*Ocotea* spp.	2.76	*Trattinickia serratifolia*	1.66
Eschweilera apiculata	1.68	*Omedia perebia*	2.25	*Trattinickia rhofolia*	1.45
Eschweilera spp.	2.47	*Ormosia* spp.	2.24	*Trattinnickia rhoifolia*	1.32
Eugenia sp.	1.17	*Ormosia* spp.	2.19	*Trema micrantha*	2.88

Species	N (%)	Species	N (%)	Species	N (%)
Fagaria spp.	1.50	*Parinaria guianensis*	1.51	*Vatairea paraensis*	1.77
Fusaea spp.	1.52	*Parkia multijuga*	2.67	*Vismia guianensis*	1.87
Glycydendron amazonicum	1.35	*Platymiscium guianensis*	3.07	*Xilopia aromatica*	2.56
Goupia glabra	2.19	*Poesilanta perfusa*	2.57	*Xilopia paraensis*	2.09

APPENDIX 12.2: SPECIES USED IN CANOPY SPECTRAL SEPARABILITY ANALYSES OF FIGURE 12.14 FOR HAWAIIAN FOREST SPECIES

Species	Chl-a (μg/cm^2)	Chl-b (μg/cm^2)	Carotenoids (μg/cm^2)	N (%)	P (%)	SLA (cm^2/g)	H$_2$O (%)
Acacia koa	50.5	17.2	15.8	2.68	0.11	43.2	54.9
Casuarina equisetifolia	27.3	9.7	9.3	1.37	0.07	73.8	66.4
Cecropia obtusifolia	27.4	9.2	8.1	2.35	0.14	118.2	70.4
Cryptomeria japonica	29.1	9.9	9.1	1.02	0.28	46.1	63.3
Diospyros sandwicensis	34.4	13.1	11.3	1.09	0.08	60.4	52.6
Eucalyptus deglupta	14.4	3.8	6.1	0.96	0.09	65.8	51.9
Eucalyptus globulus	41.1	14.7	12.8	1.46	0.12	51.9	55.8
Falcataria moluccana	37.7	10.4	9.5	2.94	0.09	108.6	56.6
Ficus benjamina	17.0	5.1	7.6	1.10	0.09	68.4	57.9
Ficus elastica	38.5	14.1	12.5	1.30	0.15	51.2	71.9
Ficus microcarpa	41.5	14.8	14.7	1.35	0.10	65.6	50.4
Ficus microcarpa	43.9	17.1	14.0	1.40	0.10	69.0	53.8
Fraxinus uhdei	22.9	7.6	10.5	1.26	0.15	71.5	55.0
Grevillea robusta	30.7	11.5	12.5	1.10	0.06	59.1	47.4
Juniperus bermudiana	28.4	11.7	8.5	0.82	0.11	72.6	61.2
Macaranga mappa	33.6	13.2	10.1	1.39	0.08	110.3	64.2
Magnolia grandiflora	36.3	13.0	13.2	1.12	0.12	50.9	53.5
Mangifera indica	18.8	5.8	6.7	0.83	0.11	68.6	56.4
Melastoma candidum	29.8	12.2	7.7	1.55	0.08	159.7	68.5
Metrosideros polymorpha	40.4	15.3	12.4	0.69	0.06	38.9	50.5
Morella faya	53.4	20.3	18.2	1.71	0.05	57.2	51.6
Myrsine lessertiana	43.1	15.7	13.5	1.34	0.12	78.4	66.9
Nestegis sandwicensis	33.8	13.4	11.9	1.08	0.08	64.2	51.4

Species	Chl-a (µg/cm²)	Chl-b (µg/cm²)	Carotenoids (µg/cm²)	N (%)	P (%)	SLA (cm²/g)	H₂O (%)
Pisonia brunoniana	31.4	12.7	11.6	0.94	0.08	70.9	56.1
Podocarpus neriifolius	18.2	6.7	7.3	1.19	0.27	76.7	64.3
Psidium cattleianum	43.6	18.7	12.9	1.02	0.05	64.3	59.0
Sapindus saponaria	50.8	18.9	16.8	3.34	0.20	127.7	63.3
Schefflera actinophylla	37.5	15.6	12.0	0.94	0.10	48.7	60.3
Schinus molle	44.0	14.8	14.9	1.85	0.20	76.3	61.8
Tibouchina granulosa	22.8	7.5	8.9	1.20	0.14	81.7	68.3
Trema orientalis	39.4	12.3	12.3	2.66	0.16	108.2	66.7

13 Tropical Remote Sensing—Opportunities and Challenges

John A. Gamon

CONTENTS

13.1 INTRODUCTION

Historically, the potential of remote sensing for ecological studies remained limited for a variety of reasons, including the insufficient spatial, spectral, or temporal resolution of most remote sensing data. As the chapters of this book illustrate, the fine spectral detail of hyperspectral remote sensing, or imaging spectrometry, presents a wealth of possibilities for expanding our understanding of tropical ecosystems. However, the potential of imaging spectrometry is limited by the shortage of hyperspectral data, and the sheer volume and complexity of hyperspectral data when they are available. This chapter will focus on two applications of this untapped potential: (1) assessment of biodiversity, and (2) evaluation of biosphere–atmosphere interactions—both areas where imaging spectrometry can provide new opportunities. To make progress in the face of such challenges, we will need to develop an informatics framework that encompasses the full complexity of hyperspectral data and the complexity of the task at hand. Thus, this chapter also outlines some basic considerations of the cyberinfrastructure needed to accommodate this task.

13.2 BIODIVERSITY

Biodiversity occurs at many scales and includes genetic diversity, species diversity (species richness), functional diversity, and ecosystem diversity [1,2]. The tropics, broadly defined as the zones falling between 23.5 N and 23.5 S degrees latitude, contain a large variety of terrestrial ecosystems, far greater than is commonly realized.

Tropical ecosystem types include desert, alpine tundra, evergreen forests, seasonally deciduous ("dry") forests, grassland, and savanna, as well as a variety of shoreline or marine ecosystems, including coral reefs and mangrove forests. To these broad ecosystem types, we must now add a complex mix of disturbed urban and agricultural ecosystems, as the tropics contain some of the most densely populated regions of the world [3]. Tropical ecosystems are among the world's hotspots of species richness and endemism [4]. Many of the forested ecosystems are disappearing or being degraded at rapid rates, and a chief function of remote sensing has been to document this loss [5–7].

In the past, remote assessment of biodiversity has been largely indirect [8,9] because satellite sensors have generally lacked the spectral and spatial resolution needed to capture patterns of species richness directly. Additionally, optical remote sensing of biodiversity is generally limited to what can be detected from above and is heavily weighted by the signal returning from top surface layer—for example, the upper canopy layers of a forest. With the advent of new imaging spectrometers, we now have several more direct pathways for linking remote sensing to some measure of biodiversity. Some of these alternate measures of diversity may be more accessible from remote sensing than species richness, per se. For example, hyperspectral sensors typically have sufficiently narrow bandwidths to resolve individual absorption features in reflectance spectra [10,11] (see also chapter 12). Particularly when combined with spatial detail (small pixel sizes) and temporal resolution (multitemporal imagery), hyperspectral sensors provide a rich array of tools for direct assessment of vegetation functional and structural diversity from remote sensing.

This detection of biodiversity with remote sensing rests on the central hypothesis that optical diversity (variation in optical patterns detected by remote sensing) is related to biodiversity at some level. Thus, instead of detecting species, per se, the hypothesis states that remote sensing detects functional and structural properties that vary with species or functional groups. There is an emerging body of evidence in support of this hypothesis (e.g., Gamon et al. [12], Clark, Roberts, and Clark [13], and chapter 12), although understanding the exact reasons for this remains a critical research challenge. A related prediction is that various levels of biodiversity (e.g., genetic diversity, species diversity, and functional-type diversity) are related to each other in some coherent way, a phenomenon that has been called *surrogacy* in the biodiversity literature [14]. If the hypothesis holds true, there should be consistent links between optical diversity detectable from surface canopy layers and some independent measure of biodiversity.

Fundamental to this hypothesis is the observation that hyperspectral reflectance detects features associated with biochemical, physiological, and structural variability. There is now a large body of literature discussing the ability of hyperspectral tools to assess the biochemical composition of vegetation [10]. Furthermore, critical measures of physiological function, including pigment composition [15,16], water content [17–19], and relative photosynthetic rates or light use efficiency [11,20], have been explored with some success by hyperspectral reflectance for some time. Particularly when applied at a fine spatial resolution (small pixel size), hyperspectral reflectance also provides a powerful tool for assessing the structural complexity of vegetation (chapter 12).

To test the biodiversity hypothesis, we are primarily limited by the shortage of hyperspectral data, lack of biodiversity information, and the complex, multidimensional structure of these data, which leaves remote sensing of biodiversity as one of the primary theoretical and technical challenges in tropical ecology. Given the critical concern that we are losing biodiversity in the tropics and the need to identify areas of high diversity and endemism for conservation priorities, remote sensing of biodiversity remains a critical, unmet need.

13.3 BIOSPHERE–ATMOSPHERE EXCHANGE AND RELEVANCE TO CLIMATE CHANGE

Tropical forests are often called the "lungs" of the world, for their gas exchange has a disproportionate impact on global climate and atmospheric composition [21,22]. The tropics strongly affect climate and atmospheric composition for several reasons. Combined with their high productivity, the large area of tropical forests ensures that tropical ecosystems are responsible for a disproportionate share of the global carbon fixation and respiration [21].

Remote sensing has a critical role to play in assessing the interaction of tropical ecosystems, atmosphere, and climate. Remote sensing is often used to define the "surface conditions" in global circulation models (GCMs), and thus plays a critical role in assessing the interaction of tropical ecosystems, atmosphere, and climate. Unfortunately, most global satellite products cannot fully represent the diversity of cover types and ecosystem states found in the tropics. For example, current global models of carbon flux generally treat the tropical forests as a single biome type [23], even though functionally and phenologically distinct types are present.

Simulations with GCM models incorporating more realistic surface properties for tropical regions have demonstrated that subtle changes in the rates of gas exchange, degree of land cover, or extent of fragmentation can have measurable affects on climate and atmospheric composition. Land use change associated with deforestation and land conversion alters albedo microclimate, thus altering the balance between sensible and latent heat exchange and affecting cloud formation and weather patterns. Modeling studies have found that deforestation can have large ramifications for regional climate. More recently, Werth and Avissar [24] concluded that tropical land conversion can affect climate in distant regions of the world. Furthermore, the impact of *physiological status* of vegetation on atmospheric processes has only recently been explored, indicating a measurable effect of physiology on climate [25,26]. Thus, improved characterization of tropical surface conditions possible with imaging spectrometry, including functional type and physiological activity, should lead to more realistic climate predictions.

It might seem that the invisible processes of biosphere–atmosphere gas exchange would be inaccessible with remote sensing. However, remote sensing provides many ways to approach this problem indirectly. For example, remote sensing of carbon dioxide uptake is often based on the "light-use efficiency" (LUE) model (also called a radiation-use efficiency model), which states that the photosynthetic carbon uptake is a function of the photosynthetically active radiation absorbed (APAR) and the *efficiency* (ε) with which that absorbed radiation is converted to biomass (fixed carbon). When summed or integrated over a given time period (typically a year), this can be used to assess net primary production (NPP), the net carbon gained by photosynthetic vegetation:

$$NPP = \varepsilon \times \Sigma APAR \qquad (13.1)$$

This model, driven at least in part from remote sensing inputs, is now the basis for global assessment of NPP [23,27]. Each term in the light-use efficiency model can be derived from remote sensing [11], and could presumably be improved by the greater spectral detail offered by hyperspectral instruments if these instruments were more fully available.

While surface–atmosphere flux models driven by remote sensing are based on well-understood principles, *validation* of these models remains largely incomplete, primarily due to the shortage of data and the limitations of the data sets used for validation. The models themselves or their validation often requires field measurements (ground truthing), including flux monitoring, meteorological measurements, and field optical sampling—all of which are limited, particularly in many parts of the tropics. For example, the coverage of meteorological stations is insufficient to fully characterize the surface conditions at the resolution sampled by satellite data, causing a mismatch between the spatial scales of different model inputs [23]. Eddy covariance, which is often considered the "gold standard" of ecosystem carbon and water vapor flux measurements, has its own limitations and sources of error [28]. Until recently, there have only been a small number of eddy covariance measurements in the tropics, making it difficult to validate remote sensing estimates of ecosystem gas exchange, although this is now beginning to change (e.g., Malhi and Grace [21]). Other methods of validation, including inversion modeling applied to atmospheric gas sampling, are beginning to provide new insights into the global patterns of carbon exchange, and strengthen the notion that tropical ecosystems play a disproportionate role in the global carbon cycle [29].

The limited validation of remote sensing flux models using eddy covariance and other field measurements indicates that a number of errors remain in the absolute values or seasonal patterns of flux estimates, and these errors are now partly understood and arise from a variety of sources [30–35]. What is also clear is that hyperspectral sensors offer several solutions to the problem of estimating biosphere–atmosphere gas exchange. For example, the ability of hyperspectral sensors to detect the level and activity of photosynthetic and photoprotective plant pigments offers a variety of methods to assess both the absorption and efficiency terms of the light-use efficiency model [11,33]. The ability to remotely detect the depth of water absorption features, and thus the moisture status of leaves and canopies and subsequent physiological responses [17–19,36,37], offers further information on physiological state that may be able to improve our estimates of both carbon and water vapor fluxes.

The rich array of physiological and structural tools available with imaging spectrometry allows us to greatly improve the mapping of vegetation structure and function. For example, different types of vegetation process carbon, water, and nutrients at different rates, and contain different levels of water, pigments, and other compounds; these differences are detectable with imaging spectrometry, provided the spectral and spatial resolution is fine enough. Additionally, the fine spectral and spatial resolution available with some imaging spectrometers, particularly when combined with light detection and ranging (LiDAR) or other remote sensing approaches, can better resolve the three-dimensional structure of vegetation (chapters 3 and 12). Since vegetation structure is both a product of, and an influence on, gas exchange processes, improved estimation of structure is another way that imaging spectrometry can improve our estimates of biosphere–atmosphere gas exchange.

13.4 CYBERINFRASTRUCTURAL NEEDS

The examples of biodiversity monitoring or surface–atmosphere gas exchange estimation illustrate the need for a cyberinfrastructure that can assist with these complex but critical challenges. At its core, this encompasses the informatics challenges of linking remote sensing to ecological variables we wish to assess. In its broadest sense, cyberinfrastructure includes the topics of field sampling, sensor networks, data communication and storage, and visualization and analytical tools [38], all of which require a considerable investment if we are to realize the full potential of hyperspectral remote sensing in the tropics for assessing biodiversity or surface–atmosphere gas exchange.

Informatics for ecology and environmental science—or "eco-informatics" as it is sometimes called—is among the most complex informatics challenges we face. Its complexity derives both from the nature and scale of the scientific questions addressed, and from a variety of technical issues. For example, the topic of biodiversity is itself a challenging one, both in terms of its multiple definitions and its theoretical implications [1,2]. At a technical level, attempting to sample diversity at the level of the entire tropics is a formidable challenge that continues to elude taxonomists and field ecologists. By definition, any particular method of sampling is necessarily limited to a particular level of diversity and may miss other critical forms of diversity. However, remote sensing has many characteristics that can provide a fresh new approach to this problem and also facilitate the development of a suitable cyberinfrastructure.

From an informatics standpoint, the benefits of remote sensing lie in its uniform data format that is readily searchable and retrievable. Thus, within a particular instrument, we have a built-in data standard. However, among instruments, there is no single standard, and a central informatics challenge remains, developing data management tools for handling multiple standards from multiple sensors and for relating them to each other in a coherent way. For example, spectral response, spatial resolution, and temporal sampling frequency vary between sensors, with sensor age and sampling protocols, and this variation must be characterized and understood if we are to integrate data from different sensors. This is essentially a problem of scaling in time, space, and spectral space, and developing the theory and tools for scaling will be a critical part of any informatics solution involving remote sensing [39].

Perhaps a bigger informatics challenge lies in relating remotely sensed signals to the ecological states and processes we wish to understand. This task is further complicated by the inherent complexity of ecological data, which are themselves rarely adequately standardized to the level required by typical databases [40]. In fact, because most ecological data are inherently diverse in format and purpose, many attempts to derive ecological databases focus more on the metadata than the actual data themselves [41].

Fortunately, hyperspectral remote sensing offers a partial solution to the nonstandard nature of ecological data. From an informatics perspective, the key to successful utilization of hyperspectral data for ecological goals lies in recognizing that these data—reflectance spectra comprising numerical arrays—follow highly structured and repeatable patterns, much like the pattern in a DNA sequence that comprises a genetic code. Similarly to the way that DNA sequences represent particular functional traits (or "genes"), the patterns in hyperspectral data represent specific variables (levels of pigments, nitrogen, water, etc.) that are indicators of ecological states and processes. Thus, reflectance spectra provide a powerful, readily searchable, *proxy* data set that can be related to a particular goal or ecological variable through proper calibration or validated models. This strength of hyperspectral remote sensing as a structured proxy data set has barely been tapped.

To make full use of this potential, a number of technical issues must be tackled that have not yet been fully addressed by the ecological and remote sensing communities but that are the topics of research in informatics theory. A few of these challenges are outlined here. We must develop intelligent databases [42] that readily accommodate the multiple data structures of different sensors. One feature of these databases should be easily usable tools for comparing different spectral, spatial, and temporal responses of different sensors. In principle, many of these tools exist in the form of routines used in image processing or remote sensing software, but these routines typically require considerable information about the sensor and technical skill on the part of the user. Additionally, these databases must be inherently searchable and able to link to ecological databases, many of which do not yet fully exist. Databases of flux measurements for different ecosystems, including some tropical sites, are now available [43], but linking these databases to remote sensing and ecological data sets requires considerable knowledge and technical skill and is hardly an automated process. Databases of biodiversity also exist [40], but more work is needed to structure these in a way that can be easily compared to remote sensing databases. A more appropriate way to proceed might be to develop standardized metrics of optical diversity from remote sensing and use these as tools to develop and prioritize the biodiversity databases.

13.5 CONCLUSION

Hyperspectral remote sensing is well suited for addressing many of the most crucial questions in tropical ecology; biodiversity assessment and surface–atmosphere fluxes are two examples. For all of this to occur, we need better access to hyperspectral sensors, a sustained campaign of field monitoring and validation, and the cyberinfrastructure suitable for handling these large, complex data sets. To achieve these goals, we need to have better communication among the different scientific communities, including field ecologists, remote sensing scientists, atmospheric scientists, and computer scientists, than we have generally had in the past. The separate pieces exist; they must be simply and intelligently assembled into a coherent whole.

REFERENCES

1. DeLong, D.C., Defining biodiversity, *Wildlife Society Bulletin*, 24, 738, 1996.
2. Gaston, K.J., *Biodiversity: An introduction*, 2nd ed. Blackwell Publishing, Malden, MA, 2004.
3. Lepers, E., Lambin, E.F., Janetos, A.C., et al., A synthesis of information on rapid land-cover change for the period 1981–2000, *BioScience*, 55, 115, 2005.
4. Myers, N., Mittermeier, R.A., Mittermeier, C.G., et al., Biodiversity hotspots for conservation priorities, *Nature*, 403, 853, 2000.
5. Skole, D. and Tucker, C., Tropical deforestation and habitat fragmentation in the Amazon: Satellite data from 1978 to 1988, *Science*, 260, 1905, 1993.
6. Asner, G.P., Knapp, D.E., Broadbent, E.N., et al., Selective logging in the Brazilian Amazon, *Science*, 310, 480, 2005.
7. Morton, D.C., DeFries, R.S., Shimabukuro, Y.E., et al., Rapid assessment of annual deforestation in the Brazilian Amazon using MODIS data, *Earth Interactions*, 9, 1, 2005.
8. Nagendra, H., Using remote sensing to assess biodiversity, *International Journal of Remote Sensing*, 22, 2377, 2001.
9. Turner, D.P., Spector, S., Gardiner, M., et al., Remote sensing for biodiversity and conservation, *Trends in Ecology and Evolution*, 18, 306, 2003.
10. Ustin, S.L., Roberts, D.A., Gamon, J.A., et al., Using imaging spectroscopy to study ecosystem processes and properties, *BioScience*, 54, 523, 2004.
11. Gamon, J.A., Qui, H.-L., and Sanchez-Azofeifa, G.A., Ecological applications of remote sensing at multiple scales, in *Handbook of functional plant ecology*, 2nd ed., F.I. Pugnaire and F. Valladares, eds. CRC Press, Boca Raton, FL, 2007, 655.
12. Gamon, J.A., Kitajima, K., Mulkey, S.S., et al., Diverse optical and photosynthetic properties in a neotropical forest during the dry season: Implications for remote estimation of photosynthesis, *Biotropica*, 37, 547, 2005.
13. Clark, M.L., Roberts, D.A., and Clark, D.B., Hyperspectral discrimination of tropical rain forest species at leaf to crown scales, *Remote Sensing of Environment*, 96, 375, 2005.
14. Magurran, A.E., *Measuring biological diversity*. Blackwell Publishing, Malden, MA, 2004.
15. Sims, D.A. and Gamon, J.A., Relationships between leaf pigment content and spectral reflectance across a wide range of species, leaf structures and developmental stages, *Remote Sensing of Environment*, 81, 337, 2002.
16. Gitelson, A.A., Vina, A., Ciganda, V., et al., Remote estimation of canopy chlorophyll content in crops, *Geophysical Research Letters*, 32, L08403, doi: 10.1029/2005 GL022688, 2005.
17. Peñuelas, J., Filella, I., Biel, C., et al., The reflectance at the 950–970-nm region as an indicator of plant water status, *International Journal of Remote Sensing*, 14, 1887, 1993.
18. Serrano, L., Ustin, S.L., Roberts, D.A., et al., Deriving water content of chaparral vegetation from AVIRIS data, *Remote Sensing of Environment*, 74, 570, 2000.
19. Sims, D.A. and Gamon, J.A., Estimation of vegetation water content and photosynthetic tissue area from spectral reflectance: A comparison of indices based on liquid water and chlorophyll absorption, *Remote Sensing of Environment*, 84, 526, 2003.
20. Gamon, J.A., Peñuelas, J., and Field, C.B., A narrow-waveband spectral index that tracks diurnal changes in photosynthetic efficiency, *Remote Sensing of Environment*, 41, 35, 1992.
21. Malhi, Y. and Grace, J., Tropical forests and atmospheric carbon dioxide, *Trends in Ecology and Evolution*, 15, 332, 2000.
22. Foley, J.A., Asner, G.P., Costa, M.H., et al., Amazonia revealed: Forest degradation and the loss of ecosystem goods and services in the Amazon Basin, *Frontiers in Ecology and the Environment*, 5, 25, 2007.

23. Running, S.W., Nemani, R.R., Heinsch, F.A., et al., A continuous satellite-derived measure of global primary production, *BioScience*, 54, 547, 2004.
24. Werth, D. and Avissar, R., The local and global effects of Amazon deforestation, *Journal of Geophysical Research-Atmospheres*, 107, No. D20, 8087, doi:10.1029/2001JD000717, 2002.
25. Sellers, P.J., Bounoua, L., Collatz, G.J., et al., Comparison of radiative and physiological effects of double atmospheric CO_2 on climate, *Science*, 271, 1402, 1996.
26. Bounoua, L., Collatz, G.J., Sellers, P.J., et al., Interactions between vegetation and climate: Radiative and physiological effects of doubled atmospheric CO_2, *Journal of Climate*, 12, 309, 1999.
27. Field, C.B., Behrenfeld, M.J., Randerson, J.T., et al., Primary production of the biosphere: Integrating terrestrial and oceanic components, *Science*, 281, 237, 1998.
28. Moncrieff, J.B., Malhi, Y., and Leuning, R., The propagation of errors in long-term measurements of land–atmosphere fluxes of carbon and water, *Global Change Biology*, 2, 231, 1996.
29. Stephens, B.B., Gurney, K.R., Tans, P.P., et al., Weak northern and strong tropical land carbon uptake from vertical profiles of atmospheric CO_2, *Science*, 316, 1732, 2007.
30. Turner, D.P., Ritts, W.D., Cohen, W.B., et al., Scaling gross primary production (GPP) over boreal and deciduous forest landscapes in support of MODIS GPP product validation, *Remote Sensing of Environment*, 88, 256, 2003.
31. Turner, D.P., Ritts, W.D., Cohen, W.B., et al., Site-level evaluation of satellite-based global terrestrial gross primary production and net primary production monitoring, *Global Change Biology*, 11, 666, 2005.
32. Fensholt, R., Sandholt, I., and Rasmussen, M.S., Evaluation of MODIS LAI, fAPAR, and the relation between fAPAR and NDVI in a semi-arid environment using in-situ measurement, *Remote Sensing of Environment*, 91, 490, 2004.
33. Rahman, A.F., Cordova, V.D., Gamon, J.A., et al., Potential of MODIS ocean bands for estimating CO_2 flux from terrestrial vegetation: A novel approach, *Geophysical Research Letters*, 31, L10503, doi:10.1029/2004GL019778, 2004.
34. Huemmrich, K.F., Privette, J.L., Mukelabai, M., et al., Time-series validation of MODIS land biophysical products in a Kalahari woodland, Africa, *International Journal of Remote Sensing*, 26, 4381, 2005.
35. Cheng, Y., Gamon, J.A., Fuentes, D.A., et al., A multiscale analysis of dynamic optical signals in a Southern California chaparral ecosystem: A comparison of field, AVIRIS, and MODIS data, *Remote Sensing of Environment*, 103, 369, 2006.
36. Asner, G.P., Nepstad, D., Cardinot, G., et al., Drought stress and carbon uptake in an Amazon forest measured with space-borne imaging spectrometry, *Proceedings of the National Academy of Sciences U.S.A.*, 101, 6039, 2004.
37. Claudio, H.C., Gamon, J.A., Cheng, Y., et al., Monitoring drought effects on vegetation water content and fluxes in chaparral with the 970 nm water band index, *Remote Sensing of Environment*, 103, 304, 2006.
38. NSF, *Cyberinfrastructure Vision for 21st Century Discovery*, NSF Cyberinfrastructure Council, ed., U.S. National Science Foundation, 2007.
39. Gamon, J.A., Rahman, A.F., Dungan, J.L., et al., Spectral network (Specnet)—What is it and why do we need it? *Remote Sensing of Environment*, 103, 227, 2006.
40. Michener, W.K. and Brunt, J.W., *Ecological data: Design, management and processing.* Blackwell Science, Oxford, U.K., 2000.
41. Jones, M.B., Berkley, C., Bojilova, J., et al., Managing scientific metadata, *IEEE Internet Computing*, 5, 59, 2001.
42. Parsaye, K., Chignelle, M., Khoshafian, S., et al., *Intelligent databases: Object oriented, deductive hypermedia technologies.* John Wiley & Sons, New York, 1989.
43. Baldocchi, D., Falge, E., Gu, L., et al., FLUXNET: A new tool to study the temporal and spatial variability of ecosystem-scale carbon dioxide, water vapor, and energy flux densities, *Bulletin of the American Meteorological Society*, 82, 2415, 2001.

Index

Printed and bound by CPI Group (UK) Ltd, Croydon, CR0 4YY

23/10/2024

01778227-0016